Vibrations, Waves, and Acoustics

D. Chattopadhyay
P. C. Rakshit

Celebrating 60 Years 1960-2020
Since Books and Allied 1960

[*For the undergraduate students of Physics (Hons/Major) (CBCS) and Engineering*]

Vibrations, Waves, and Acoustics

D. Chattopadhyay, Ph.D., D.Sc., P.R.S. (Mouat Medalist)

Former Professor and Head,
Institute of Radio Physics and Electronics,
University of Calcutta.

P. C. Rakshit, Ph. D.

Former Professor and Head,
Institute of Radio Physics and Electronics,
University of Calcutta.

BOOKS & ALLIED (P) Ltd.

No.1-E(1) 'Shubham Plaza' (1st Floor)
83/1, Beliaghata Main Road, Kolkata 700010 (India)
Tel : (033) 23740062 • 8274085530
e-mail : booksandallied1960@gmail.com • booksmktng@gmail.com

Vibrations, Waves, and Acoustics

First Published : 1998
Second Edition : 2000
Third Edition : 2002
Fourth Edition : 2005
Fifth Edition (thoroughly revised) **: 2006**
Sixth Edition : 2007
Seventh Edition : 2008
Eighth Edition : 2010
Reprinted : 2011
Revised Reprint : 2012
Reprinted : 2014, 2015, 2016, 2019
Reprinted : 2020

Publisher : Ruba Sen
BOOKS & ALLIED (P) LTD.
8/1 Chintamoni Das Lane, Kolkata 700009

Printer : EUREKA
91/A, Baithakkhana Road, Kolkata 700009

Typesetter : BOOKS & ALLIED (P) LTD.
8/1 Chintamoni Das Lane, Kolkata 700009

Project Team : Salil Das, Abhijit Ghoshal & Koely Chatterjee

ISBN 978-81-87134-14-5

Price : **₹350.00**

Preface to the Eighth Edition

This revised edition embodies some new material particularly in chapters 6, 7, 9 and 13, and several additional review questions and problems, both solved and unsolved. In its present form, the book is expected to serve better its users who are sincerely thanked for their steadfast support.

October 16, 2009
Institute of Radio Physics and Electronics
92, Acharya Prafulla Chandra Road
Kolkata 700009

D. Chattopadhyay

Preface to the Seventh Edition

In this edition, a few changes have been effected to meet the requirements of the students. Some more problems and review questions have also been included in the light of recent examination papers.

We hope that the book, in its present form, will be more useful to the readers.

July 15, 2008
Institute of Radio Physics and Electronics
92, Acharya Prafulla Chandra Road
Kolkata 700009

D. Chattopadhyay
P. C. Rakshit

Preface to the Sixth Edition

The fifth edition was sold out rather quickly, giving us yet another opportunity to overhaul the book. We have, in this edition, modified and enlarged some parts of the book to remove obscurities and include more information. A few more review questions and problems, solved and unsolved, have also been incorporated.

We offer our sincere thanks to the users of the book for their patronage.

June 30, 2007
Institute of Radio Physics and Electronics
92, Acharya Prafulla Chandra Road
Kolkata 700009

D. Chattopadhyay
P. C. Rakshit

Preface to the Fifth Edition

In this edition, the book has been updated with the incorporation of some new material in the text. More review questions and problems, both solved and unsolved, have also been included in the light of current examinations.

We hope that the book in its present form will be more helpful to its users.

D. CHATTOPADHYAY
P. C. RAKSHIT

August 15, 2006
Institute of Radio Physics and Electronics
92, Acharya Prafulla Chandra Road
Kolkata 700009

Preface to the First Edition

This book is meant for the Honours undergraduate students of Physics. The current syllabi of different universities have been consulted in selecting the contents of the book. The noteworthy feature of the book is its clarity and brevity. The physical aspects and the mathematical analyses have been logically developed herein in a systematic way. A number of problems have been worked out to train the students to apply the formulae and concepts they have learnt in the text. A variety of review questions and problems are also included for the practice of the students. Many of the problems and questions incorporated in the book have been taken from recent examination papers.

We shall deem our labour amply rewarded if the book is appreciated by the teachers and the students. Any suggestion for the improvement of the book is welcome.

D. CHATTOPADHYAY
P. C. RAKSHIT

January 1, 1998
Institute of Radio Physics and Electronics
92, Acharya Prafulla Chandra Road
Calcutta 700009

Contents

Chapter 1

Simple Harmonic Motion

1.1 Introduction

If the motion of a particle is repeated at regular intervals of time, the particle is said to execute a *periodic motion.* A particle in periodic motion traces out the same path again and again, and comes back to a given point on the path at regular intervals of time. The time taken by the particle in going once round the path is called the *time period* of the motion. The number of times the path is described by the particle in a unit time is known as the *frequency.* If T is the time period, clearly the frequency is $f = 1/T$.

Periodic motions can be *linear, circular* or *elliptical.* The examples of linear periodic motion include the motions of a pendulum, of the piston in the cylinder of a steam engine, and of a mass suspended by a spring, because such motions are confined to a straight line. The motion of the hands of a clock is a circular periodic motion, while the revolution of the earth round the sun is an elliptic periodic motion.

A particle in periodic motion performs a *vibratory* or *oscillatory* motion if it moves to and fro repeatedly over the same path at regular intervals of time. The periodic motion of a pendulum is clearly vibratory or oscillatory. A particle moving uniformly in a circle executes a periodic motion, but the motion is not vibratory or oscillatory.

The simplest type of oscillatory motion is termed the simple harmonic motion (abbreviated SHM). *If a particle moves to and fro in a straight line such that the restoring force is always directed towards a fixed point on its path and is proportional to its distance from the fixed point, the particle is said to perform a simple harmonic motion.* For small displacements, the vertical oscillations of a weight carried by an elastic string, the oscillations of a liquid column in a U-tube, the motion of a simple pendulum are simple harmonic in nature. A particle in simple harmonic motion is termed a *linear harmonic oscillator*.

If the vibratory motion of a particle of a material medium is imparted to the neighbouring particles of the medium due to elastic forces between them, *waves* are generated in the medium. When the frequency of the wave lies in the *sonic range* extending roughly from 20 Hz to 20 kHz, the energy carried by the wave produces the sensation of hearing in our ear. Ordinarily, *sound* refers to this stimulus causing auditory sensation in our ear. In a wider sense, sound refers to waves in a material medium in the audible and also in the inaudible range. Sound waves with frequencies below 20 Hz are called *infrasonics*, and those with frequencies above 20 kHz are known as *ultrasonics* and *hypersonics*.

Acoustics is the branch of physical science dealing with the generation, propagation, reception, detection, and practical applications of sound. It has various applications such as in entertainment through music, design of microphones and loudspeakers, sound recording studios, auditoriums, medical diagnosis, flaw detection in materials, etc. As the generation and propagation of sound involve vibration of physical bodies and wave motion in material media, the basic principles of acoustics rest on the theories of vibrations and waves.

1.2 Motion of a Linear Harmonic Oscillator

Let x be the displacement of a particle executing a simple harmonic motion at any instant of time t from its position of rest which is a fixed point on its path. For small displacements, the *restoring force* or the *force of restitution* F tending to bring the particle back to its position of rest is proportional to x. Thus we write

$$F = -sx, \tag{1.1}$$

where s is the constant of proportionality between force and displacement, and is referred to as the *spring constant* or the *stiffness factor*, indicating the restoring force per unit displacement. The reciprocal of s is termed the *compliance*. The negative sign in Eq. (1.1) implies that F and x are oppositely directed. Equating the restoring force to mass × acceleration, i.e., $m\frac{d^2x}{dt^2}$ (by Newton's second law of motion), we have

$$m\frac{d^2x}{dt^2} = -sx, \tag{1.2}$$

where m is the mass of the particle. Putting

$$\omega^2 = \frac{s}{m}, \tag{1.3}$$

Eq. (1.2) can be written as

$$\frac{d^2x}{dt^2} + \omega^2 x = 0. \tag{1.4}$$

Equation (1.2) or Eq. (1.4) is referred to as the *equation of motion* of the linear harmonic oscillator. As a trial solution of Eq. (1.4), we take $x = e^{nt}$. Substituting into Eq. (1.4) gives $n^2 + \omega^2 = 0$ or $n = \pm j\omega$, where $j = \sqrt{-1}$. Thus, the general solution of Eq. (1.4) can be written as

$$x = Ce^{j\omega t} + De^{-j\omega t} \tag{1.5}$$

where C and D are two complex constants. Let $C = C_1 + jC_2$ and $D = D_1 + jD_2$, where C_1, C_2, D_1 and D_2 are real constants. So,

$$x = (C_1 + jC_2)(\cos\omega t + j\sin\omega t) + (D_1 + jD_2)(\cos\omega t - j\sin\omega t) \tag{1.6}$$

where we have used Euler's relationships $e^{\pm j\theta} = \cos\theta \pm j\sin\theta$. Equation (1.6) can be written as

$$\begin{aligned} x = {} & [(C_1 + D_1)\cos\omega t + (D_2 - C_2)\sin\omega t] \\ & + j[(C_2 + D_2)\cos\omega t + (C_1 - D_1)\sin\omega t]. \end{aligned} \tag{1.7}$$

Either the real or the imaginary part on the right-hand side of Eq. (1.7) is the acceptable *physical* solution for the displacement (which is real). The real or the imaginary part in Eq. (1.7) gives x in the form

$$x = a\cos\omega t + b\sin\omega t \tag{1.8}$$

where a and b are two arbitrary *real* constants. Thus Eq. (1.8) gives the *general physical solution* of Eq. (1.4). Putting $a = A\sin\phi$ and $b = A\cos\phi$, so that $A = \sqrt{a^2 + b^2}$ and $\tan\phi = a/b$, Eq. (1.8) can be written as

$$x = A\sin(\omega t + \phi). \tag{1.9}$$

Again, putting $\phi = \theta + \frac{\pi}{2}$, Eq. (1.9) gives

$$x = A\cos(\omega t + \theta). \tag{1.10}$$

Equations (1.9) and (1.10) are *alternative forms* of the general physical solution of Eq. (1.4).

The quantity A in Eq. (1.9) or (1.10) denotes the maximum displacement of the particle from the position of rest or the mean position, and is called the *amplitude* of oscillation. If T is the *time period*, then x at time t and that at time $t + T$ are the same. As $\sin(\omega t + \phi)$ remains unchanged when the angle increases by 2π, we have $\omega T = 2\pi$. So,

$$T = \frac{2\pi}{\omega}. \tag{1.11}$$

The *frequency* of oscillation is

$$f = \frac{1}{T} = \frac{\omega}{2\pi} \tag{1.12}$$

The quantity ω is referred to as the *angular frequency*. The angles $(\omega t+\phi)$ and $(\omega t + \theta)$ in Eqs. (1.9) and (1.10), respectively are known as the *phase angles*. At time $t = 0$, the phase angle is ϕ or θ. This *initial phase angle* is often referred to as the *epoch*.

1.3 Rotating Phasors

From Euler's relationship $e^{j\omega t} = \cos\omega t + j\sin\omega t$, we find that $\cos\omega t = Re(e^{j\omega t})$ and $\sin\omega t = Im(e^{j\omega t})$, where '*Re*' and '*Im*' stand for 'real part of' and 'imaginary part of', respectively. Thus, sine and cosine functions can be generated from the complex number $e^{j\omega t}$. In Fig. 1.1, the radius vector OP of magnitude unity denotes this complex number. The components of OP along the real and the imaginary axes are $\cos\omega t$ and $\sin\omega t$, respectively. The angle ωt increases with time, so that the radius vector OP rotates in the counterclockwise direction with the angular velocity ω.

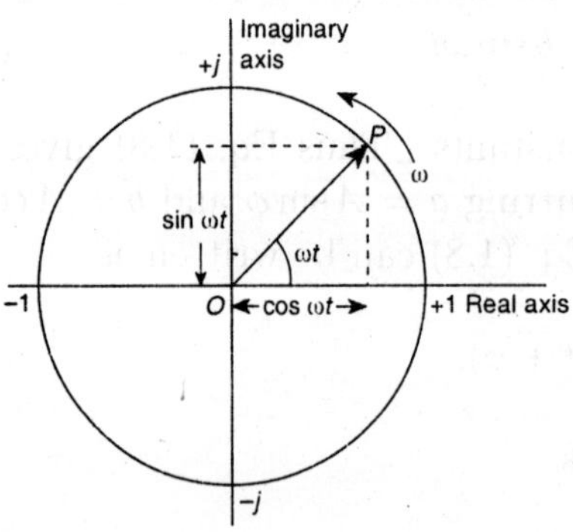

Fig.1.1 Graphical representation of $e^{j\omega t}$

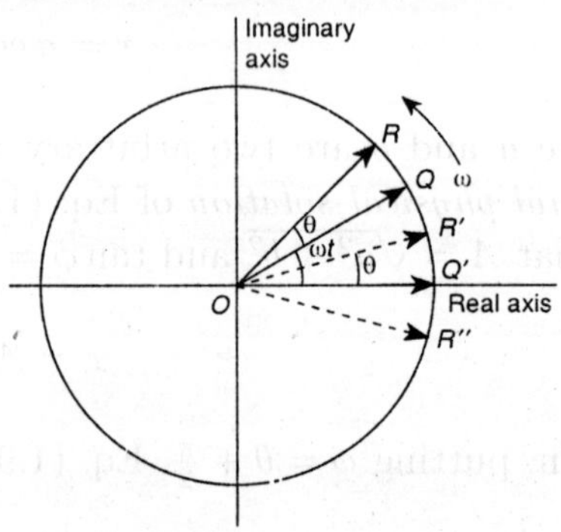

Fig.1.2 Phasor representation related to SHM

For the complex number $e^{-j\omega t}$, the angle is $-\omega t$. Therefore, the complex number $e^{-j\omega t}$ is represented by a unit radius vector which rotates with time in the clockwise direction with the angular velocity ω. The radius vectors denoting complex numbers are termed *phasors*.

The displacement $x = A\cos\omega t = Re(Ae^{j\omega t})$ is represented by the phasor OQ in Fig. 1.2, where $|OQ| = A$. The phasor OR represents the displacement $x = A\cos(\omega t + \theta) = Re[Ae^{j(\omega t+\theta)}]$. The phasors OQ and OR rotate with time in the counterclockwise direction with the angular velocity ω, but the angle θ between them does not change. OQ' and OR' are the positions of OQ and OR, respectively at $t = 0$. If θ is positive, OR leads OQ whereas if θ is negative, OR lags OQ. If θ is negative, OR'' is the position of OR at $t = 0$.

Since the projection of OR on the real axis is $A\cos(\omega t + \theta)$ and that on the imaginary axis is $A\sin(\omega t + \theta)$, *the simple harmonic motion can be looked upon as the projection of a uniform circular motion on any diameter.*

1.4 Evaluation of Constants

The general solution of the second order differential equation [Eq. (1.4)] given by Eq. (1.8) involves two unknown arbitrary constants a and b which can be determined from two *boundary conditions*, as shown below.

(a) Suppose that the particle is displaced from its mean position by x_0 and then released. Thus the boundary conditions are (i) at $t = 0$, $x = x_0$ and (ii) at $t = 0$, $\dot{x} = 0$, the velocity of the particle at the instant of release being zero. Putting the boundary condition (i) in Eq. (1.8), we obtain $x_0 = a$. Thus $x = x_0\cos\omega t + b\sin\omega t$, and $\dot{x} = \frac{dx}{dt} = -x_0\omega\sin\omega t + b\omega\cos\omega t$. The boundary condition (ii) gives $0 = b\omega$ or, $b = 0$. Therefore, the displacement of the particle is given by

$$x = x_0\cos\omega t. \tag{1.13}$$

(b) If the particle is given an impulse at time $t = 0$ when it is at the position of rest, the boundary conditions are (i) at $t = 0$, $x = 0$ and (ii) at $t = 0$, $\dot{x} = \dot{x}_0$. Using the boundary condition (i) in Eq. (1.8) we obtain $a = 0$, so that $x = b\sin\omega t$, and $\dot{x} = b\omega\cos\omega t$. The boundary condition (ii) now gives $b = \dot{x}_0/\omega$. In this case, the solution is

$$x = \frac{\dot{x}_0}{\omega}\sin\omega t. \tag{1.14}$$

(c) We now consider the general boundary condition:

(i) at $t = 0$, $x = x_0$ and (ii) at $t = 0$, $\dot{x} = \dot{x}_0$. The boundary condition (i), used in Eq. (1.8), gives $x_0 = a$, so that $x = x_0\cos\omega t + b\sin\omega t$. Hence, $\dot{x} = -x_0\omega\sin\omega t + b\omega\cos\omega t$. The boundary condition (ii) gives $\dot{x}_0 = b\omega$, or, $b = \dot{x}_0/\omega$. Therefore, the solution is

$$x = x_0\cos\omega t + \frac{\dot{x}_0}{\omega}\sin\omega t. \tag{1.15}$$

If the displacement and the velocity of the oscillating particle are x_0 and $\dot{x}_0$, respectively, at $t = \tau$ instead of at $t = 0$, the time t in Eq. (1.15) is to be replaced by $t - \tau$. In this case, the solution is

$$x = x_0 \cos\omega(t-\tau) + \frac{\dot{x}_0}{\omega}\sin\omega(t-\tau). \tag{1.16}$$

Observation

Equations (1.8), (1.9) and (1.10) give three forms of the general physical solution for a particle in simple harmonic motion. Some more forms can be given, as shown below.

Putting $D = C^*$, where C^* is the complex conjugate of C in Eq. (1.5), we have $C_1 = D_1$ and $C_2 = -D_2$. Then the imaginary part of Eq. (1.7) goes to zero, and x reduces to the form of Eq. (1.8). Thus the general physical solution of Eq. (1.4) can be written as

$$x = Ce^{j\omega t} + C^* e^{-j\omega t}. \tag{1.17}$$

In the solution $x = Ce^{j\omega t}$, putting $C = C_1 + jC_2$, we have

$$x = (C_1\cos\omega t - C_2\sin\omega t) + j(C_2\cos\omega t + C_1\sin\omega t).$$

Both the real and the imaginary parts of this equation are of the form of Eq. (1.8) and can be taken as the general physical solution; thus

$$\begin{aligned} x &= Re(Ce^{j\omega t}) \\ \text{or,}\quad x &= Im(Ce^{j\omega t}), \end{aligned} \tag{1.18}$$

where the complex number C contains two arbitrary real constants C_1 and C_2.

In the same fashion, starting with $x = Ce^{-j\omega t}$, we can show that

$$\begin{aligned} x &= Re(Ce^{-j\omega t}) \\ \text{or,}\quad x &= Im(Ce^{-j\omega t}) \end{aligned} \tag{1.19}$$

represents the general physical solution of Eq. (1.4).

1.5 Energy of a Linear Harmonic Oscillator

Let x be the displacement of the particle at any instant of time t. The force of restitution opposing the displacement is sx. If the particle is displaced further by dx, the work done against the restoring force is $sx\,dx$, and this is the increase of the potential energy for the displacement dx. So, the total potential energy for a displacement x is

$$E_p = \int_0^x sx\,dx = \frac{1}{2}sx^2. \tag{1.20}$$

The instantaneous kinetic energy of the particle is

$$E_k = \frac{1}{2}m\left(\frac{dx}{dt}\right)^2. \tag{1.21}$$

The total energy E is the sum of the kinetic and the potential energies:

$$E = E_k + E_p = \frac{1}{2}m\left(\frac{dx}{dt}\right)^2 + \frac{1}{2}sx^2. \tag{1.22}$$

If there is no dissipation of energy, the total energy remains constant. Therefore, $\frac{dE}{dt} = 0$, or,

$$\frac{d}{dt}\left[\frac{1}{2}m\left(\frac{dx}{dt}\right)^2 + \frac{1}{2}sx^2\right] = 0$$

$$\text{or,}\quad m\frac{dx}{dt}\cdot\frac{d^2x}{dt^2} + sx\frac{dx}{dt} = 0$$

$$\text{or,}\quad m\frac{d^2x}{dt^2} + sx = 0, \tag{1.23}$$

since $\frac{dx}{dt} \neq 0$. Equation (1.23) is the equation of motion, obtained here from the *principle of conservation of energy*. Taking the general physical solution

$$x = A\cos(\omega t + \theta), \tag{1.24}$$

we have for the instantaneous kinetic energy

$$E_k = \frac{1}{2}m\left(\frac{dx}{dt}\right)^2 = \frac{1}{2}m\omega^2A^2\sin^2(\omega t + \theta). \tag{1.25}$$

Clearly, the maximum kinetic energy is

$$E_{km} = \frac{1}{2}m\omega^2A^2. \tag{1.26}$$

The instantaneous potential energy is

$$E_p = \frac{1}{2}sA^2\cos^2(\omega t + \theta) = \frac{1}{2}m\omega^2A^2\cos^2(\omega t + \theta), \tag{1.27}$$

using Eq. (1.3). The maximum potential energy is

$$E_{pm} = \frac{1}{2}m\omega^2A^2. \tag{1.28}$$

Equations (1.26) and (1.28) show that the maximum kinetic and potential energies are equal.

The *average kinetic energy* is

$$\bar{E}_k = \frac{1}{T}\left[\frac{1}{2}m\int_0^T\left(\frac{dx}{dt}\right)^2 dt\right] = \frac{m}{2T}\omega^2A^2\int_0^T \sin^2(\omega t+\theta)dt$$
$$= \frac{1}{4}m\omega^2A^2 = \frac{1}{2}E_{km}. \quad (1.29)$$

Also, the *average potential energy* is

$$\bar{E}_p = \frac{1}{T}\left[\frac{1}{2}s\int_0^T x^2dt\right] = \frac{s}{2T}A^2\int_0^T \cos^2(\omega t+\theta)dt$$
$$= \frac{1}{4}sA^2 = \frac{1}{4}m\omega^2A^2 = \frac{1}{2}E_{pm}. \quad (1.30)$$

Equations (1.29) and (1.30) show that the average kinetic and potential energies are half the corresponding maximum energies.

Adding Eqs. (1.25) and (1.27), we find that $E = E_k + E_p = \frac{1}{2}m\omega^2A^2$, which is a constant, equal to the maximum kinetic (or potential) energy. As the particle oscillates, its energy changes between kinetic and potential forms, one increasing at the expense of the other.

Note that the restoring force is derivable from the potential energy : $F = -sx = -\frac{dE_p}{dx}$. After a complete cycle, the net work done by the restoring force is zero, since the particle returns to the same point along a closed path, i.e., to the state of the same potential energy. The restoring force is thus a *conservative force.*

1.6 Typical Harmonic Oscillators

A few elementary examples of simple harmonic motion have already been cited. Some more useful illustrations are given below.

(i) Cantilever loaded at the free end: We consider a horizontal cantilever of length l, at the free end of which a vertical load F acts. If Y is Young's modulus of the material of the cantilever and I is the geometrical moment of inertia of the cross section about the neutral plane, the deflection y at a distance x from the fixed end is

$$YIy = \frac{1}{2}Flx^2 - \frac{1}{6}Fx^3. \quad (1.31)$$

At the point of loading, $x = l$ and the depression is δ, say. So, Eq. (1.31) gives

$$YI\delta = \frac{1}{3}Fl^3$$
$$\text{or,} \quad F = \frac{3YI\delta}{l^3}. \quad (1.32)$$

Substituting for F from Eq. (1.32) into Eq. (1.31), we obtain

$$y = \frac{\delta}{2}\left[\frac{3x^2}{l^2} - \frac{x^3}{l^3}\right]. \tag{1.33}$$

Therefore, the velocity of a particle of the beam at a distance x from the fixed end is

$$\dot{y} = \frac{dy}{dt} = \frac{\dot{\delta}}{2}\left[\frac{3x^2}{l^2} - \frac{x^3}{l^3}\right]. \tag{1.34}$$

Let μ be the mass per unit length of the cantilever. Then its kinetic energy is

$$E_k' = \int_0^l \frac{1}{2}\mu\left[\frac{3x^2}{l^2} - \frac{x^3}{l^3}\right]^2 \frac{\dot{\delta}^2}{4}dx = \frac{33}{280}\mu l\dot{\delta}^2. \tag{1.35}$$

If M is the mass of the load at the end of the cantilever, its kinetic energy is $\frac{1}{2}M\dot{\delta}^2$. So, the total kinetic energy of the system is

$$E_k = \frac{1}{2}M\dot{\delta}^2 + E_k'. \tag{1.36}$$

The potential energy of the beam, being the work done in straining it, is

$$E_p = \int_0^\delta F\,d\delta = \frac{3YI}{l^3}\int_0^\delta \delta\,d\delta = \frac{3YI}{2l^3}\delta^2. \tag{1.37}$$

using Eq. (1.32). For a conservative system, the total energy $E_k + E_p$ is a constant. Thus, adding Eqs. (1.36) and (1.37), we obtain

$$\left(\frac{1}{2}M + \frac{33}{280}\mu l\right)\dot{\delta}^2 + \frac{3YI}{2l^3}\delta^2 = \text{constant}.$$

Differentiating with respect to t gives

$$\left(M + \frac{33}{140}\mu l\right)\ddot{\delta} + \frac{3YI}{l^3}\delta = 0. \tag{1.38}$$

Comparing Eq. (1.38) with the equation for simple harmonic motion [Eq. (1.4)], we recognise at once that the cantilever oscillates simple harmonically with a time period

$$T = 2\pi\sqrt{\left(M + \frac{33}{140}\mu l\right)\frac{l^3}{3YI}}. \tag{1.39}$$

(ii) Torsional oscillation: Let a rigid body be suspended from one end of a thin vertical wire, the other end of which is attached to a fixed support. The axis of the wire passes through the centre of gravity of the body. The body is slightly rotated about the axis of suspension and then released, so that it executes angular oscillations about the axis. Such an oscillatory motion is referred to as the *torsional oscillation* or *angular simple harmonic motion*.

As the wire is twisted by applying a torque to the body, the wire tends to untwist itself and thus exerts a restoring torque on the body. Within elastic limit, the restoring torque is proportional to the angle of twist. If c is the restoring torque per unit angle of twist and θ is the instantaneous angle of twist, the instantaneous restoring torque is $c\theta$. Let I be the moment of inertia of the body about the axis of rotation. Then the equation of the torsional oscillation of the body is

$$I\frac{d^2\theta}{dt^2} = -c\theta$$

$$\text{or,} \quad \frac{d^2\theta}{dt^2} + \omega^2\theta = 0, \tag{1.40}$$

where $\omega^2 = c/I$. The solution of Eq. (1.40) is

$$\theta = \theta_0 \sin(\omega t + \phi). \tag{1.41}$$

Thus θ varies simple harmonically with time with a period

$$T = \frac{2\pi}{\omega} = 2\pi\sqrt{\frac{I}{c}}. \tag{1.42}$$

(iii) Bifilar suspension: Suppose that a rigid rod AB be suspended by two parallel threads CD and EF of equal length l (Fig. 1.3). The threads are attached at equal distances on either side of the centre of gravity G of the rod AB. Such a system is known as the *bifilar suspension*.

Let the rod be rotated through a small angle in the horizontal plane and then released. Suppose that θ is the instantaneous angle made by the rod AB with its position of rest and that α is the corresponding angle made by each thread with its equilibrium position. If T_0 is the tension in each thread, the restoring forces acting on the rod are the components $T_0 \sin\alpha$ for the two threads. The restoring torque is $2dT_0 \sin\alpha$, where $2d$ is the distance CE between the threads. If M is the mass of the rod, we have $T_0 = Mg/2$. As the angles θ and α are small, we have $\sin\alpha \simeq \alpha$ and $\sin\theta \simeq \theta$. Also, $CC' = EE' = \theta d = \alpha l$. Thus the restoring torque can be written as $2dT_0\alpha = Mg\alpha d = Mgd^2\theta/l$. If I is the moment of inertia of the rod about a vertical axis through G, the equation of its torsional oscillation is

$$I\frac{d^2\theta}{dt^2} = -\frac{Mgd^2}{l}\theta. \tag{1.43}$$

The motion is thus simple harmonic having the time period

$$T = 2\pi\sqrt{\frac{Il}{Mgd^2}}. \tag{1.44}$$

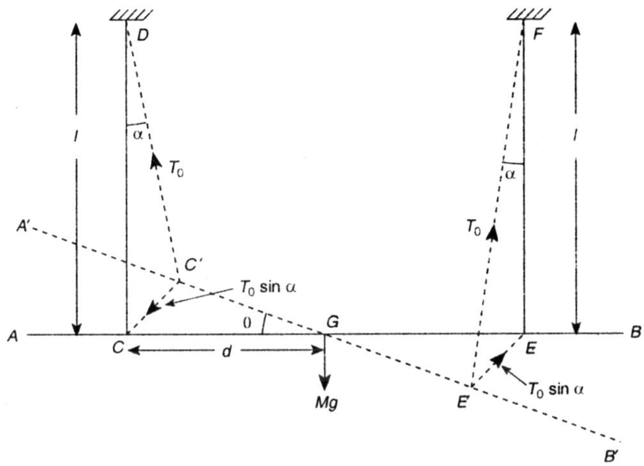

Fig. 1.3 Bifilar suspension

1.7 Superposition of Simple Harmonic Motions

(i) *Simple harmonic motions of the same frequency but different phases and amplitudes along the same straight line:*

Let the two simple harmonic motions be given by

$$x_1 = a_1 \cos(\omega t + \alpha_1) \tag{1.45}$$

$$\text{and} \quad x_2 = a_2 \cos(\omega t + \alpha_2), \tag{1.46}$$

where x_1, x_2 are the instantaneous displacements; a_1, a_2 are the amplitudes; ω is the angular frequency; and α_1, α_2 are the initial phase angles. The instantaneous resultant displacement is

$$\begin{aligned} x &= x_1 + x_2 = a_1 \cos(\omega t + \alpha_1) + a_2 \cos(\omega t + \alpha_2) \\ &= (a_1 \cos\alpha_1 + a_2 \cos\alpha_2)\cos\omega t - (a_1 \sin\alpha_1 + a_2 \sin\alpha_2)\sin\omega t. \end{aligned} \tag{1.47}$$

Let $A\cos\phi = a_1\cos\alpha_1 + a_2\cos\alpha_2$ and $A\sin\phi = a_1\sin\alpha_1 + a_2\sin\alpha_2$, where A and ϕ are constants given by

$$\begin{aligned} A^2 &= (a_1\cos\alpha_1 + a_2\cos\alpha_2)^2 + (a_1\sin\alpha_1 + a_2\sin\alpha_2)^2 \\ &= a_1^2 + a_2^2 + 2a_1a_2\cos(\alpha_1 - \alpha_2) \end{aligned} \tag{1.48}$$

$$\text{and} \quad \tan\phi = \frac{a_1\sin\alpha_1 + a_2\sin\alpha_2}{a_1\cos\alpha_1 + a_2\cos\alpha_2}. \tag{1.49}$$

Therefore, Eq. (1.47) can be written as

$$x = A\cos(\omega t + \phi), \tag{1.50}$$

which shows that the resultant motion is simple harmonic having the same frequency as that of the component vibrations.

The above analysis can be readily extended to include more than two oscillations. Let the individual vibrations be represented by $a_1\cos(\omega t+\alpha_1)$, $a_2\cos(\omega t+\alpha_2)$, $a_3\cos(\omega t+\alpha_3)$, and so on. The resultant motion is

$$\begin{aligned} x &= a_1\cos(\omega t+\alpha_1) + a_2\cos(\omega t+\alpha_2) + a_3\cos(\omega t+\alpha_3) + \cdots \\ &= (a_1\cos\alpha_1 + a_2\cos\alpha_2 + a_3\cos\alpha_3 + \cdots)\cos\omega t \\ &\qquad -(a_1\sin\alpha_1 + a_2\sin\alpha_2 + a_3\sin\alpha_3 + \cdots)\sin\omega t \\ &= A\cos\phi\cos\omega t - A\sin\phi\sin\omega t \\ &= A\cos(\omega t+\phi), \end{aligned} \tag{1.51}$$

where

$$A\sin\phi = a_1\sin\alpha_1 + a_2\sin\alpha_2 + a_3\sin\alpha_3 + \cdots = \sum_i a_i\sin\alpha_i$$

and $$A\cos\phi = a_1\cos\alpha_1 + a_2\cos\alpha_2 + a_3\cos\alpha_3 + \cdots = \sum_i a_i\cos\alpha_i,$$

so that

$$A^2 = \left(\sum_i a_i\sin\alpha_i\right)^2 + \left(\sum_i a_i\cos\alpha_i\right)^2 \tag{1.52}$$

and $$\tan\phi = \sum_i a_i\sin\alpha_i \Big/ \sum_i a_i\cos\alpha_i. \tag{1.53}$$

Equation (1.51) shows that the resultant oscillation is simple harmonic in nature having a frequency equal to that of each component oscillation.

Observation

Simple harmonic motions can be readily added resorting to the phasor diagram. To find the sum of the simple harmonic motions $a_1\cos(\omega t+\alpha_1)$, $a_2\cos(\omega t+\alpha_2)$, $a_3\cos(\omega t+\alpha_3), \cdots$, we draw the rotating vectors at time $t = 0$, as shown in Fig. 1.4, one after the other, a_1 making angle α_1 with the real axis, a_2 the angle α_2, and so on. In Fig. 1.4, the angles $\alpha_1, \alpha_2, \alpha_3$ and α_4 are positive (leading angles), whereas the angle α_5 is negative (lagging angle). The resultant phasor is obtained by joining the starting point to the final point of the phasors. By the laws of addition of phasors the square of the magnitude A of the resultant phasor is given by Eq. (1.52). The inclination ϕ of the resultant phasor to the real axis is given by Eq. (1.53).

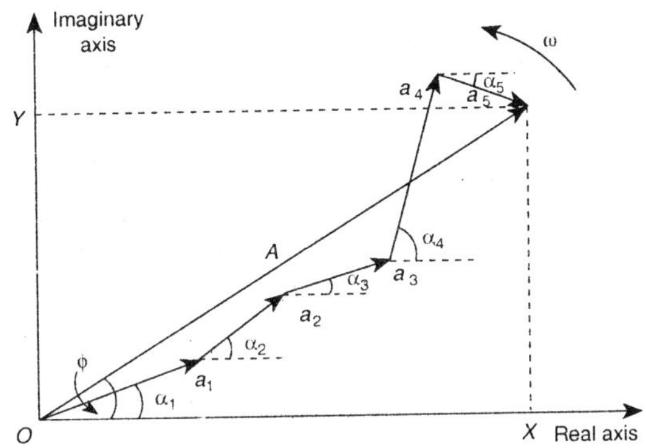

Fig. 1.4 Composition of simple harmonic motions using phasors

We have shown the phasors in one position only, i.e., at $t = 0$ in Fig. 1.4. With increasing time, all the phasors rotate with the same angular frequency ω. Thus the entire phasor diagram rotates as a rigid structure with the angular velocity ω.

The individual motions being consine functions, are the real parts of the phasors. Hence the resultant motion is given by the real part of the resultant phasor, i.e., $x = A\cos(\omega t + \phi)$. If instead, the individual motions are sine functions, their combined motion is the imaginary part of the resultant phasor, i.e., $x = A\sin(\omega t + \phi)$.

(ii) *Two simple harmonic motions of slightly different frequencies along the same straight line (Beats)*:

Let the two motions have the angular frequencies ω and $\omega - \Delta\omega$, respectively. If α_1 and α_2 are the initial phases, the two motions can be expressed by $x_1 = a_1\cos(\omega t + \alpha_1)$ and $x_2 = a_2\cos[(\omega - \Delta\omega)t + \alpha_2]$, respectively. Putting $\alpha_2' = \alpha_2 - (\Delta\omega)t$, we have for the resultant displacement

$$\begin{aligned} x = x_1 + x_2 &= a_1\cos(\omega t + \alpha_1) + a_2\cos(\omega t + \alpha_2') \\ &= A\cos(\omega t + \phi), \end{aligned} \tag{1.54}$$

where

$$A^2 = a_1^2 + a_2^2 + 2a_1a_2\cos(\alpha_1 - \alpha_2') \tag{1.55}$$

$$\text{and} \quad \tan\phi = \frac{a_1\sin\alpha_1 + a_2\sin\alpha_2'}{a_1\cos\alpha_1 + a_2\cos\alpha_2'}. \tag{1.56}$$

Equations (1.55) an (1.56) are obtained from Eqs. (1.48) and (1.49), respectively, on replacing α_2 by α_2'. We have $\alpha_1 - \alpha_2' = \alpha_1 - \alpha_2 + (\Delta\omega)t$. So, if $\alpha_1 - \alpha_2 + (\Delta\omega)t = (2p + 1)\pi$, where $p = 0, 1, 2, 3$, etc., Eq. (1.55) gives $A = a_1 - a_2$. On the other hand, if $\alpha_1 - \alpha_2 + (\Delta\omega)t = 2p\pi$, we have $A = a_1 + a_2$. Equation (1.54) shows that the resultant vibration has the angular frequency ω. The amplitude A of the resultant vibration is not a constant. Equation (1.55) reveals that A changes with time between $(a_1 - a_2)$ and $(a_1 + a_2)$ with an angular frequency $\Delta\omega$. If $a_1 = a_2 = a$, the amplitude varies periodically between 0 and $2a$. Note that the frequency of the amplitude variation is the *difference* between the two component frequencies.

Thus when two simple harmonic motions of slightly different frequencies superimpose, the amplitude of the resultant vibration changes regularly with time between a maximum and a minimum. This phenomenon is referred to as *beats*. For a clear realization of beats, the frequency difference $\Delta\omega$ should be small and the amplitudes a_1 and a_2 should be equal, or nearly so. The time between two successive maxima or minima of amplitude is $2\pi/\Delta\omega$ or, $1/\Delta f$ second, where Δf is the frequency difference in Hz between the component vibrations. So, the number of beats per second i.e., the number of maxima or minima per second is Δf.

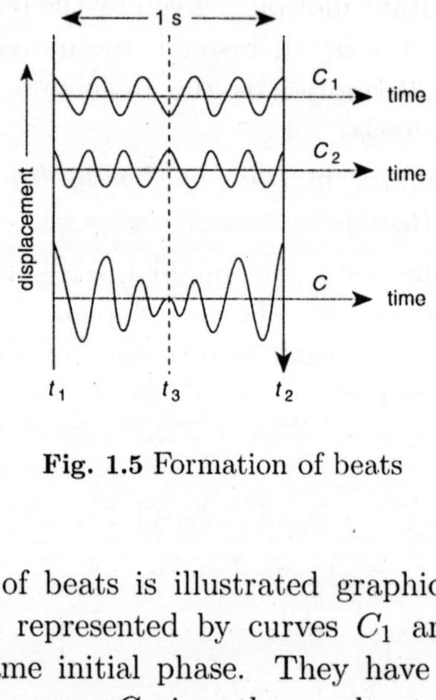

Fig. 1.5 Formation of beats

The phenomenon of beats is illustrated graphically in Fig. 1.5. The component vibrations represented by curves C_1 and C_2, have the same amplitude and the same initial phase. They have the frequencies 5 and 6Hz, respectively. The curve C gives the resultant vibration. At times t_1 and t_2, the two simple harmonic motions are in phase so that the resultant displacements at t_1 and t_2 are a maximum. At time t_3, the two component vibrations superimpose in opposite phase; so the resultant displacement is a minimum at t_3.

Two consecutive maxima (or two consecutive minima) give one beat. Figure 1.5 depicts that one beat is produced in one second. Thus, the number of beats equals the difference of frequencies of the two combining vibrations.

(iii) *Two simple harmonic motions of the same frequency at right angles to each other*:

Let the two simple harmonic vibrations acting along the x-and the y-axes be represented respectively by $x = a_1 \cos(\omega t + \alpha_1)$ and $y = a_2 \cos(\omega t + \alpha_2)$. We have

$$\begin{aligned} y &= a_2 \cos[\omega t + \alpha_1 - (\alpha_1 - \alpha_2)] \\ &= a_2 \cos(\omega t + \alpha_1)\cos(\alpha_1 - \alpha_2) + a_2 \sin(\omega t + \alpha_1)\sin(\alpha_1 - \alpha_2) \\ &= a_2 \frac{x}{a_1}\cos(\alpha_1 - \alpha_2) + a_2\sqrt{1 - \frac{x^2}{a_1^2}}\sin(\alpha_1 - \alpha_2) \\ &= \frac{a_2}{a_1}x\cos(\alpha_1 - \alpha_2) + \frac{a_2}{a_1}\sqrt{a_1^2 - x^2}\sin(\alpha_1 - \alpha_2) \end{aligned}$$

or, $\sqrt{a_1^2 - x^2}\sin(\alpha_1 - \alpha_2) = \dfrac{a_1}{a_2}y - x\cos(\alpha_1 - \alpha_2)$.

Squaring and rearranging, we obtain

$$a_1^2 \sin^2(\alpha_1 - \alpha_2) = x^2 + \frac{a_1^2}{a_2^2}y^2 - 2\frac{a_1}{a_2}xy\cos(\alpha_1 - \alpha_2)$$

$$\text{or,} \quad \frac{x^2}{a_1^2} + \frac{y^2}{a_2^2} - 2\frac{xy}{a_1 a_2}\cos(\alpha_1 - \alpha_2) = \sin^2(\alpha_1 - \alpha_2). \tag{1.57}$$

Equation (1.57) represents an ellipse confined inside a rectangle of sides $2a_1$ and $2a_2$ (Fig. 1.6). The major axis of the ellipse makes an angle δ with the x-axis, where

$$\tan 2\delta = \frac{2a_1 a_2}{a_1^2 - a_2^2}\cos(\alpha_1 - \alpha_2). \tag{1.58}$$

We now specialise to the following cases:

Case 1. If $\alpha_1 - \alpha_2 = 0$, i.e., the two simple harmonic motions are *in phase*, Eq. (1.57) simplifies to

$$\frac{x}{a_1} - \frac{y}{a_2} = 0, \tag{1.59}$$

which is a straight line passing through the origin with a slope $\frac{a_2}{a_1}$ (Fig. 1.7).

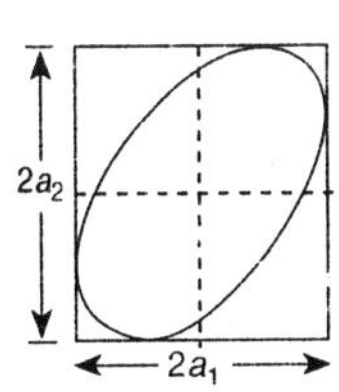

Fig.1.6 An ellipse representing Eq. (1.57)

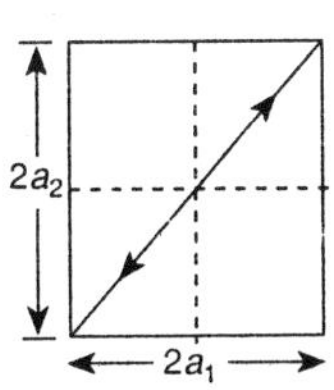

Fig.1.7 A straight line of slope a_2/a_1

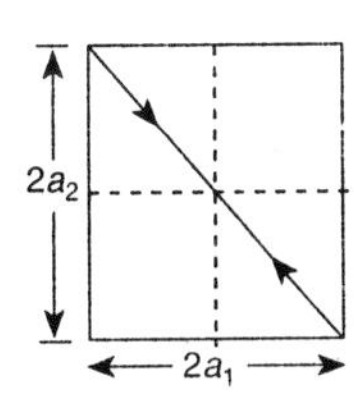

Fig.1.8 A straight line of slope $-a_2/a_1$

Case 2. Let $\alpha_1 - \alpha_2 = \pi$, i.e., the two motions are in *opposite phase.* Then Eq. (1.57) reduces to

$$\frac{x}{a_1} + \frac{y}{a_2} = 0, \tag{1.60}$$

which is also a straight line passing through the origin with a slope $-\frac{a_2}{a_1}$ (Fig. 1.8).

Case 3. If $\alpha_1 - \alpha_2 = \frac{\pi}{2}$, i.e., the two motions are in *quadrature*, we have from Eq. (1.57)

$$\frac{x^2}{a_1^2} + \frac{y^2}{a_2^2} = 1, \tag{1.61}$$

which gives an ellipse of semi-axes a_1 and a_2, the axes of the ellipse coinciding with the x and y-axes (Fig. 1.9).

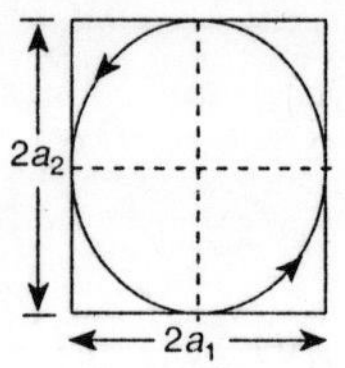

Fig.1.9 An ellipse representing Eq. (1.61)

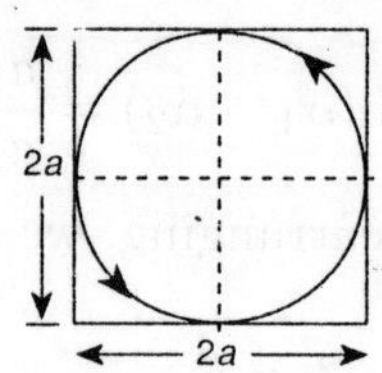

Fig.1.10 A circle obeying Eq. (1.62)

Case 4. If $a_1 = a_2 = a$ and $\alpha_1 - \alpha_2 = \frac{\pi}{2}$, then Eq. (1.57) reduces to

$$x^2 + y^2 = a^2, \tag{1.62}$$

which is a circle of radius a and centre at the origin. Here the resultant motion is a uniform circular motion of angular frequency ω (Fig. 1.10).

(iv) *Two simple harmonic motions of slightly different frequencies at right angles to each other* :

Let the two simple harmonic motions along the x and the y-axes, respectively be given by $x = a_1 \cos(\omega t + \alpha_1)$ and $y = a_2 \cos[(\omega - \Delta\omega)t + \alpha_2]$, where $\Delta\omega$ is small. Putting $\alpha_2' = \alpha_2 - (\Delta\omega)t$, we have $y = a_2 \cos(\omega t + \alpha_2')$. Proceeding as before, we find that the resultant motion is given by Eq. (1.57) with α_2 replaced by $\alpha_2' = \alpha_2 - (\Delta\omega)t$, i.e., by

$$\frac{x^2}{a_1^2} + \frac{y^2}{a_2^2} - \frac{2xy}{a_1 a_2} \cos[\alpha_1 + (\Delta\omega)t - \alpha_2] = \sin^2[\alpha_1 + (\Delta\omega)t - \alpha_2]. \tag{1.63}$$

The phase angle $\delta = \alpha_1 + (\Delta\omega)t - \alpha_2$ in Eq. (1.63) varies slowly with time. The resultant motion is similar to that when the two frequencies are equal, but owing to the gradually changing phase, the pattern changes with time, being always confined within the rectangle of sides $2a_1$ and $2a_2$. When

$\alpha_2 = \alpha_1 + (\Delta\omega)t$, the two component motions are in phase and the pattern is the straight line of Fig. 1.7. As t increases, the pattern opens out into an ellipse (Fig. 1.6), and when $\alpha_1 + (\Delta\omega)t - \alpha_2 = \frac{\pi}{2}$ it takes the shape of Fig. 1.9. With further increase of time, the pattern becomes an oblique ellipse inclined to the left, and shrinks to the straight line of Fig. 1.8 when $\alpha_1 + (\Delta\omega)t - \alpha_2 = \pi$. As t increases further, the pattern repeats the shapes in the reverse order. The sequence of the pattern is shown in Fig. 1.11.

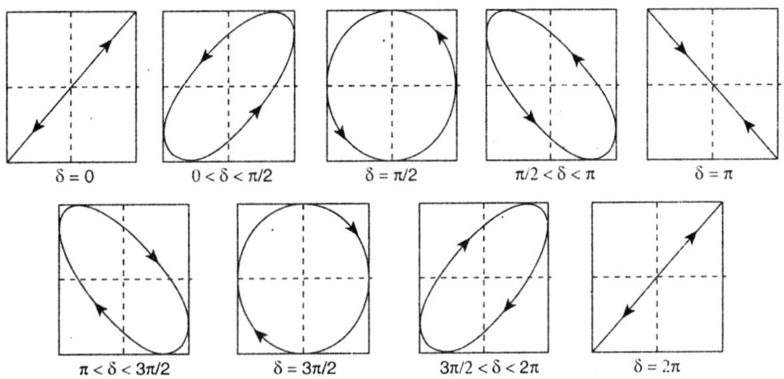

Fig. 1.11 Sequence of paths as δ increases from 0 to 2π

The time interval between two identical patterns can be found as follows. Let at any two successive instants of time t_1 and $t_2 (> t_1)$, Eq. (1.63) remain unchanged. Then $\alpha_1 + (\Delta\omega)t_2 - \alpha_2 = 2\pi + \alpha_1 + (\Delta\omega)t_1 - \alpha_2$ or, $(\Delta\omega)(t_2 - t_1) = 2\pi$,

$$\text{or,} \quad t_2 - t_1 = \frac{2\pi}{\Delta\omega}. \tag{1.64}$$

This shows that the pattern changes with a frequency $\Delta f = \frac{\Delta\omega}{2\pi}$. This allows the frequency difference Δf between the component vibrations to be determined. If the pattern is repeated in every τ seconds, we have $\Delta f = \frac{1}{\tau}$. If the frequency of one component is known, the other can be immediately found.

(v) *Two simple harmonic motions of commensurate frequencies at right angles to each other* :

(a) Suppose that two simple harmonic vibrations of frequencies in the ratio 1 : 2 and differing in initial phase by ϕ are impressed simultaneously on a particle along the x and the y-axes, respectively. The component displacements can then be written as $x = a_1 \cos \omega t$ and $y = a_2 \cos(2\omega t + \phi)$. We have

$$\begin{aligned} \frac{y}{a_2} &= \cos 2\omega t \cos\phi - \sin 2\omega t \sin\phi \\ &= (2\cos^2 \omega t - 1)\cos\phi - 2\sin\omega t \cos\omega t \sin\phi \end{aligned}$$

$$= \left(2\frac{x^2}{a_1^2} - 1\right)\cos\phi - 2\sqrt{1 - \frac{x^2}{a_1^2}} \cdot \frac{x}{a_1}\sin\phi$$

$$\text{or,} \qquad \frac{y}{a_2} - \left(2\frac{x^2}{a_1^2} - 1\right)\cos\phi = -2\frac{x}{a_1}\sqrt{1 - \frac{x^2}{a_1^2}}\sin\phi.$$

Squaring we obtain

$$4\frac{x^2}{a_1^2}\left(1 - \frac{x^2}{a_1^2}\right)\sin^2\phi = \left[\frac{y}{a_2} - \left(\frac{2x^2}{a_1^2} - 1\right)\cos\phi\right]^2 \tag{1.65}$$

If $\phi = 0$, Eq. (1.65) becomes

$$\left(\frac{y}{a_2} - \frac{2x^2}{a_1^2} + 1\right)^2 = 0, \tag{1.66}$$

which gives two coincident parabolas (Fig. 1.12), each of which is represented by

$$x^2 = \frac{a_1^2}{2a_2}(y + a_2).$$

If $\phi = \frac{\pi}{2}$, Eq. (1.65) reduces to

$$\frac{y^2}{a_2^2} + \frac{4x^2}{a_1^2}\left(\frac{x^2}{a_1^2} - 1\right) = 0. \tag{1.67}$$

Equation (1.67) is one of fourth degree in x and represents a curve containing two loops (Fig. 1.13). If the phase difference ϕ is changed, the shape of the loops will change.

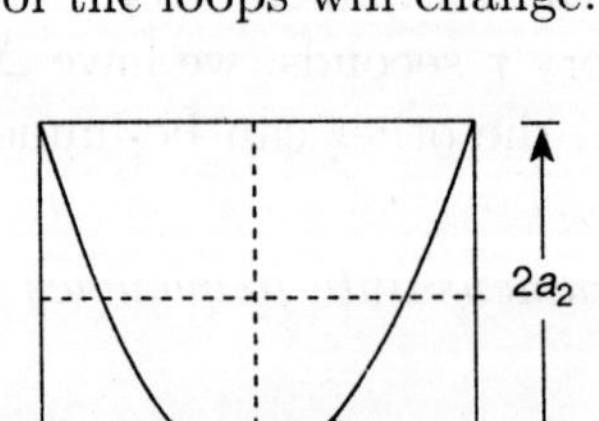

Fig. 1.12 Coincident parabolas represented by Eq. (1.66)

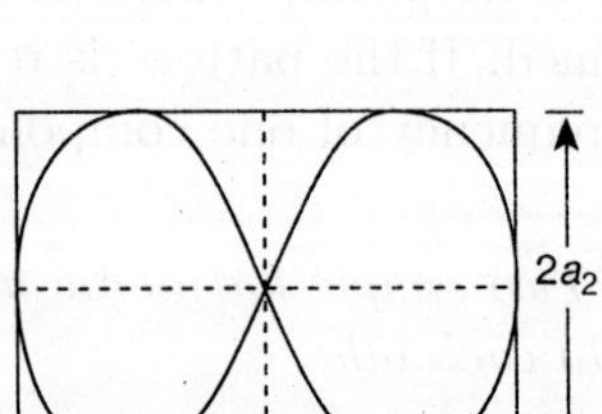

Fig. 1.13 A curve containing two loops [see Eq. (1.67)]

(b) Suppose that two simple harmonic oscillations of frequencies in the ratio of 1 : 3 and having an initial phase difference ϕ act simultaneously on a particle along the x and the y-axes, respectively. The components displacements are then given by $x = a_1 \cos\omega t$ and $y = a_2 \cos(3\omega t + \phi)$. We have

$$\begin{aligned}
\frac{y}{a_2} &= \cos 3\omega t \cos\phi - \sin 3\omega t \sin\phi \\
&= (4\cos^3\omega t - 3\cos\omega t)\cos\phi - (3\sin\omega t - 4\sin^3\omega t)\sin\phi \\
&= \left(4\frac{x^3}{a_1^3} - 3\frac{x}{a_1}\right)\cos\phi - \left[3\left(1-\frac{x^2}{a_1^2}\right)^{1/2} - 4\left(1-\frac{x^2}{a_1^2}\right)^{3/2}\right]\sin\phi \\
&= \left(4\frac{x^3}{a_1^3} - 3\frac{x}{a_1}\right)\cos\phi - \left(1-\frac{x^2}{a_1^2}\right)^{1/2}\left(4\frac{x^2}{a_1^2} - 1\right)\sin\phi.
\end{aligned}$$

Transposing and squaring we get

$$\left(1-\frac{x^2}{a_1^2}\right)\left(4\frac{x^2}{a_1^2}-1\right)^2 \sin^2\phi = \left[\frac{y}{a_2} - \left(\frac{4x^3}{a_1^3} - \frac{3x}{a_1}\right)\cos\phi\right]^2. \tag{1.68}$$

If $\phi = 0$, Eq. (1.68) gives

$$\left(\frac{y}{a_2} + \frac{3x}{a_1} - \frac{4x^3}{a_1^3}\right)^2 = 0. \tag{1.69}$$

Equation (1.69) gives two coincident cubic curves (Fig. 1.14). If $\phi = \frac{\pi}{2}$, Eq. (1.68) reduces to

$$\left(1-\frac{x^2}{a_1^2}\right)\left(\frac{4x^2}{a_1^2}-1\right)^2 - \frac{y^2}{a_2^2} = 0. \tag{1.70}$$

Equation (1.70) is of sixth degree in x and represents a curve having three loops (Fig. 1.15). As the phase difference ϕ is changed, the shape of the loops will change.

In general, if the ratio of the frequencies is N, the curve representing the path of the particle will contain N loops.

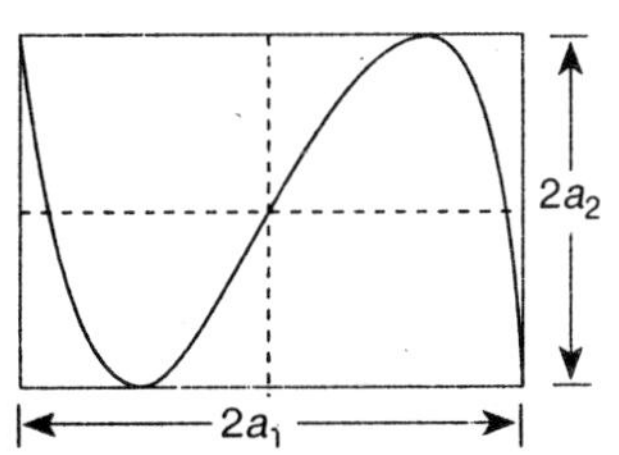

Fig. 1.14 Two coincident cubic curves

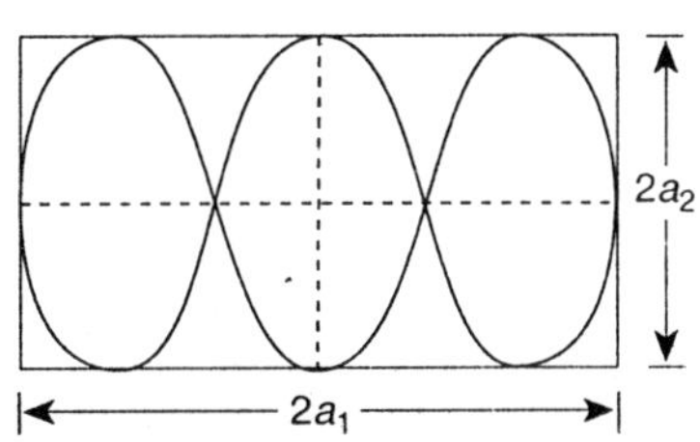

Fig. 1.15 Path of the particle having three loops

1.8 Lissajous Figures

The curves formed by the superposition of two simple harmonic motions at right angles to each other are referred to as *Lissajous figures.* The shape of the Lissajous figures is determined by the relative amplitudes, frequencies, and the initial phases of the component simple harmonic motions. The curves shown in Figs. 1.11 through 1.15 are examples of Lissajous figures.

Let the simple harmonic motion along the x-axis be given by $x = a_1 \cos(\omega_1 t + \alpha_1)$ and that along the y-axis by $y = a_2 \cos(\omega_2 t + \alpha_2)$. If the x- and the y-values are repeated after a time interval T, we have

$$x = a_1 \cos(\omega_1 t + \alpha_1) = a_1 \cos\left[\omega_1(t+T) + \alpha_1\right]$$

$$\text{and} \quad y = a_2 \cos(\omega_2 t + \alpha_2) = a_2 \cos\left[\omega_2(t+T) + \alpha_2\right],$$

giving $\omega_1 T = 2m\pi$ and $\omega_2 T = 2n\pi$, where m and n are positive integers. Thus, $\frac{\omega_1}{\omega_2} = \frac{m}{n}$. As ω_1/ω_2 is the ratio of two integers, it is a rational number. Hence, *the trajectory resulting from the superposition of the two motions is repeated only if the ratio of the frequencies of the two component simple harmonic motions is a rational number.*

The ratio of the frequencies of the component vibrations can be readily found from the Lissajous figures. Draw a horizontal and a vertical line to intersect the curve and find the number of cuts each line makes with the curves. The ratio of the number of cuts made by the horizontal line to that made by the vertical line is the ratio of the vertical to the horizontal frequency. The ratio of the sides of the rectangle enclosing the figure gives the ratio of the amplitudes of the component oscillations.

Lissajous figures can be produced in a number of ways. A suitable method is to use the cathode-ray oscilloscope (abbreviated CRO). The heart of the CRO is the cathode ray tube (CRT), the main components of which are the electron gun, the horizontal and the vertical deflecting systems, and the fluorescent screen (Fig. 1.16). The electron gun produces a narrow beam of high-speed electrons and focusses the beam on the fluorescent screen. A

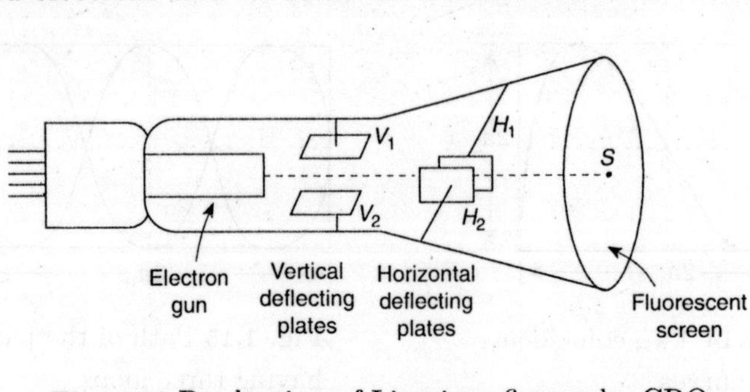

Fig. 1.16 Production of Lissajous figures by CRO

visible spot S is formed on the screen where the beam strikes. On its way from the electron gun to the fluorescent screen, the electron beam passes through a set of parallel deflecting plates. The first set of plates (V_1, V_2) deflects the beam vertically and the second set (H_1, H_2), horizontally. The deflection is achieved by applying voltage between the deflecting plates. A voltage across the vertical deflecting plates V_1 and V_2 deflects the spot vertically on the screen. On the other hand, a voltage across the horizontal deflecting plates H_1 and H_2 moves the spot horizontally.

To display a Lissajous figure on the CRO screen, the two simple harmonic vibrations are first converted into sinusoidal voltages by means of two microphones. These voltages are amplified and then fed to the horizontal and the vertical deflecting plates of the CRT. The spot on the fluorescent screen moves under the simultaneous action of the two sinusoidal voltages at right angles and thus traces out the desired Lissajous figure, which is visible.

1.9 Generation of a Uniform Circular Motion from Two Simple Harmonic Motions at Right Angles

Let two simple harmonic motions $x = a \sin \omega t$ and $y = a \cos \omega t$ acting along the x and y-axes be simultaneously impressed on a particle. Since $\cos \omega t = \sin\left(\omega t + \frac{\pi}{2}\right)$, the y-motion *leads* the x-motion by $\frac{\pi}{2}$. We have

$$x^2 + y^2 = a^2(\sin^2 \omega t + \cos^2 \omega t) = a^2$$

which represents a circle of radius a and centre at the origin. Thus the particle has a uniform circular motion of angular frequency ω. At time $t = 0$, we have $x = 0$ and $y = a$. So the particle is at A (Fig. 1.17) at $t = 0$. At $t = \frac{T}{4}$, where T is the time period, we have $\omega t = \omega T/4 = \frac{\pi}{2}$. Hence at $t = \frac{T}{4}$, $x = a$ and $y = 0$, so that the particle is at B. Thus the particle describes the circle in the *clockwise direction*.

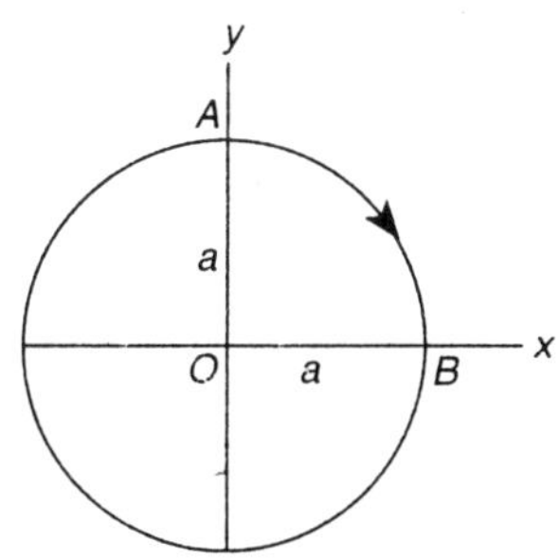

Fig. 1.17 Circular motion from two SHMs

If the motions are $x = a \sin \omega t$ and $y = -a \cos \omega t$, the result is again a uniform circular motion, but the circle is traced in the *counterclockwise direction*. Here the y-motion *lags* the x-motion by $\frac{\pi}{2}$, for $-\cos \omega t = \sin\left(\omega t - \frac{\pi}{2}\right)$.

Thus two perpendicular simple harmonic motions of the same amplitude and frequency, and differing in phase by $\frac{\pi}{2}$, combine to produce a circular motion. The sense of the circular motion depends on which SHM leads the other.

1.10 Combination of Two Opposite Circular Motions

Consider a uniform circular motion of radius a and angular frequency ω described in the clockwise sense. This motion can be resolved into two perpendicular simple harmonic motions given by $x = a\sin\omega t$ and $y = a\cos\omega t$. Consider another uniform circular motion of the same radius and angular frequency described in the counterclockwise sense. This circular motion has two perpendicular simple harmonic components $x = a\sin\omega t$ and $y = -a\cos\omega t$. If the two circular motions superimpose, the resultant motion is $x = 2a\sin\omega t$ and $y = 0$. Thus, the particle executes a simple harmonic motion along the x-axis of angular frequency ω and amplitude equal to the diameter of either circular motion.

1.11 Solved Problems

Problem 1. *A particle oscillating simple harmonically has the displacements x_1 and x_2 when its velocities are v_1 and v_2, respectively. Calculate the amplitude and the time period.*

Solution. Let the SHM be given by $x = a\sin(\omega t - \theta)$, where a is the amplitude, ω is the angular frequency, and θ is the initial phase. The velocity is

$$v = \frac{dx}{dt} = \omega a\cos(\omega t - \theta).$$

Thus, $\sin(\omega t - \theta) = \frac{x}{a}$ and $\cos(\omega t - \theta) = \frac{v}{\omega a}$.
Squaring and adding, we get

$$\frac{x^2}{a^2} + \frac{v^2}{a^2\omega^2} = 1.$$

Using the given conditions, we obtain

$$\frac{x_1^2}{a^2} + \frac{v_1^2}{a^2\omega^2} = 1 \quad \text{and} \quad \frac{x_2^2}{a^2} + \frac{v_2^2}{a^2\omega^2} = 1.$$

Solving for ω and a, we get

$$\omega = \left(\frac{v_1^2 - v_2^2}{x_2^2 - x_1^2}\right)^{1/2} \quad \text{and} \quad a = \left(\frac{x_2^2 v_1^2 - x_1^2 v_2^2}{v_1^2 - v_2^2}\right)^{1/2}.$$

The time period is

$$T = \frac{2\pi}{\omega} = 2\pi\left(\frac{x_2^2 - x_1^2}{v_1^2 - v_2^2}\right)^{1/2}.$$

Problem 2. *A particle in SHM is at its mean position at time $t = 0$. At time $t = 0.1\,\text{s}$ and $t = 0.2\,\text{s}$, the displacements are $1\,\text{cm}$. and $1.5\,\text{cm}$., respectively. Determine the frequency and the amplitude of oscillation.*

Solution. The displacement is given, in general, by $x = A\cos\omega t + B\sin\omega t$ where A and B are constants, and ω is the angular frequency. At $t = 0$, $x = 0$. So, $A = 0$, and $x = B\sin\omega t$.

Using the other conditions, we have

$$1 = B\sin(0.1 \times \omega) \quad \text{and} \quad 1.5 = B\sin(0.2 \times \omega)$$

$$\text{or,} \quad 1.5 = \frac{\sin(\omega/5)}{\sin(\omega/10)} = \frac{2\sin(\omega/10)\cos(\omega/10)}{\sin(\omega/10)} = 2\cos(\omega/10)$$

$$\text{or,} \quad \cos(\omega/10) = \frac{1.5}{2} = 0.75 = \cos 41.4°.$$

So, $\frac{\omega}{10} = 41.4° = 0.723$ rad, or $\omega = 7.23$ rad/s.
The frequency is $f = \frac{\omega}{2\pi} = 1.15$ Hz.
The amplitude is

$$B = \frac{1}{\sin(0.1 \times \omega)} = \frac{1}{\sin 41.4°} = 1.51\,\text{cm}.$$

Problem 3. *A simple harmonic oscillator of mass $0.2\,\text{g}$ has an amplitude $4\,\text{cm}$. Its velocity at zero displacement is $1\,\text{m/s}$. Find the frequency and the energy of oscillation.*

Solution. With usual notations, the displacement is $x = a\sin(\omega t + \theta)$. As $a = 4$ cm., we have $x = 4\sin(\omega t + \theta)$. The velocity is $\frac{dx}{dt} = 4\omega\cos(\omega t + \theta)$. At zero displacement, $\sin(\omega t + \theta) = 0$, whence $(\omega t + \theta) = 0$, so that $\cos(\omega t + \theta) = 1$. Then

$$\frac{dx}{dt} = 4\omega = 100 \text{ cm/s}, \quad \text{or,} \quad \omega = 25 \text{ rad/s}.$$

The frequency is $f = \frac{\omega}{2\pi} = \frac{25}{2\pi} = 3.98$ Hz.
The energy of the oscillator is equal to its maximum kinetic energy, given by

$$\frac{1}{2} \times 0.2 \times 100^2 = 10^3 \text{ erg}.$$

Problem 4. *A particle of mass M is attached to the mid-point of a string of length L stretched vertically. The ends of the string are fixed to rigid supports. Determine the time period of small horizontal oscillations of the mass, neglecting the weight of the string.*

Solution. Let the particle of mass M attached to the mid-point P of the vertical string be slightly displaced horizontally by x. Let the tensions in the two halves of the string be T_1 and T_2, respectively (Fig. 1.18). We assume that the tensions remain constant as the particle oscillates. Resolving the forces on the particle vertically, we have

$$T_1 \cos\theta = Mg + T_2 \cos\theta,$$

since θ is small, $\cos\theta \simeq 1$, so that

$$T_1 - T_2 = Mg.$$

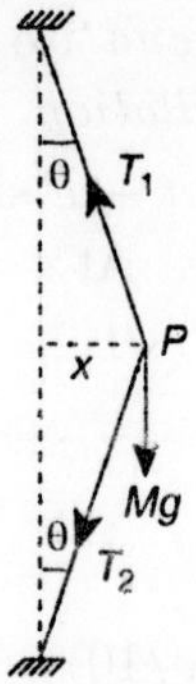

Fig. 1.18 Figure for problem 4

The force of restitution opposite to x is $T_1 \sin\theta + T_2 \sin\theta = (T_1 + T_2)\theta$, since $\sin\theta \simeq \theta$ when θ is small. We have $\theta = x/(L/2) = 2x/L$. Hence the equation of the SHM is

$$M\frac{d^2x}{dt^2} + 2(T_1 + T_2)\frac{x}{L} = 0 \quad \text{or,} \quad \frac{d^2x}{dt^2} + \omega^2 x = 0,$$

where $\omega^2 = \frac{2(T_1+T_2)}{ML}$.

The time period is $T = \frac{2\pi}{\omega} = \pi\sqrt{\frac{2ML}{T_1+T_2}}$.

Problem 5. *A sphere of mass M capable of producing an undamped simple harmonic motion along the x-axis is at rest. The stiffness constant is s. Another sphere of mass m moving with a velocity u_0 hits the first sphere elastically such that the line joining the centres is along the x-axis. Find the displacement of the first sphere at any subsequent instant of time in terms of M, m, u_0 and s.* **[C.U. 1986]**

Solution. Let v_1 and v_2 be the x-velocities of the spheres of mass M and m, respectively, after the collision. The principle of conservation of linear momentum gives

$$mu_0 = mv_1 + Mv_2. \tag{i}$$

The collision being elastic, the relative velocity of one sphere with respect to the other before the collision is equal and opposite to the relative velocity after the collision. So, $u_0 - 0 = v_2 - v_1$ or, $v_1 = v_2 - u_0$. Therefore, from Eq. (i), we get

$$mu_0 = m(v_2 - u_0) + Mv_2 \quad \text{or,} \quad v_2 = \frac{2mu_0}{m + M}. \tag{ii}$$

The equation of motion of the first sphere of mass M is

$$M\frac{d^2x}{dt^2} + sx = 0 \quad \text{or,} \quad \frac{d^2x}{dt^2} + \omega^2 x = 0,$$

where

$$\omega = \sqrt{\frac{s}{M}}. \tag{iii}$$

The solution of the equation is $x = A\cos\omega t + B\sin\omega t$. At time $t = 0$, $x = 0$, so that $A = 0$. Hence $x = B\sin\omega t$, and $\frac{dx}{dt} = \omega B\cos\omega t$. At $t = 0$, we have $\frac{dx}{dt} = v_2$. So, $B = \frac{v_2}{\omega}$. Thus, the displacement is given by

$$x = \frac{2mu_0}{m+M}\sqrt{\frac{M}{s}}\sin\sqrt{\frac{s}{M}}t,$$

using Eqs. (ii) and (iii).

Problem 6. *Two tuning forks A and B of approximately the same frequency are used to obtain Lissajous figures. It is observed that the figures go through a cycle of changes in 20 s. On slightly loading B, the figures go through one cycle in 15 s. If A has a frequency 400 Hz, find the frequency of B before and after loading.* **[C.U. 1983]**

Solution. The frequency of the pattern change before loading is $\Delta f_1 = \frac{1}{20} = 0.05$ Hz. The same after loading is $\Delta f_2 = \frac{1}{15} = 0.067$ Hz. As the loading decreases the frequency of B and the frequency difference between A and B increases when B is loaded, clearly the frequency of B is originally less than that of A. So, the frequency of B before loading is $400 - 0.05 = 399.95$ Hz and that after loading is $400 - 0.067 \simeq 399.93$ Hz.

Problem 7. *A weightless spring is suspended vertically from a rigid support. Two equal masses are attached to the free end of the spring so that the spring elongates by x_0. If one mass drops off, find the motion of the remaining mass.*

Solution. If s is the spring constant and each mass is m, we have $sx_0 = 2mg$, or $x_0 = 2mg/s$. After one mass drops off, let at time t, the displacement of the remaining mass from its equilibrium position be x. The elongation of the spring at this equilibrium position is $\frac{x_0}{2}$. The restoring force is $-sx$, so that the motion is simple harmonic, the equation being

$$m\frac{d^2x}{dt^2} + sx = 0 \quad \text{or,} \quad \frac{d^2x}{dt^2} + \omega^2 x = 0,$$

where $\omega = \sqrt{s/m}$. The amplitude of the SHM is $x_0 - \frac{x_0}{2} = \frac{x_0}{2} = mg/s = g/\omega^2$. The solution is $x = A\cos\omega t + B\sin\omega t$. At $t = 0$, $x = \frac{x_0}{2}$, so that $A = \frac{x_0}{2}$. Also, at $t = 0$, $\frac{dx}{dt} = 0$, whence $B = 0$. Hence,

$$x = \frac{x_0}{2}\cos\omega t = \frac{x_0}{2}\cos\sqrt{\frac{2g}{x_0}}t.$$

Problem 8. *Three simple harmonic motions of the same frequency act on a particle simultaneously in the same direction. Their amplitudes are 1.0, 1.5, and 2.0 cm., respectively. The phase angle of the second with respect to the* *first is 60°, and that of the third with respect to the second is 30°. Obtain the resultant amplitude and phase angle relative to the first.* **[Burd. U. 1999]**

Solution. The resultant motion is

$$\begin{aligned} x &= \cos\omega t + 1.5\cos(\omega t + 60°) + 2\cos(\omega t + 90°) \\ &= \cos\omega t + 1.5\cos 60° \cos\omega t - 1.5\sin 60° \sin\omega t - 2\sin\omega t \\ &= \left(1 + \frac{3}{4}\right)\cos\omega t - \left(\frac{3\sqrt{3}}{4} + 2\right)\sin\omega t \\ &= A\cos\phi\cos\omega t - A\sin\phi\sin\omega t = A\cos(\omega t + \phi) \end{aligned}$$

where

$$A\cos\phi = 1 + \frac{3}{4} = \frac{7}{4} \quad \text{and} \quad A\sin\phi = \frac{3\sqrt{3}}{4} + 2 = \frac{8 + 3\sqrt{3}}{4}.$$

Hence,

$$A^2 = \left(\frac{7}{4}\right)^2 + \left(\frac{8 + 3\sqrt{3}}{4}\right)^2 = 13.95 \quad \text{or,} \quad A = 3.73 \text{ cm.}$$

$$\text{and} \quad \tan\phi = \frac{8 + 3\sqrt{3}}{7} = 1.885 \quad \text{or,} \quad \phi = 62°.$$

So, the resultant amplitude is 3.73 cm. and the phase angle relative to the first component is 62°.

Problem 9. *A thin rod of length l and material density ρ_1 is pivoted at its lowest point A in a stationary, homogeneous and nonviscous liquid of density $\rho_2 (> \rho_1)$ [Fig. 1.19]. The rod can rotate freely in a vertical plane about a horizontal axis through A. Show that if the rod is displaced by a small angle θ from its equilibrium position and then released, it performs a simple harmonic motion with the angular frequency $[3g(\rho_2 - \rho_1)/(2\rho_1 l)]^{1/2}$, where g is the acceleration due to gravity.*

Solution. If α is the cross-sectional area of the rod, its volume is $V = \alpha L$. The mass of the rod is $m = V\rho_1 = \alpha l\rho_1$. When the rod is displaced by the angle θ, its weight acting vertically downward through the centre of gravity O is $W = mg = \alpha l\rho_1 g$. The force of buoyancy acting vertically upward through O is

$$F_B = V\rho_2 g = \alpha l\rho_2 g.$$

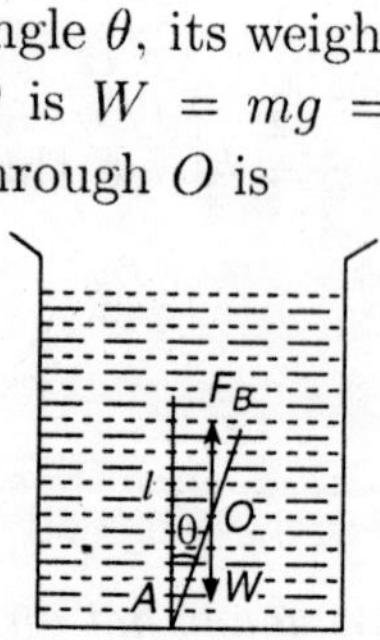

Fig. 1.19 Figure for problem 9

Since $\rho_2 > \rho_1$, the net force acting vertically upwards through O is

$$F = F_B - W = \alpha l(\rho_2 - \rho_1)g.$$

The torque about A is

$$\Gamma = -F\frac{l}{2}\sin\theta = -\frac{\alpha l^2}{2}(\rho_2 - \rho_1)g\sin\theta$$

The negative sign signifies the restoring nature of the torque. For small θ, $\sin\theta \simeq \theta$ and so

$$\Gamma = -\frac{\alpha l^2}{2}(\rho_2 - \rho_1)g\theta.$$

The moment of inertia of the rod about A is

$$I = \frac{1}{3}ml^2 = \frac{1}{3}\alpha\rho_1 l^3.$$

The equation of motion of the rod is

$$I\frac{d^2\theta}{dt^2} = \Gamma$$

$$\text{or,} \quad \frac{d^2\theta}{dt^2} = \frac{\Gamma}{I} = -\frac{3g(\rho_2 - \rho_1)}{2\rho_1 l}\theta = -\omega^2\theta$$

$$\text{where} \quad \omega^2 = \frac{3g(\rho_2 - \rho_1)}{2\rho_1 l}.$$

Thus the motion of the rod is simple harmonic with the angular frequency

$$\omega = \left[\frac{3g(\rho_2 - \rho_1)}{2\rho_1 l}\right]^{1/2}.$$

Problem 10. *Find the result of superposition of n number of simple harmonic motions having the same amplitude a and the same angular frequency ω, but equal successive phase advancement δ.*

Solution. Let the simple harmonic motions be denoted by $x_1 = ae^{j\omega t}$, $x_2 = ae^{j(\omega t+\delta)}, \cdots, x_n = ae^{j[\omega t+(n-1)\delta]}$. The resultant motion is

$$x = x_1 + x_2 + \cdots + x_n = ae^{j\omega t}\left[1 + e^{j\delta} + e^{j2\delta} + \cdots + e^{j(n-1)\delta}\right]$$

$$= ae^{j\omega t}\frac{1 - e^{jn\delta}}{1 - e^{j\delta}} = ae^{j\omega t}\frac{e^{jn\delta/2}}{e^{j\delta/2}}\cdot\frac{e^{-jn\delta/2} - e^{jn\delta/2}}{e^{-j\delta/2} - e^{j\delta/2}}$$

$$= a\frac{\sin(n\delta/2)}{\sin(\delta/2)}e^{j[\omega t+(n-1)\delta/2]}, \quad \text{since } \sin\theta = \frac{1}{j2}\left(e^{j\theta} - e^{-j\theta}\right)$$

Taking the real part as the physical solution, we get

$$x = a\frac{\sin(n\delta/2)}{\sin(\delta/2)}\cos\left[\omega t + (n-1)\delta/2\right]$$

Thus the resultant motion is simple harmonic having the amplitude $a\sin(n\delta/2)/\sin(\delta/2)$ and the phase angle $(n-1)\delta/2$.

[It may be mentioned that this result is useful in finding the diffraction pattern due to a plane diffraction grating in optics.]

Problem 11. *The potential energy of a diatomic molecule is*

$$E_p = -\frac{A}{x^6} + \frac{B}{x^{12}}$$

where x is the distance between the constituent atoms, and A and B are positive quantities. Determine the equilibrium distance x_0 and the time period of small oscillation. Assume that one atom is heavy and at rest.

Solution. At equilibrium, the potential energy E_p is a minimum. So,

$$\frac{dE_p}{dx} = \frac{6A}{x^7} - \frac{12B}{x^{13}} = 0,$$

$$\text{giving} \quad x = x_0 = \left(\frac{2B}{A}\right)^{1/6}.$$

The restoring force on the vibrating atom of mass m is $F = -\dfrac{dE_p}{dx} = -\dfrac{6A}{x^7} + \dfrac{12B}{x^{13}}$. Let $x = x_0 + x_1$, where $x_1 \ll x_0$. Hence

$$F = -\frac{6A}{x_0^7}\left(1 - \frac{7x_1}{x_0}\right) + \frac{12B}{x_0^{13}}\left(1 - \frac{13x_1}{x_0}\right)$$

expanding binomially and retaining terms up to the linear in x_1/x_0. Thus

$$F = \left(-\frac{12 \times 13}{x_0^{14}}B + \frac{6 \times 7}{x_0^8}A\right)x_1 = -36A\left(\frac{A}{2B}\right)^{4/3} x_1$$

The equation of motion is

$$m\frac{d^2x_1}{dt^2} = F$$

$$\text{or,} \quad m\frac{d^2x_1}{dt^2} = -36A\left(\frac{A}{2B}\right)^{4/3} x_1.$$

$$\text{or} \quad \frac{d^2x_1}{dt^2} + \omega^2 x_1 = 0$$

$$\text{where } \omega = \left[(36A/m)(A/2B)^{4/3}\right]^{1/2} = 3\left(\frac{4A^7}{m^3B^4}\right)^{1/6}.$$

Clearly, the motion is simple harmonic with the period

$$T = \frac{2\pi}{\omega} = \frac{2\pi}{3}\left(\frac{m^3B^4}{4A^7}\right)^{1/6}.$$

Problem 12. *The position vector of a particle of mass m is $\vec{r}(t) = (a\cos\omega t)\hat{i} + (a\sin\omega t)\hat{j}$, where 'a' is a constant, and $\hat{i}$ and $\hat{j}$ are unit vectors along the x-axis and y-axis, respectively. Show that the force acting on the particle is a no-work force.*

Solution. The x-component of $\vec{r}$ is $x = a\cos\omega t$, and the y-component is $y = a\sin\omega t$. The path described by the particle is the circle represented by

$$x^2 + y^2 = a^2 \tag{i}$$

The velocity of the particle is

$$\vec{v} = \dot{\vec{r}} = (-a\omega\sin\omega t)\hat{i} + (a\omega\cos\omega t)\hat{j}$$

The acceleration is

$$\vec{f} = \dot{\vec{v}} = -\omega^2\left[(a\cos\omega t)\hat{i} + (a\sin\omega t)\hat{j}\right] = -\omega^2\vec{r}$$

The force acting on the particle is

$$\vec{F} = m\vec{f} = -m\omega^2\vec{r} = -m\omega^2(\hat{i}x + \hat{j}y)$$

If $d\vec{r} = \hat{i}dx + \hat{j}dy$ is the displacement, the work done by the force is

$$\begin{aligned} dW &= \vec{F}\cdot d\vec{r} = -m\omega^2(\hat{i}x + \hat{j}y)\cdot(\hat{i}dx + \hat{j}dy) \\ &= -m\omega^2(x\,dx + y\,dy) \end{aligned} \tag{ii}$$

From (i), $x\,dx + y\,dy = 0$. Hence, from (ii), $dW = 0$. So, $\vec{F}$ is a no-work force.

Problem 13. *A spring stretches* $0.03\,\text{m}$ *when subjected to a force of* $6\,\text{N}$. *A* $0.5\,\text{kg}$ *mass is attached to the free end of the spring suspended vertically from a rigid support. The mass is given an initial displacement of* $0.015\,\text{m}$ *and an initial velocity of* $0.4\,\text{m/s}$. *Find the period and amplitude of the motion.*

[C.U. 2008]

Solution. A force $6\,\text{N}$ stretches the spring by $0.03\,\text{m}$. So,, the spring constant is $s = 6/0.03 = 200\,\text{N/m}$. The suspended mass $m = 0.05$ kg executes a simple harmonic motion with an angular velocity $\omega = \sqrt{s/m} = \sqrt{200/0.5} = 20$ rad/s. The time period is

$$T = 2\pi/\omega = 2 \times 3.1416/20 = 0.31416\,\text{s}$$

If A is the amplitude and α is the initial phase, the displacement x at time t is given by $x = A\sin(\omega t + \alpha)$. Given, $x = 0.015$ m at $t = 0$. Hence

$$A\sin\alpha = 0.015 \tag{i}$$

Also, the velocity at time t is

$$\frac{dx}{dt} = A\omega\cos(\omega t + \alpha).$$

Given, $\frac{dx}{dt} = 0.4$ m/s at $t = 0$. Therefore,

$$A\cos\alpha = \frac{0.4}{\omega} = \frac{0.4}{20} = 0.02 \quad \text{(ii)}$$

From Eqs. (i) and (ii), we get

$$A = \left[(0.015)^2 + (0.02)^2\right]^{1/2} = 0.025\,\text{m}$$

REVIEW QUESTIONS

1. Define simple harmonic motion and set up the differential equation for it. Solve the differential equation and give the different forms of the physical solution. Show that the solution is periodic in nature.

2. (a) How are simple harmonic motions related to rotating phasors?

   (b) Show that the restoring force on a simple harmonic oscillator is conservative.

3. Write the forms of the simple harmonic motion for the following boundary conditions: (i) at $t = 0$, $x = a$ and $\dot{x} = 0$, (ii) at $t = 0$, $x = 0$ and $\dot{x} = u$, (iii) at $t = \tau$, $x = x_\tau$ and $\dot{x} = \dot{x}_\tau$.

4. Derive the differential equation for the simple harmonic motion from the energy principle. Show that the average kinetic and potential energies of a particle in SHM are half the total energy.

5. Two simple harmonic motions act simultaneously on a particle at right angles to each other. Show that the path of the particle will be an ellipse when the two motions have the same period but different amplitudes and initial phases. What happens when the phase difference between the motions is (i) zero, (ii) π, and (iii) $\frac{\pi}{2}$?

6. Discuss the formation of Lissajous figures by the superposition of two simple harmonic motions when (i) the periods are the same but the amplitudes and phases are different; (ii) the periods are in the ratio 1 : 2, and the amplitudes and phases are different. **[C.U. 1983]**

7. What are Lissajous figures? How can you demonstrate them experimentally? Two simple harmonic motions of nearly the same frequency and acting at right angles to each other are superimposed. Show how the resulting pattern changes with time.

8. What is the relation between a simple harmonic motion and a uniform circular motion? How can you generate a uniform circular motion from two mutually perpendicular simple harmonic motions? Also show that a simple harmonic motion is equivalent to two opposite circular motions.

9. Consider a light horizontal plate of rectangular cross-section fixed to a rigid support at one end and a mass m placed on the other end. The free end is depressed and let go. Establish the differential equation of SHM of the mass and hence find an expression for the time period. **[C.U. 1989]**

10. What are beats? Give an analytical description of the phenomenon of beats and show that the beat frequency is equal to the difference between the frequencies of the component oscillations. **[C.U. 1996, 99, 2007, '08; urd. U. 1998, 2000]**

11. A particle is simultaneously under two simple harmonic motions at right angles to each other. The two motions are represented by $x = a\sin\omega t$ and $y = b\sin(\omega t + \delta)$. (i) Show that the resultant motion is an ellipse given by $\frac{x^2}{a^2} + \frac{y^2}{b^2} - \frac{2xy}{ab}\cos\delta = \sin^2\delta$. (ii) What is the locus of the particle when $\delta = \pi/2$ and $a = b$? **[C.U. 2002]**

12. A particle moving in the xy-plane has the displacement $x = \cos\omega t$ along the x-axis and the displacement $y = \cos\alpha\omega t$ along the y-axis. Show that the trajectory is repeated only if α is a rational number. **[C.U. 2004]**

PROBLEMS

1. A particle of mass 0.1g is projected from its equilibrium position with an initial velocity of 10 cm/s. It momentarily comes to rest in 1.5s. Write its equation of motion, and calculate the stiffness constant and the total energy.

[**Ans.** $x = \frac{30}{\pi}\sin\frac{\pi}{3}t$ cm, $0.011\pi^2$ dyne/cm, 5 erg]

2. A cylinder of mass M and cross sectional area A floats vertically in a liquid of density ρ. If it is depressed slightly and then released, find the frequency of vertical oscillations. **[C.U. 1980]**

[**Ans.** $\frac{1}{2\pi}\sqrt{\frac{A\rho g}{M}}$]

3. The equation of motion of a particle in the x-direction is $x = 4\sin^2(\omega t - \frac{\pi}{6})$ cm. Show that the particle is in SHM, and find the mean position, the frequency, and the amplitude of oscillation.

[**Ans.** 2 cm, $\frac{\omega}{\pi}$ Hz, 2 cm]

4. Two tuning forks of slightly different frequencies are used to produce Lissajous figures. A circular figure is found to appear in every 10s. If one tuning fork has the frequency 300 Hz, what is the frequency of the other?

[**Ans.** 300.1 Hz or 299.9 Hz]

5. A Lissajous figure is produced by superposing a vertical and a horizontal simple harmonic motion. The pattern has 4 intersections with the vertical and 6 intersections with the horizontal. If the horizontal frequency is 2 kHz, what is the vertical frequency?

[**Ans.** 3 kHz]

6. Two vibrations, at right angles to each other, are described by the equations $x = 3\cos 4\pi t$ and $y = 3\cos(8\pi t + \frac{\pi}{3})$, where x and y are expressed in centimetres and t in seconds. Construct the Lissajous figure of the combined motion. **[Burd. U. 1994]**

[**Ans.** The equation of the Lissajous figure containing two loops is $16x^4 - 24x^2y - 144x + 36y^2 + 108y + 81 = 0$]

7. Two simple harmonic motions act in the same direction simultaneously on a particle. The SHMs have the same frequency 5 Hz; amplitudes of 0.005 m and 0.002 m respectively; and the second component leads the first by 45°. Determine the amplitude of the resultant displacement and its phase with respect to the first component. Give the expression for the resultant displacement as a function of time. **[Burd. U. 1997]**

[**Ans.** 0.006568 m, 12.43°, $x = 0.006568\cos(10\pi t + 12.43°)$ m]

8. A particle of mass m executes a simple harmonic motion along x-axis with an angular frequency ω and an amplitude a. For the instantaneous displacement x, the particle has the instantaneous linear momentum p_x. Show that p_x versus x plot is an ellipse.

[HINTS. The total energy of the oscillator is $E = \frac{1}{2}m\omega^2a^2$. The instantaneous kinetic energy is $E_k = p_x^2/(2m)$, and the instantaneous potential energy is $E_p = \frac{1}{2}m\omega^2x^2$. Since $E_k + E_p = E$, we have $p_x^2/(m\omega a)^2 + x^2/a^2 = 1$. Hence the result.]

9. A particle vibrates harmonically with a frequency 0.25 Hz. At the initial moment, it is in the equilibrium position moving at a speed of 10 cm/s. Write its equation of motion. **[C.U. 2009]**

[**Ans.** $x = \frac{20}{\pi}\sin 0.5\pi t$ cm]

Chapter 2

Free and Forced Vibrations

2.1 Free Vibration

When vibrations are excited in a body and the body is left to itself after removing the exciting force, the body continues to oscillate with its own characteristic frequency. Such oscillations are known as *free* or *natural vibrations* of the body. The frequency of free vibrations, termed the *natural frequency*, depends on the mass, shape, size, and the elastic properties of the body. The reciprocal of the natural frequency is called the *free* or the *natural period* of vibration. Ideally, the body vibrates with its natural frequency for an indefinite time with a constant amplitude. In practice, however, the amplitude decays with time and finally the body comes to rest at its mean position. The decay of the vibrations is caused by the frictional resistance, known as *damping*. To overcome the frictional resistance, the vibrating body must expend energy continuously. Consequently, the energy of the body decreases, so that the amplitude of oscillations diminishes with time. Such vibrations of decaying amplitude are referred to as *resisted* or *damped vibrations*. The motion of a simple pendulum is an example of a *damped simple harmonic motion*, because the amplitude gradually decreases with time and the bob finally comes to rest. Here damping is caused by the friction of the air and at the point of suspension. The frequency of the damped oscillation is slightly less than the frequency of the undamped natural oscillations.

2.2 Analytical Treatment of Damped Vibrations

Let x be the instantaneous displacement of a particle of mass m performing a damped motion from its initial position of rest. The following forces act on the particle :

(a) The *force of restitution* or the *restoring force* which is proportional to the displacement x and tends to bring the particle back to its initial position. This force is $-sx$, where s is the restoring force per unit displacement and is called the *stiffness factor*.

(b) The *retarding* or the *resisting force* caused by friction. This force is proportional to the instantaneous velocity of the particle and is given by $-k\frac{dx}{dt}$, where k is the resisting force per unit velocity. The negative sign indicates that the force opposes the motion. The system is subjected to viscous friction here.

The net force acting on the particle is $-sx - k\frac{dx}{dt}$. This must be equated to mass $\times$ acceleration, i.e., $m\frac{d^2x}{dt^2}$, giving for the equation of motion

$$m\frac{d^2x}{dt^2} = -sx - k\frac{dx}{dt}$$

$$\text{or,} \qquad \frac{d^2x}{dt^2} + 2b\frac{dx}{dt} + \omega_0^2 x = 0, \tag{2.1}$$

where $2b = k/m$ and $\omega_0^2 = s/m$. The quantity $2b$ denotes the resisting force per unit mass per unit velocity and the quantity ω_0^2 represents the restoring force per unit mass per unit displacement. Note that $\omega_0 (= \sqrt{s/m})$ is the undamped angular frequency of the particle (see Chapter 1).

To solve the above second order differential equation [Eq. (2.1)], we take $x = e^{\alpha t}$ as a trial solution. Then $\frac{dx}{dt} = \alpha e^{\alpha t} = \alpha x$ and $\frac{d^2x}{dt^2} = \alpha^2 e^{\alpha t} = \alpha^2 x$. Substituting in Eq. (2.1) gives

$$\alpha^2 + 2b\alpha + \omega_0^2 = 0$$

$$\text{or,} \qquad \alpha = -b \pm \sqrt{b^2 - \omega_0^2}.$$

Thus, the general solution of Eq. (2.1) can be written as

$$\begin{aligned} x &= A_1 e^{(-b+\sqrt{b^2-\omega_0^2})t} + A_2 e^{(-b-\sqrt{b^2-\omega_0^2})t} \\ &= e^{-bt}\left(A_1 e^{\sqrt{b^2-\omega_0^2}\,t} + A_2 e^{-\sqrt{b^2-\omega_0^2}\,t}\right), \end{aligned} \tag{2.2}$$

where A_1 and A_2 are two arbitrary constants to be determined from initial conditions.

Case 1. Let $b > \omega_0$, i.e., damping is large. Suppose that x_0 is the displacement and $\dot{x}_0$ is the velocity at time $t = 0$. Putting $t = 0$, we obtain from Eq. (2.2)

$$x_0 = A_1 + A_2. \tag{2.3}$$

We have from Eq. (2.2)

$$\frac{dx}{dt} = -be^{-bt}\left(A_1 e^{b_1 t} + A_2 e^{-b_1 t}\right) + e^{-bt}\left(b_1 A_1 e^{b_1 t} - b_1 A_2 e^{-b_1 t}\right) \tag{2.4}$$

where $b_1 = \sqrt{b^2 - \omega_0^2}$. Putting $t = 0$ in Eq. (2.4), we get

$$\dot{x}_0 = -b(A_1 + A_2) + b_1(A_1 - A_2). \tag{2.5}$$

Equations (2.3) and (2.5) give

$$A_1 = \frac{x_0}{2}\left(1 + \frac{b + \dot{x}_0/x_0}{b_1}\right) \quad \text{and} \quad A_2 = \frac{x_0}{2}\left(1 - \frac{b + \dot{x}_0/x_0}{b_1}\right).$$

Substituting these in Eq. (2.2) and inserting the value of b_1, we obtain

$$x = \frac{x_0}{2}e^{-bt}\left[\left(1 + \frac{b + \dot{x}_0/x_0}{\sqrt{b^2 - \omega_0^2}}\right)e^{\sqrt{b^2-\omega_0^2}\,t} + \left(1 - \frac{b + \dot{x}_0/x_0}{\sqrt{b^2 - \omega_0^2}}\right)e^{-\sqrt{b^2-\omega_0^2}\,t}\right]. \tag{2.6}$$

The variation of x with t is depicted by curve 1 in Fig. 2.1. The motion of the particle is clearly *nonoscillatory* or *aperiodic*. The displacement falls off and asymptotically approaches zero with increasing time. This is the *overdamped motion* and is exhibited by a pendulum moving in a highly viscous liquid or by an aperiodic moving coil galvanometer.

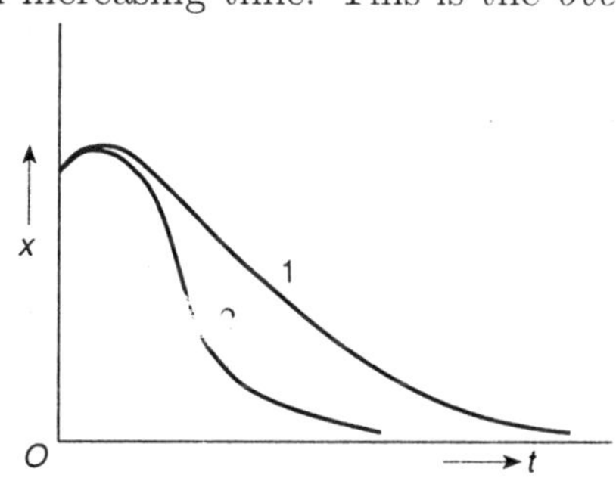

Fig. 2.1 Overdamped (curve 1) and critically damped (curve 2) motions

Case 2. Let $b = \omega_0$. To obtain the solution in this case, we let $b \to \omega_0$ in Eq. (2.6). Thus, we take $\sqrt{b^2 - \omega_0^2} = \delta$ in Eq. (2.6), where δ is a small quantity. Expanding and neglecting the terms higher than the linear in δt, Eq. (2.6) gives

$$x = \frac{x_0}{2}e^{-bt}\left[\left(1 + \frac{b + \dot{x}_0/x_0}{\delta}\right)(1 + \delta t) + \left(1 - \frac{b + \dot{x}_0/x_0}{\delta}\right)(1 - \delta t)\right]$$
$$= \frac{x_0}{2}e^{-bt}\left[2 + 2\left(b + \frac{\dot{x}_0}{x_0}\right)t\right] = x_0 e^{-bt}\left[1 + \left(b + \frac{\dot{x}_0}{x_0}\right)t\right], \tag{2.7}$$

which is the desired solution. Equation (2.7) shows that here also the motion is aperiodic, but x approaches zero quicker than the overdamped case. The motion is now said to be *critically damped*, the variation of x with t being displayed by curve 2 in Fig. 2.1.

If an initial velocity $\dot{x}_0$ is imparted to the particle while at rest ($x_0 = 0$), Eq. (2.7) reduces to $x = \dot{x}_0 te^{-bt}$. Then $\frac{dx}{dt} = \dot{x}_0 e^{-bt}(1 - bt)$. The particle momentarily comes to rest when $\frac{dx}{dt} = 0$ or, $t = \frac{1}{b} = \frac{1}{\omega_0}$, which is independent of initial velocity. At this stage, the particle attains the maximum displacement which is $x_m = \dot{x}_0/(e\omega_0)$. Note that x_m depends on the initial velocity $\dot{x}_0$.

Equation (2.7) can be *alternatively* derived as follows :

Since $b = \omega_0$, Eq. (2.2) gives $x = (A_1 + A_2)e^{-bt} = Ae^{-bt}$, which contains only one arbitrary constant A and is thus not a general solution. So, let $x = e^{-bt}f(t)$ be the general solution of Eq.(2.1) when $b = \omega_0$. Substitution in Eq. (2.1) gives

$$e^{-bt}\frac{d^2f}{dt^2} = 0.$$

As $e^{-bt} \neq 0$, $\frac{d^2f}{dt^2} = 0$, giving $f = A_1 + A_2t$, where A_1 and A_2 are constants to be determined from the initial conditions. The solution is now

$$x = (A_1 + A_2t)e^{-bt}.$$

As $x = x_0$, at $t = 0$, we have $x_0 = A_1$, so that

$$x = (x_0 + A_2t)e^{-bt} \tag{2.7A}$$

Also,
$$\frac{dx}{dt} = [-bx_0 + A_2(1 - bt)]\, e^{-bt}.$$

If, at $t = 0$, $\frac{dx}{dt} = \dot{x}_0$, we get

$$\dot{x}_0 = -bx_0 + A_2$$

whence
$$A_2 = bx_0 + \dot{x}_0$$

Substituting for A_2 in Eq. (2.7A) we obtain

$$x = x_0e^{-bt}\left[1 + \left(b + \frac{\dot{x}_0}{x_0}\right)t\right],$$

which agrees with Eq. (2.7).

Case 3. Let $b < \omega_0$, i.e., the damping is small. Now, $\sqrt{b^2 - \omega_0^2} = j\sqrt{\omega_0^2 - b^2}$, where $j = \sqrt{-1}$. Hence Eq. (2.2) reduces to

$$\begin{aligned} x &= e^{-bt}\left(A_1e^{j\sqrt{\omega_0^2-b^2}t} + A_2e^{-j\sqrt{\omega_0^2-b^2}t}\right) \\ &= e^{-bt}\left[(A_1 + A_2)\cos\sqrt{\omega_0^2 - b^2}\,t + j(A_1 - A_2)\sin\sqrt{\omega_0^2 - b^2}\,t\right]. \end{aligned} \tag{2.8}$$

The constants A_1 and A_2 are, in general, complex numbers and either the real or the imaginary part of Eq. (2.8) is the physical solution. Both parts can be written as

$$x = e^{-bt}\left(A\cos\sqrt{\omega_0^2 - b^2}\,t + B\sin\sqrt{\omega_0^2 - b^2}\,t\right), \tag{2.9}$$

where A and B are arbitrary real constants to be determined from the initial conditions. Let the initial displacement of the particle be x_0 and the initial velocity be $\dot{x}_0$. Putting $x = x_0$ at $t = 0$ in Eq. (2.9), we get $x_0 = A$. Taking $p = \sqrt{\omega_0^2 - b^2}$ for brevity, we obtain from Eq. (2.9)

$$\frac{dx}{dt} = -be^{-bt}(A\cos pt + B\sin pt) + pe^{-bt}(-A\sin pt + B\cos pt).$$

Since $\frac{dx}{dt} = \dot{x}_0$ at $t = 0$, we have

$$\dot{x}_0 = -bA + pB = -bx_0 + pB$$

$$\text{or,} \quad B = \frac{\dot{x}_0 + bx_0}{p}.$$

Substituting the values of A and B, we obtain from Eq. (2.9)

$$x = x_0 e^{-bt}\left(\cos\sqrt{\omega_0^2 - b^2}\,t + \frac{b + \dot{x}_0/x_0}{\sqrt{\omega_0^2 - b^2}}\sin\sqrt{\omega_0^2 - b^2}\,t\right) \tag{2.10}$$

$$= C\,e^{-bt}\cos\left(\sqrt{\omega_0^2 - b^2}\,t - \theta\right), \tag{2.11}$$

where $C\cos\theta = x_0$ and $C\sin\theta = \frac{bx_0 + \dot{x}_0}{\sqrt{\omega_0^2 - b^2}}$, so that

$$C = \left(\frac{x_0^2\omega_0^2 + \dot{x}_0^2 + 2bx_0\dot{x}_0}{\omega_0^2 - b^2}\right)^{1/2} \tag{2.12}$$

$$\text{and} \quad \tan\theta = \frac{b + \dot{x}_0/x_0}{\sqrt{\omega_0^2 - b^2}}. \tag{2.13}$$

If the particle is initially displaced to x_0 and then released, $\dot{x}_0 = 0$. Then Eq. (2.11) reduces to

$$x = \frac{x_0\omega_0}{\sqrt{\omega_0^2 - b^2}}e^{-bt}\cos\left(\sqrt{\omega_0^2 - b^2}\,t - \theta'\right), \tag{2.14}$$

where

$$\tan\theta' = \frac{b}{\sqrt{\omega_0^2 - b^2}}. \tag{2.15}$$

If instead, the particle is given an initial velocity $\dot{x}_0$ while at rest, then $x_0 = 0$ and Eq. (2.11) becomes

$$x = \frac{\dot{x}_0}{\sqrt{\omega_0^2 - b^2}}e^{-bt}\sin\sqrt{\omega_0^2 - b^2}\,t. \tag{2.16}$$

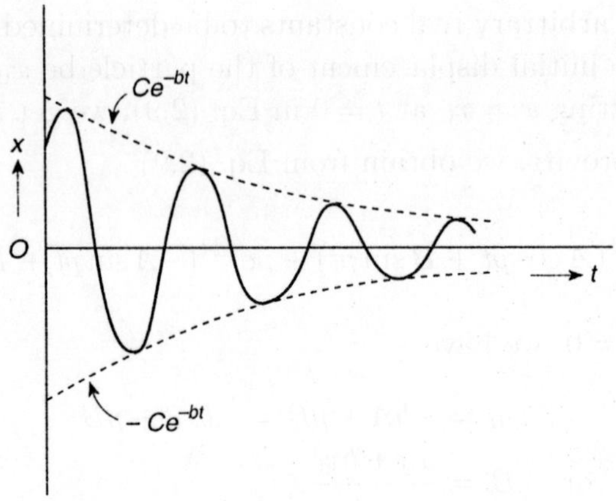

Fig. 2.2 Underdamped oscillatory motion

Equation (2.11) shows that the motion is oscillatory with an angular frequency $\omega = \sqrt{\omega_0^2 - b^2}$ and an amplitude Ce^{-bt}. Since the amplitude decays exponentially with time, the motion is *damped oscillatory.* This is the *underdamped* case, the variation of x with time being displayed in Fig. 2.2. The rate of decrease of the amplitude is determined by b, called the *decay constant.* The free *undamped* motion, obtained by setting $b = 0$ in Eq. (2.11), is

$$x = C\cos(\omega_0 t - \theta).$$

This is a simple harmonic motion with a constant amplitude. When damping is present, the amplitude is no longer a constant but decays exponentially with time. Also, the damping reduces the angular frequency from ω_0 to $\omega(= \sqrt{\omega_0^2 - b^2})$. However, if $b \ll \omega_0$, the reduction of the frequency is negligibly small.

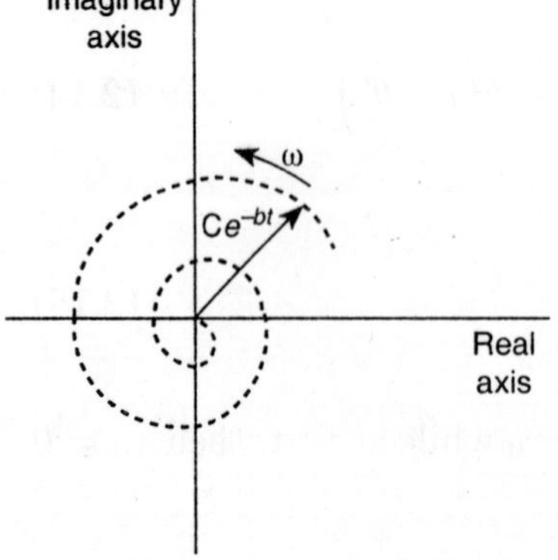

Fig. 2.3 Rotating phasor for underdamped oscillatory motion

The underdamped oscillatory motion is represented by a phasor of amplitude Ce^{-bt} rotating with the angular velocity ω (Fig. 2.3). As time passes, the length of the phasor decreases, so that its tip describes a *shrinking spiral.* The projection of the rotating phasor on the real (or the imaginary) axis gives the desired motion.

A pendulum having a massive bob oscillating in air gives an example of the underdamped oscillatory motion. The damping is increased if the bob is placed in a liquid.

The time in which the amplitude decreases to $1/e$ of its initial value is called the *time constant* or the *modulus of decay* or the *relaxation time* of the damped motion. If the initial amplitude is a and τ is the time constant, then $ae^{-b\tau} = a/e$ or, $b\tau = 1$. Thus, $\tau = 1/b = 2m/k$. This shows that the time constant is larger for a heavier mass and a smaller damping. The time constant shows the rapidity with which the motion is damped out by the resistive force.

When the motion is under damped oscillatory, the amplitude falls off with time. Let $x_1, x_2, \cdots, x_n$ be the successive maximum displacements on the opposite sides of the mean position. Then,

$$\frac{x_1}{x_2} = \frac{x_2}{x_3} = \frac{x_3}{x_4} = \cdots = e^{bT/2} = x \text{ (say)} \tag{2.17}$$

where T is the time period. The quantity x is termed the *decrement*. The *logarithmic decrement* is defined by

$$\lambda = \ln x = \frac{bT}{2}. \tag{2.18}$$

Experimentally one can find T and λ. Then b can be determined from Eq. (2.18). From the knowledge of b and T, one can find the undamped natural angular frequency ω_0 from the relationship $T = 2\pi/\sqrt{\omega_0^2 - b^2}$. Again $\omega_0 = \sqrt{s/m}$ and $b = k/2m$. If m is known, we can at once determine s and k from the values of ω_0 and b.

Note that the angular frequency ω has the unit rad/s whereas the decay constant b has the unit neper/s.

2.3 Energy Relations

Let x be the displacement of a particle of mass m executing a damped motion at time t. The potential energy E_p of the particle at this instant is the work done against the force of restitution. Thus,

$$E_p = \int_0^x sx\, dx = \frac{1}{2}sx^2.$$

The instantaneous kinetic energy is

$$E_k = \frac{1}{2}m\left(\frac{dx}{dt}\right)^2.$$

The sum of the kinetic and potential energies is

$$E = E_k + E_p = \frac{1}{2}m\left(\frac{dx}{dt}\right)^2 + \frac{1}{2}sx^2. \tag{2.19}$$

The instantaneous rate of dissipation of the energy E due to the resisting force $k\frac{dx}{dt}$ is the product of this force and the velocity. Thus,

$$-\frac{dE}{dt} = k\left(\frac{dx}{dt}\right)^2. \tag{2.20}$$

Equation (2.19) gives $\frac{dE}{dt} = m\frac{dx}{dt}\cdot\frac{d^2x}{dt^2} + sx\frac{dx}{dt}$.

Putting this in Eq. (2.20), we obtain

$$m\frac{d^2x}{dt^2} + k\frac{dx}{dt} + sx = 0,$$

which is the differential equation of the damped motion, obtained here from the energy principle.

Starting from the maximum displacement x_1 on one side of the swing, the particle after the time period T has the maximum displacement x_2 (say) on the same side. At x_1 and x_2, the particle is momentarily at rest and its energy is entirely potential. The loss of energy in one time period is thus

$$\Delta E = \frac{1}{2}s(x_1^2 - x_2^2) = \frac{1}{2}sx_1^2\left(1 - \frac{x_2^2}{x_1^2}\right) = E(1 - e^{-4\lambda}),$$

where $E\left(= \frac{1}{2}sx_1^2\right)$ is the energy at the beginning of the cycle and λ is the logarithmic decrement. The average fractional loss of energy per cycle is

$$\frac{\Delta E}{E} = 1 - e^{-4\lambda} \simeq 4\lambda, \tag{2.21}$$

if λ is small.

The efficiency of a damped oscillator to store energy is measured by a *figure of merit* or a *quality factor*, Q. It is defined as

$$Q = 2\pi \times \frac{\text{maximum energy stored per cycle}}{\text{energy lost or dissipated per cycle}} = 2\pi \times \frac{E}{\Delta E} = \frac{\pi}{2\lambda} \tag{2.21A}$$

with the aid of Eq. (2.21). As $\lambda = bT/2$ by Eq. (2.18), and $T = 2\pi/\omega_0$ for small damping, Eq. (2.21A) gives

$$Q = \frac{\omega_0}{2b} = \frac{\omega_0 m}{k} = \frac{\omega_0 \tau}{2} \tag{2.21B}$$

The higher the Q, the less the damping, and the higher the efficiency of the oscillator in storing energy.

Observation

We have seen that the restoring force on a harmonic oscillator is conservative. This force is derivable from the potential energy of the oscillator : $F = -sx = -\frac{dE_p}{dx}$. The work done by this force over a full cycle is zero, because during the time interval when the oscillator is moving towards the equilibrium position, work is done by the force; and during an equal time interval when the oscillator is moving away from its equilibrium position, an equal amount of work is done against the force. The total energy of the oscillator remains constant, and the undamped oscillator is said to be a *conservative system.*

For a damped oscillator, the viscous friction comes into play when the system is in motion, and *always* acts opposite to the velocity. Work has to be always done *against* the resisting force so that this work done over a full cycle is not zero. Energy must be drained off the oscillator to perform this work which is dissipated as heat. The energy of a damped oscillator therefore does not remain constant; rather it decays with time. The damped oscillator is hence a *nonconservative system.*

2.4 Forced Vibration and Resonance

A vibrating system gradually loses its amplitude since energy is dissipated due to frictional forces. To maintain the system in vibration, energy must be supplied from outside. If an external periodic force is applied to the vibrating system, the system tends to vibrate with its own natural frequency. But the applied driving force tries to impress its own frequency on the vibrating system. Initially, the system vibrates with both the frequencies. The natural vibrations die out in course of time due to the prevailing resisting forces and the system finally in the steady state, vibrates with the frequency of the driving force with a constant amplitude. Such vibration where the system oscillates with a frequency the same as that of an externally impressed periodic force is known as the *forced vibration* of the system.

Resonance is a particular case of forced vibration when the frequency of the driving periodic force equals the natural frequency of the system. Here the system oscillates with a considerably large amplitude. In the state of steady oscillations, the energy supplied by the external source fully replenishes the energy drained away from the system in overcoming the frictional forces. Resonant vibrations are also called *sympathetic vibrations,* and the system is said to *resonate* with the impressed periodic force.

Usually, a feeble sound is heard from a vibrating tuning fork. But if the stem of the vibrating fork is pressed on a wooden table, the sound becomes considerably louder. The vibrating fork induces a forced vibration of the table which, in turn, excites the vibrations of a large volume of the surrounding air at a frequency equal to that of the tuning fork. Thus the sound becomes louder. This, however, does not violate the principle of conservation of energy. The fork vibrates for a longer time if it is not pressed on the table. When it is placed in contact with the table, its energy is drained very quickly to set the table and the large volume of air into vibration. Thus, though the sound becomes louder, it lasts for a shorter time, keeping the total energy constant.

Hollow wooden enclosures are used in stringed musical instruments. The vibrations of the strings cause forced vibration of the enclosure and the air contained in it. This intensifies the emitted sound.

Some musical instruments like esraj, sitar etc. have, in addition to the main string, a few side strings tuned to different notes. When a note is played on the main string, a side string tuned to that particular note is excited to resonance. The intensity and the sweetness of the emitted note are thus increased.

2.5 Analytical Treatment of Forced Vibration

Let a particle of mass m, capable of executing a damped simple harmonic motion, be subjected to an external simple harmonic force of constant amplitude and frequency. Suppose that x is the displacement of the particle from its mean position at time t. The forces acting on the particle are the following:

(i) The force of restitution or the restoring force $-sx$ tending to bring the particle back to its mean position, s being the *stiffness factor*.

(ii) The retarding or the resisting force $-k\frac{dx}{dt}$, k being the resisting force per unit velocity.

(iii) The driving periodic force $F\cos\omega t$, where F is the amplitude and ω is the angular frequency of the force.

The net force in the positive x-direction is $F\cos\omega t - sx - k\frac{dx}{dt}$. This force must be equated to mass × acceleration, i.e., $m\frac{d^2x}{dt^2}$. Hence the equation of motion of the particle is

$$m\frac{d^2x}{dt^2} = F\cos\omega t - sx - k\frac{dx}{dt}$$

$$\text{or,}\quad \frac{d^2x}{dt^2} + 2b\frac{dx}{dt} + \omega_0^2 x = \frac{F}{m}\cos\omega t, \tag{2.22}$$

where $2b = k/m$, and $\omega_0^2 = s/m$. The quantity ω_0 is the *undamped natural angular frequency* of the particle and b is the *decay constant*.

The general solution of the inhomogeneous second order differential equation [Eq. (2.22)] contains a particular solution, called the *particular integral*, which does not have any arbitrary constant and a *complementary function*, which is the general solution of the homogeneous equation obtained by setting the driving force to zero. The complementary function contains two arbitrary constants which are determined by the initial conditions.

When the natural motion of the particle is oscillatory, the complementary function can be written as [see Eq. (2.11)]

$$x_1 = Ce^{-bt}\cos\left(\sqrt{\omega_0^2 - b^2}\,t - \theta\right), \tag{2.23}$$

where C and θ are two arbitrary constants.

To find the particular integral, we suppose that x_r is the displacement for the driving force $F\cos\omega t$. If, on the other hand, the driving force is $F\sin\omega t$, let x_i be the corresponding displacement. The equations of motions for the two cases are respectively,

$$\frac{d^2x_r}{dt^2} + 2b\frac{dx_r}{dt} + \omega_0^2 x_r = \frac{F}{m}\cos\omega t \tag{2.24}$$

and

$$\frac{d^2x_i}{dt^2} + 2b\frac{dx_i}{dt} + \omega_0^2 x_i = \frac{F}{m}\sin\omega t. \tag{2.25}$$

Multiplying Eq. (2.25) by $j(=\sqrt{-1})$ and adding to Eq. (2.24) we obtain

$$\frac{d^2S}{dt^2} + 2b\frac{dS}{dt} + \omega_0^2 S = \frac{F}{m}e^{j\omega t}, \tag{2.26}$$

where we have put $S = x_r + jx_i$ and have used Euler's relationship $e^{j\omega t} = \cos\omega t + j\sin\omega t$.

Since ultimately the system yields to the driving force and oscillates with its frequency, the particular integral for Eq. (2.26) is taken as $S = S_0 e^{j\omega t}$, where S_0 is a complex quantity. Then $\frac{dS}{dt} = j\omega S_0 e^{j\omega t}$ and $\frac{d^2S}{dt^2} = -\omega^2 S_0 e^{j\omega t}$. Substituting these results in Eq. (2.26) gives

$$S_0 = \frac{F/m}{(\omega_0^2 - \omega^2) + j2\omega b} = \frac{(F/m)e^{-j\phi}}{\left[(\omega_0^2 - \omega^2)^2 + 4b^2\omega^2\right]^{1/2}}, \tag{2.27}$$

where

$$\tan\phi = \frac{2\omega b}{\omega_0^2 - \omega^2}. \tag{2.28}$$

Hence

$$S = S_0 e^{j\omega t} = \frac{(F/m)e^{j(\omega t - \phi)}}{\left[(\omega_0^2 - \omega^2)^2 + 4b^2\omega^2\right]^{1/2}}. \tag{2.29}$$

The real part x_r of S given by Eq. (2.29) represents the solution when the driving force is $F \cos \omega t$, i.e., the real part of the excitation $Fe^{j\omega t}$. The imaginary part x_i gives the solution when the driving force is $F \sin \omega t$, i.e., the imaginary part of $Fe^{j\omega t}$. Note that although for mathematical convenience, complex numbers are used, they do not give the physical solutions.

The physical excitations are $F \cos \omega t$ or $F \sin \omega t$, and the physical solutions are extracted from the complex solution by taking its real or imaginary part. Thus for the driving force $F \cos \omega t$, the particular integral is

$$x_r = \frac{(F/m)\cos(\omega t - \phi)}{\left[(\omega_0^2 - \omega^2)^2 + 4b^2\omega^2\right]^{1/2}} = \frac{F\cos(\omega t - \phi)}{\omega|Z_m|} \tag{2.30}$$

where

$$|Z_m| = \left[\left(\frac{m}{\omega}\omega_0^2 - m\omega\right)^2 + 4b^2m^2\right]^{1/2} = \left[k^2 + \left(\omega m - \frac{s}{\omega}\right)^2\right]^{1/2}. \tag{2.31}$$

The complete solution is obtained by adding the complementary function [Eq. (2.23)] and the particular integral [Eq. (2.30)]; thus

$$x = x_1 + x_r = Ce^{-bt}\cos\left(\sqrt{\omega_0^2 - b^2}\, t - \theta\right) + \frac{F}{\omega|Z_m|}\cos(\omega t - \phi), \tag{2.32}$$

where the constants C and θ are determined from the initial conditions, i.e., the values of the displacement x and the velocity dx/dt at time $t = 0$.

The first term on the right hand side of Eq. (2.32) gives the natural damped simple harmonic motion of a particle. The second term gives the forced vibration, i.e., the response of the particle to the exciting force. Initially, both the terms are operative and their superposition gives an irregular motion. After a few time constants, the natural vibration dies out since its amplitude falls off exponentially with time with the decay constant b. Thus, finally only the forced vibration given by the second term persists. This is the *steady state motion.* The part of the complete solution other than the steady state solution is called the *transient solution.* So, the first term in Eq. (2.32) represents the transient solution. Note that in the steady state, the particle oscillates simple harmonically with a constant amplitude $F/(\omega Z_m)$ with the period of the impressed force, but lags the impressed force by the angle ϕ.

If the driving angular frequency ω and the natural damped angular frequency $\sqrt{\omega_0^2 - b^2}$ are nearly equal, their superposition will produce beats when transient persists. Such beats, called the *transient beats*, disappear in a few time constants. The smaller the decay constant b, the longer the life of the beats.

The transients appear when the excitation is impressed or removed. Thus, in musical instruments, they occur every time a note is started or stopped. The transients are important in the fidelity response of microphones and loudspeakers, and also in the reverberation of sound in halls. In most cases, the transients have undesirable effects.

2.6 Mechanical and Electrical Analogues

The equation of the forced vibration of a mechanical system in complex form is

$$m\frac{d^2S}{dt^2} + k\frac{dS}{dt} + sS = Fe^{j\omega t}. \tag{2.33}$$

The emf equation of a series LCR ac circuit in complex form is

$$L\frac{d^2q}{dt^2} + R\frac{dq}{dt} + \frac{q}{C} = Ve^{j\omega t}. \tag{2.34}$$

Equations (2.33) and (2.34) are similar, allowing analogies between the mechanical and the electrical quantities. It is readily seen that the mass m is analogous to the self inductance L, the resistive force per unit velocity k to the electrical resistance R, the compliance $1/s$ to the electrical capacitance C, the force F to the voltage V, the displacement S to the charge q, and the velocity $\frac{dS}{dt}$ to the electric current $\frac{dq}{dt}$. From this analogy, k is termed the *mechanical resistance*.

For the ac circuit, the steady-state current is given by

$$i = \frac{Ve^{j\omega t}}{Z} \tag{2.35}$$

where

$$Z = R + j\left(\omega L - \frac{1}{\omega C}\right). \tag{2.36}$$

The quantity Z is called the *impedance* of the circuit. It has the magnitude $|Z| = \left[R^2 + \left(\omega L - \frac{1}{\omega C}\right)^2\right]^{1/2}$ and the phase angle ϕ, where $\tan\phi = \left(\omega L - \frac{1}{\omega C}\right)/R$. The quantity $X = \left(\omega L - \frac{1}{\omega C}\right)$ is called the *reactance*; ωL is the *inductive reactance* and $\frac{1}{\omega C}$ is the *capacitive reactance*. The physical current lags the driving voltage by the angle ϕ.

For the mechanical system, Eq. (2.29) gives for the steady-state displacement

$$S = \frac{Fe^{j\omega t}}{m\left[(\omega_0^2 - \omega^2) + j2\omega b\right]} = \frac{Fe^{j\omega t}}{(s - m\omega^2) + j\omega k}.$$

The steady-state velocity is

$$v = \frac{dS}{dt} = \frac{j\omega F e^{j\omega t}}{(s - m\omega^2) + j\omega k} = \frac{F e^{j\omega t}}{k + j\left(\omega m - \frac{s}{\omega}\right)}.$$

$$= \frac{F e^{j\omega t}}{Z_m}, \tag{2.37}$$

where

$$Z_m = k + j\left(\omega m - \frac{s}{\omega}\right). \tag{2.38}$$

In analogy with the electrical case, Z_m is called the *mechanical impedance*. Its magnitude is $|Z_m| = \left[k^2 + \left(\omega m - \frac{s}{\omega}\right)^2\right]^{1/2}$ and the phase angle is α, where $\tan\alpha = \left(\omega m - \frac{s}{\omega}\right)/k$. The physical velocity lags the driving force by α. The quantity $X_m = \omega m - \frac{s}{\omega}$ is known as the *mechanical reactance*; ωm is the *inertial reactance* and s/ω is the *reactance of compliance*, in analogy with the electrical quantities.

From Eq. (2.37) the mechanical impedance can be defined as the ratio of the driving force to the velocity of the driven system. Its complex nature implies that the force and the velocity are not in phase. The magnitude of the mechanical impedance is the ratio of the maximum driving force to the resulting maximum velocity.

The mechanical impedance is often expressed in *mechanical ohm*. It has the dimensions force/velocity, i.e., mass/time. In the CGS system, its unit is g/s and its SI unit is kg/s.

For ready reference, the mechanical quantities and their electrical analogues are given in Table 2.1.

TABLE 2.1

Analogous mechanical and electrical quantities

Mechanical quantities	*Electrical analogues*
force (F)	voltage (V)
displacement (S)	charge (q)
velocity (v)	current (i)
mass (m)	inductance (L)
mechanical resistance (k)	resistance (R)
compliance ($1/s$)	capacitance (C)
mechanical impedance (Z_m)	impedance (Z)
mechanical reactance (X_m)	reactance (X)
inertial reactance (ωm)	inductive reactance (ωL)
reactance of compliance (s/ω)	capacitive reactance ($1/\omega C$)

2.7 Amplitude and Velocity Resonances

In forced vibration, the steady-state displacement and velocity amplitudes of the driven system depend on the frequency of the driving force. When the displacement amplitude is a maximum for some frequency of the driver, we have the phenomenon of *amplitude (or displacement) resonance.* On the other hand, when the velocity amplitude attains a maximum for a certain frequency of the driving force, we have *velocity resonance.*

(A) *Amplitude (or displacement) resonance*: Equation (2.30) shows that the steady-state displacement amplitude of forced vibration is

$$A = \frac{F}{\left[\omega^2 k^2 + (\omega^2 m - s)^2\right]^{1/2}}. \tag{2.39}$$

The amplitude A is a maximum for that value of ω for which the denominator on the right-hand side of Eq. (2.39) is a minimum. Therefore, for this value of the angular frequency ω of the forcing function we have

$$\frac{d}{d\omega}\left[\omega^2 k^2 + (\omega^2 m - s)^2\right] = 0$$

or, $$2\omega k^2 + 4(\omega^2 m - s)\omega m = 0$$

or, $$k^2 + 2(\omega^2 m - s)m = 0$$

or, $$\omega^2 = \frac{s}{m} - \frac{k^2}{2m^2}. \tag{2.40}$$

Since $\omega_0^2 = s/m$, where ω_0 is the undamped natural angular frequency of the forced system and $2b = k/m$, where b is the decay constant, Eq. (2.40) reduces to

$$\omega^2 = \omega_0^2 - 2b^2. \tag{2.41}$$

Substituting for ω^2 from Eq. (2.40) into Eq. (2.39) gives for the maximum displacement amplitude

$$A_m = \frac{F}{k\left(\frac{s}{m} - \frac{k^2}{4m^2}\right)^{1/2}} = \frac{F/m}{2b(\omega_0^2 - b^2)^{1/2}}. \tag{2.42}$$

Thus, amplitude resonance occurs for the angular frequency of the forcing system given by Eq. (2.41), the displacement amplitude at amplitude resonance being expressed by Eq. (2.42). Note that the angular frequency for amplitude resonance $\sqrt{\omega_0^2 - 2b^2}$ is less than both ω_0, the undamped natural angular frequency, and $\sqrt{\omega_0^2 - b^2}$, the angular frequency for the damped vibration of the forced system.

When the amplitude is a maximum, the potential energy of the driven system $\frac{1}{2}sx^2$ is a maximum. Thus, the amplitude resonance occurs simultaneously with the *maximum potential energy* of the driven system with respect to the variation of frequency of the driver.

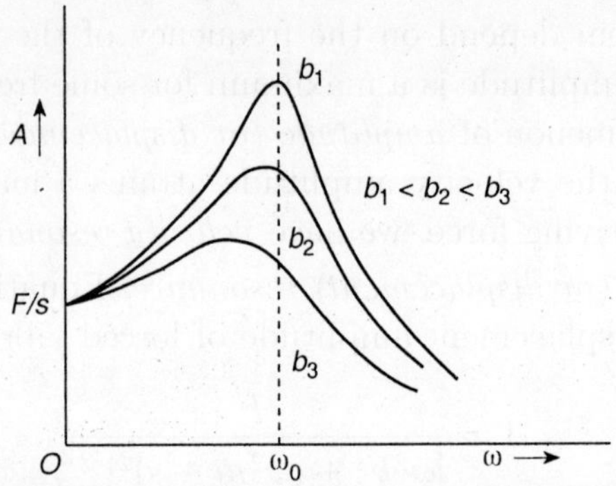

Fig. 2.4 Variation of the displacement amplitude with the angular frequency of the driver for different values of the decay constant

The variation of the displacement amplitude with the angular frequency ω of the forcing system for three different values of the decay constant, viz., b_1, b_2 and b_3, is shown in Fig. 2.4. At $\omega = 0$, we have $A = F/s$ for all values of damping. As damping decreases, the angular frequency at which A is a maximum moves towards the undamped angular frequency ω_0. Also, the maximum value of A increases with diminishing damping. In the absence of damping ($b = 0$), the amplitude A becomes infinite at $\omega = \omega_0$.

(B) *Velocity resonance*: If the forcing function is $F \cos \omega t$, the steady-state velocity for the forced vibration is obtained by taking the real part of Eq. (2.37). Thus, the physical velocity is

$$v = \frac{F\cos(\omega t - \alpha)}{\sqrt{k^2 + \left(\omega m - \frac{s}{\omega}\right)^2}}. \tag{2.43}$$

The velocity amplitude is

$$V = \frac{F}{\left[k^2 + \left(\omega m - \frac{s}{\omega}\right)^2\right]^{1/2}}. \tag{2.44}$$

As ω is varied, V attains its maximum value

$$V_m = \frac{F}{k}, \tag{2.45}$$

when $\omega m - s/\omega = 0$, i.e., when the mechanical reactance is zero. Hence, velocity resonance occurs when $\omega m = s/\omega$

$$\text{or,} \quad \omega^2 = \frac{s}{m} = \omega_0^2$$

$$\text{or,} \quad \omega = \omega_0, \tag{2.46}$$

i.e., when the forcing frequency equals the undamped natural frequency of the forced system. The kinetic energy of the forced system being $\frac{1}{2}mv^2$, the velocity resonance occurs simultaneously with the *maximum kinetic energy* of the driven system as the frequency of the driver is varied. Hence the velocity resonance is sometimes referred to as the *energy resonance*. The maximum kinetic energy is

$$E_m = \frac{1}{2}mV_m^2 = \frac{mF^2}{2k^2}. \tag{2.47}$$

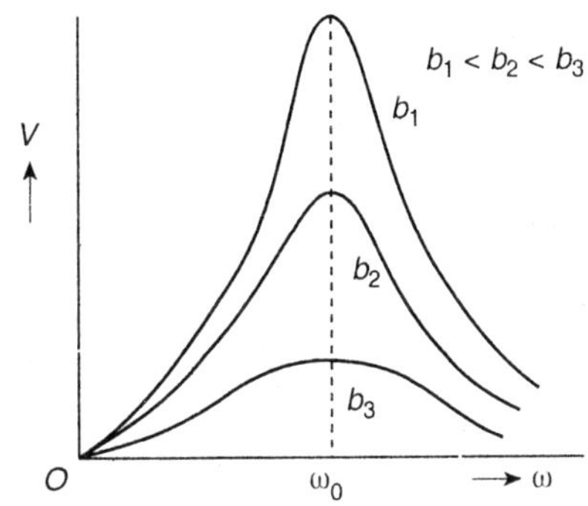

Fig. 2.5 Plot of the velocity amplitude of the forcing system versus angular frequency of the driver for different values of the decay constant

Figure 2.5 gives the variation of the velocity amplitude with the angular frequency of the driver for three different values of the decay constant, namely, b_1, b_2, and b_3. At $\omega = 0$, $V = 0$. The maximum value of V occurs at $\omega = \omega_0$ for all values of b. The maximum velocity amplitude increases as k, i.e., b decreases. For zero damping, the velocity amplitude is infinite at $\omega = \omega_0$.

Observations

(1) In a series LCR circuit, resonance is said to occur when the reactance vanishes. From this analogy, the *velocity resonance* for which the mechanical reactance disappears is referred to simply as *resonance*. Thus, ordinarily, resonance implies velocity resonance and one says that *resonance occurs at that frequency of the driver for which the mechanical reactance is zero*. The mechanical impedance at resonance is purely resistive, being equal to k.

The amplitude (or displacement) resonance is analogous to the capacitor charge resonance in a series LCR circuit. When damping is very small, both the amplitude and the velocity resonances occur at practically the same frequency which is the natural frequency of the system.

(2) It is not desirable for a group of soldiers to march on a bridge; for, their concerted footsteps will then cause forced vibrations of the bridge. If

the frequency of the rhythmic footsteps of the soldiers coincides with the natural frequency of vibration of the bridge, the bridge vibrates vigorously due to resonance, and hence may collapse. That is why *soldiers are asked to break steps while crossing a bridge.* Also, the bridge is provided with sufficient damping for safety's sake.

2.8 Power Relations in Forced Vibration and Resonance

The power of the driver is the rate at which it does work. In the steady-state, the instantaneous power of the driver in forced vibration is

$$P = (F\cos\omega t)\frac{dx}{dt} = \frac{F^2}{|Z_m|}\cos\omega t\cos(\omega t - \alpha), \tag{2.48}$$

where the expression for $v\left(=\frac{dx}{dt}\right)$, given by Eq. (2.43), is used. Hence

$$\begin{aligned} P &= \frac{F^2}{|Z_m|}\left(\cos^2\omega t\cos\alpha + \sin\omega t\cos\omega t\sin\alpha\right) \\ &= \frac{F^2}{|Z_m|}\left(\cos^2\omega t\cos\alpha + \frac{1}{2}\sin 2\omega t\sin\alpha\right). \end{aligned} \tag{2.49}$$

The average power over a complete cycle is

$$P_{av} = \frac{1}{T}\int_0^T P\,dt \tag{2.50}$$

where $T(= 2\pi/\omega)$ is the period of the driving force. Since over the cycle, the average values of $\cos^2\omega t$ and $\sin 2\omega t$ are $1/2$ and zero, respectively, we obtain from Eq. (2.49)

$$P_{av} = \frac{F^2}{2|Z_m|}\cos\alpha. \tag{2.51}$$

We have $\tan\alpha = \frac{X_m}{k}$, where $X_m \left(= \omega m - \frac{s}{\omega}\right)$ is the mechanical reactance. Hence

$$\cos\alpha = \frac{k}{(k^2 + X_m^2)^{1/2}} = \frac{k}{|Z_m|}. \tag{2.52}$$

Substituting for $\cos\alpha$ in Eq. (2.51) we get

$$P_{av} = \frac{F^2 k}{2|Z_m|^2}. \tag{2.53}$$

Note that although the instantaneous power, given by Eq. (2.49), can be both positive or negative depending on time t, the average power, given by Eq. (2.53), is always positive.

At resonance, we have $|Z_m| = k$, so that the average power at resonance is

$$(P_{av})_r = \frac{F^2}{2k}. \tag{2.54}$$

The *effective* or the *root mean square* (rms) value of the exciting simple harmonic force is $\left[\frac{1}{T}\int_0^T (F\cos\omega t)^2 dt\right]^{1/2} = F/\sqrt{2}$. Using Eq. (2.44) the rms value of the steady-state velocity is found to be $F/(\sqrt{2}|Z_m|)$. Thus, Eq. (2.51) can be rewritten as

$$P_{av} = F_{rms}V_{rms}\cos\alpha, \tag{2.55}$$

where F_{rms} and V_{rms} are the rms values of the force and the velocity, respectively. Due to the phase angle α between the force and the velocity, the average power is less than the product $F_{rms}V_{rms}$ by the factor $\cos\alpha$, which is known as the *power factor*. Eq. (2.52) gives the expression for the power factor. At resonance, $|Z_m| = k$, so that $\cos\alpha = 1$. Thus, *resonance and unity power factor are synonimous.* Obviously, at resonance the force and the velocity are in phase.

We shall now show that the average power supplied by the driver is dissipated in overcoming the resistive force. The resistive force is $k\frac{dx}{dt}$, so that the instantaneous rate at which work is done against it, is

$$\frac{dW}{dt} = k\left(\frac{dx}{dt}\right)^2 = \frac{kF^2}{|Z_m|^2}\cos^2(\omega t - \alpha). \tag{2.56}$$

The average value of $\cos^2(\omega t - \alpha)$ is 1/2 over a full cycle. Therefore, the average rate at which work is done against the resisting force is

$$\left(\frac{dW}{dt}\right)_{av} = \frac{kF^2}{2|Z_m|^2}. \tag{2.57}$$

Equations (2.53) and (2.57) show that $P_{av} = \left(\frac{dW}{dt}\right)_{av}$, i.e., the average power supplied by the driving force is consumed in overcoming the resistive force.

2.9 Sharpness of Resonance

Equation (2.53) can be rewritten as

$$P_{av} = \frac{F^2 k}{2\left[k^2 + \left(\omega m - \frac{s}{\omega}\right)^2\right]}. \tag{2.58}$$

If P_{av} is plotted against the angular frequency ω of the driver, clearly, P_{av} will attain its maximum value of $F^2/(2k)$ at the resonant frequency $\omega = \omega_0 = \sqrt{s/m}$, and will fall off on either side of this frequency (Fig. 2.6).

When k is small, the resonance peak of P_{av} is high, and vice versa. The smaller the damping, the greater the drop of P_{av} as ω differs a little from ω_0. In other words, resonance is *sharp* when damping is small. For large damping, the resonance is *broad* or *flat*. This is clearly understood by comparing curves (a) and (b) of Fig. 2.6, the former curve being for a smaller value of k than the latter. The response curve of the human ear-drum resembles more closely the curve (b), i.e., the curve for a larger damping.

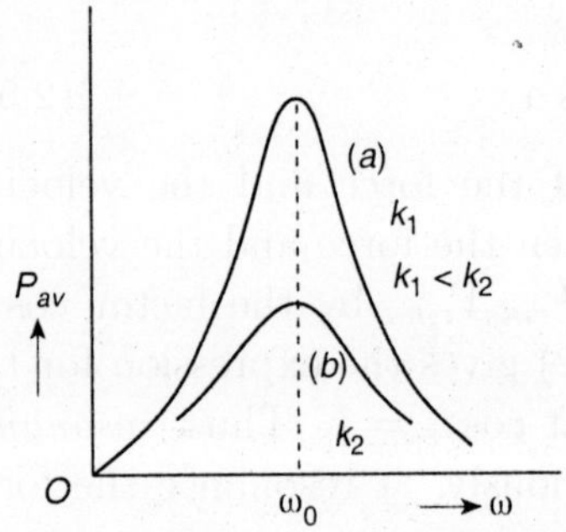

Fig. 2.6 Plot of P_{av} versus ω

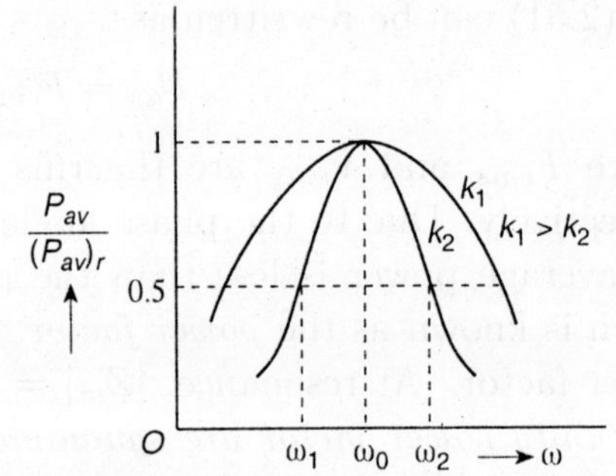

Fig. 2.7 Plot of $P_{av}/(P_{av})_r$ versus ω

The sharpness of resonance gives the rapidity with which the average power P_{av} supplied by the driver drops off as ω differs from its value at resonance. The higher sharpness of resonance for low damping is clearly revealed in Fig. 2.7 where the normalised power $P_{av}/(P_{av})$ is plotted against ω for two values of k, namely, k_1 and k_2.

Equation (2.58) yields

$$\frac{P_{av}}{(P_{av})_r} = \frac{k^2}{k^2 + m^2\left(\omega - \frac{\omega_0^2}{\omega}\right)^2} = \frac{4b^2}{4b^2 + \omega_0^2\left(\frac{\omega}{\omega_0} - \frac{\omega_0}{\omega}\right)^2}$$

$$= \frac{4b^2}{4b^2 + \omega_0^2\Delta^2}, \tag{2.59}$$

$$\text{where} \quad \Delta = \frac{\omega}{\omega_0} - \frac{\omega_0}{\omega}. \tag{2.60}$$

$\frac{P_{av}}{(P_{av})_r} = 1$ at $\omega = \omega_0$, i.e., $\Delta = 0$. Below or above resonance, $\Delta \neq 0$ and $\frac{P_{av}}{(P_{av})_r} < 1$. The sharpness of resonance is quantitatively defined as the reciprocal of $|\Delta|$ at which $P_{av}/(P_{av})_r = 1/2$. Figure 2.7 shows that for a given value of k or b, there are two values of ω, namely, ω_1 and ω_2 for which the average power P_{av} is half its value at resonance. These frequencies are called the *half-power frequencies*. ω_1 being less than ω_0, is termed the *lower* half-power frequency, whereas ω_2 being higher than ω_0, is known as the *upper* half-power frequency.

At the half-power frequencies, Eq. (2.59) gives

$$\frac{1}{2} = \frac{4b^2}{4b^2 + \omega_0^2\Delta^2}$$

$$\text{or,} \quad \omega_0^2\Delta^2 = 4b^2 \tag{2.61}$$

$$\text{or,} \quad \omega_0\Delta = \pm 2b$$

$$\text{or,} \quad \omega - \frac{\omega_0^2}{\omega} = \pm 2b$$

$$\text{or,} \quad \omega^2 \mp 2b\omega - \omega_0^2 = 0. \tag{2.62}$$

The two roots of this quadratic equation gives the half-power frequencies ω_1 and ω_2. Thus, accepting only the positive roots, we have

$$\omega_1 = \sqrt{b^2 + \omega_0^2} - b \tag{2.63}$$

$$\text{and} \quad \omega_2 = \sqrt{b^2 + \omega_0^2} + b. \tag{2.64}$$

Equation (2.61) gives for the sharpness of resonance

$$S_r = \frac{1}{|\Delta|} = \frac{\omega_0}{2b}. \tag{2.65}$$

From Eqs. (2.63) and (2.64), we obtain $2b = \omega_2 - \omega_1$. Hence Eq. (2.65) reduces to

$$S_r = \frac{\omega_0}{\omega_2 - \omega_1}.$$

In analogy with resonance in electric circuits, the quantity $\omega_0/(\omega_2 - \omega_1)$ is termed the *quality factor*, denoted by Q. The frequency range $(\omega_2 - \omega_1)$ is called the *half-power bandwidth* or simply the *bandwidth*. Thus,

$$Q = S_r = \frac{\omega_0}{\omega_2 - \omega_1}. \tag{2.66}$$

Obviously, Q is a measure of the sharpness of resonance. The smaller the damping, the smaller the bandwidth. Then Q is higher, implying that the resonance is sharper. A sharp resonance implies that the response of the driven system to the driving force is markedly *selective*: the response is vigorous at the resonant frequency and poor at off-resonant frequencies.

Observations

(1) From Eq. (2.65), we obtain $Q = \omega_0 m/k$, which is analogous to the expression $Q = \omega_0 L/R$ in electric circuits.

(2) From Eq. (2.18), we have for the logarithmic decrement $\lambda = bT/2 = \omega_0 T/(4Q) \simeq \pi/(2Q)$, since $\omega_0 T \simeq 2\pi$ for small damping. Hence

$$Q = \frac{\pi}{2\lambda}. \tag{2.67}$$

(3) The maximum kinetic energy at resonance is $E_m = mF^2/(2k^2)$ [see Eq. (2.47)]. The energy dissipated per cycle at resonance, obtained from Eq. (2.54) is $E_d = F^2T/(2k) = F^2\pi/(\omega_0 k)$. Thus,

$$\frac{E_m}{E_d} = \frac{mF^2}{2k^2} \times \frac{\omega_0 k}{\pi F^2} = \frac{\omega_0 m}{2\pi k} = \frac{Q}{2\pi}$$

$$\text{or,} \quad Q = 2\pi\frac{E_m}{E_d} = 2\pi\frac{\text{maximum kinetic energy at resonance}}{\text{energy dissipated per cycle at resonance}}. \tag{2.68}$$

Equation (2.68) offers an *alternative definition* of Q.

(4) For $\omega < \omega_0$, we have $\omega m < \frac{s}{\omega}$. Now the stiffness factor predominates, so that the regions of the curves in Fig. 2.5 for $\omega < \omega_0$ are said to be *stiffness-controlled.* For $\omega > \omega_0$, we have $\omega m > \frac{s}{\omega}$. The mass or the inertia predominantly controls the motion here. Hence the portions of the curves in Fig. 2.5 for $\omega > \omega_0$ are said to be *mass-controlled* or *inertia-controlled.* At resonance, $\omega m = \frac{s}{\omega}$ and the motion is *resistance-controlled.*

(5) Resonance is often used to find an unknown frequency (e.g., with the help of a sonometer) or to detect a particular frequency in a note containing a number of frequencies (e.g., with the help of a Helmholtz resonator). The resonance must be sharp in such cases. In loudspeakers or microphones, however, the resonance should be flat because a band of frequencies must be accommodated in such cases. A sharp resonance in such instruments will reject most of the desired frequencies, resulting in distortion. The resonance is made flatter here by increasing the damping.

(6) Equations (2.63) and (2.64) yield $\omega_1\omega_2 = \omega_0^2$. Thus the resonant frequency is the *geometric mean* between the half-power frequencies.

(7) Velocity resonance occurs when $\omega = \omega_0$. From Eq. (2.39) the displacement amplitude at velocity resonance (A_R) is found to be

$$A_R = \frac{F}{\omega_0 k} = \frac{QF}{m\omega_0^2} = \frac{QF}{s}.$$

(8) As $Q = \omega_0/(2b)$, the frequency for amplitude resonance can be written from Eq. (2.41) as $\omega = \omega_0\sqrt{1 - \frac{1}{2Q^2}}$.

2.10 Phase Angles in Forced Vibration

From Eq. (2.30) we find that the displacement x lags the forcing function $F\cos\omega t$ by the angle ϕ, where ϕ is given by Eq. (2.28), which can be rewritten as

$$\tan\phi = \frac{\omega k}{m\left(\frac{s}{m} - \omega^2\right)} = \frac{k}{\frac{s}{\omega} - \omega m} = -\frac{k}{X_m}, \tag{2.69}$$

where X_m $(= \omega m - \frac{s}{\omega})$ is the mechanical reactance.

Equation (2.37) shows that the velocity v lags the forcing function by the angle α where

$$\tan\alpha = \frac{X_m}{k}. \tag{2.70}$$

From Eqs. (2.69) and (2.70), we get

$$\phi = \frac{\pi}{2} + \alpha. \tag{2.71}$$

At resonance, $X_m = 0$, so that Eq. (2.70) gives $\alpha = 0$. The velocity at resonance is thus in phase with the driver. Consequently, the displacement at resonance lags the driver by $\pi/2$.

For $\omega > \omega_0$, i.e., when the motion is *mass-controlled*, we have $X_m > 0$ and hence α is positive. Thus above resonance, the velocity lags the driver by an angle which increases with increasing frequency and approaches $\pi/2$ as $\omega \to \infty$. The corresponding displacement lags the driving force by the angle ϕ which increases from $\pi/2$ and approaches π as ω increases above resonance.

For $\omega < \omega_0$, i.e., for *stiffness-controlled* motion, X_m and hence α is negative. Thus as ω decreases below resonance, the velocity leads the driving force by an angle which tends to $\pi/2$ as $\omega \to 0$. The corresponding displacement lags the driving force by an angle which decreases and approaches zero as $\omega \to 0$.

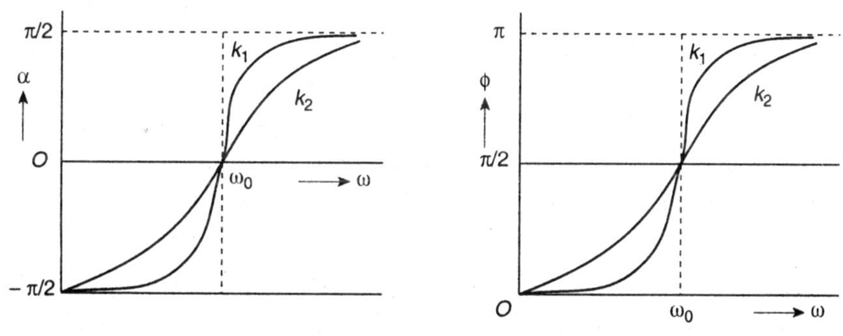

Fig. 2.8 Plots of the phase angles (a) α and (b) ϕ against ω for two values of k, viz., k_1 and $k_2 (k_1 < k_2)$

For small damping, the variations of α and ϕ with ω are very rapid in the vicinity of resonance, as shown in Figs. 2.8(a) and (b).

Comparison of amplitude and velocity resonances

Amplitude resonance	Velocity resonance
1. The amplitude of the steady-state displacement of the driven system is a maximum.	**1.** The amplitude of the steady-state velocity of the driven system is a maximum.

(*contd. on next page*)

(*contd. from previous page*)

Amplitude resonance	Velocity resonance
2. The phenomenon occurs when the angular frequency of the driving force is $\sqrt{\omega_0^2 - 2b^2}$, where ω_0 is the undamped natural angular frequency of the forced system and b is the decay constant.	**2.** Velocity resonance takes place when the forcing frequency is equal to the undamped natural frequency of the forced system.
3. With decreasing damping, the angular frequency of amplitude resonance approaches ω_0. Also, the maximum displacement amplitude increases with decreasing damping and becomes infinitely large for zero damping ($b = 0$) at ω_0.	**3.** As damping decreases, the maximum velocity amplitude increases and becomes infinitely large for zero damping ($b = 0$). For very small damping, both amplitude and velocity resonances occur practically at ω_0.
4. At amplitude resonance, the potential energy of the driven system is a maximum.	**4.** At velocity resonance, the kinetic energy of the driven system is a maximum.
5. The amplitude resonance is analogous to the capacitor charge resonance in a series LCR circuit.	**5.** The velocity resonance is analogous to true resonance, i.e., current resonance in a series LCR circuit.
6. At amplitude resonance, the mechanical reactance of the driven system does not vanish, and the power factor is less than unity.	**6.** At velocity resonance, the mechanical reactance of the driven system is zero, and the power factor is unity.
7. The displacement lags the force by an angle less than 90°. With diminishing damping, the lagging phase angle tends to 90°.	**7.** The force and the velocity are in phase.

2.11 Coupled Vibrations

In forced vibration discussed above, the frequency and the amplitude of the driving force are assumed to remain unaffected by the motion of the driven system. In other words, the energy fed back to the driver by the driven system is assumed to be negligibly small. However, if two systems of comparable masses interact, the roles of the driver and the driven system can be interchanged. A system consisting of two or more harmonic oscillators which can interact with each other to exchange energy, is termed a *coupled system*. The vibrations of such a system are known as *coupled vibrations*. The interchange of energy among the component oscillators affects the nature of vibration of each oscillator.

A simple example of coupled vibration is illustrated by the vibration of a pair of pendulums O_1A_1 and O_2A_2 of equal lengths suspended from a thick cord (Fig. 2.9). The bobs A_1 and A_2 have equal masses. The bob A_1 is made to vibrate in a direction at right angles to O_1O_2. The energy is transmitted through the cord to the bob A_2 which starts to oscillate. By Newton's third law of motion, the vibration of A_2 applies a periodic force on A_1 which is equal and opposite to that applied by A_1 on A_2. As a result of the reaction of A_2, the amplitude of A_1 decreases. If it were not so, the amplitude and energy of both the pendulums would increase simultaneously, in disagreement with the principle of conservation of energy.

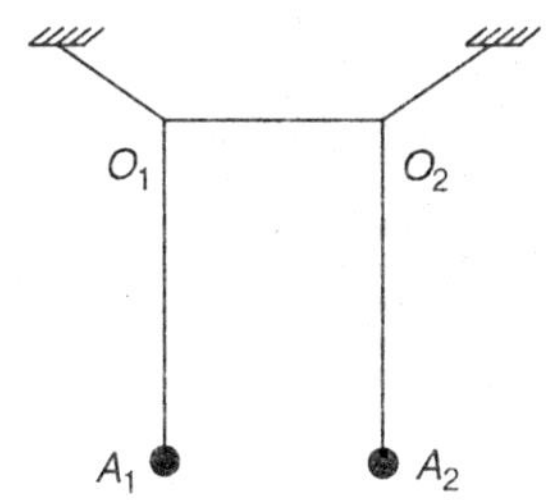

Fig. 2.9 Two coupled pendulums

The amplitude of A_1 diminishes whereas that of A_2 increases. When the amplitude of A_2 attains a maximum, A_1 stops momentarily. Then A_2 becomes the driver and A_1, the driven pendulum. The amplitude of A_1 now starts to grow and that of A_2 gradually falls to zero when A_1 attains its maximum amplitude. The process goes on, the energy exchanging continuously between A_1 and A_2. Due to frictional forces, the amplitudes fall off in successive cycles, so that finally both the bobs stop oscillating.

2.12 Mathematical Treatment of Coupled Vibrations

Consider two coupled masses m_1 and m_2 (Fig. 2.10). Let at any instant of time t, the displacements of m_1 and m_2 from their mean positions be x_1 and x_2, respectively. The net restoring force on m_1 is $-s_1x_1 - s(x_1 - x_2)$ and

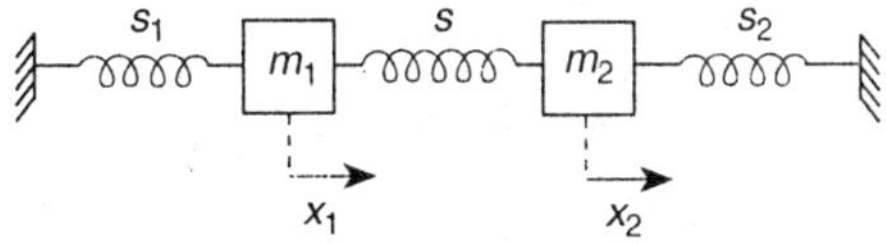

Fig. 2.10 Coupled oscillations

that on m_2 is $-s_2x_2 - s(x_2 - x_1)$. Therefore, in the absence of damping, the equations of motion of the two masses are

$$m_1 \frac{d^2x_1}{dt^2} + s_1x_1 + s(x_1 - x_2) = 0 \tag{2.72}$$

$$\text{and} \qquad m_2 \frac{d^2 x_2}{dt^2} + s_2 x_2 + s(x_2 - x_1) = 0. \tag{2.73}$$

These equations are linear and homogeneous of the second order. To solve them, we take

$$x_1 = A_1 e^{\alpha t} \quad \text{and} \quad x_2 = A_2 e^{\alpha t}, \tag{2.74}$$

where A_1 and A_2 are in general complex constants. Substituting for x_1 and x_2 in Eqs. (2.72) and (2.73) and dividing through by $e^{\alpha t}$, we get

$$(m_1\alpha^2 + s_1 + s)A_1 - sA_2 = 0 \tag{2.75}$$

$$\text{and} \qquad -sA_1 + (m_2\alpha^2 + s_2 + s)A_2 = 0. \tag{2.76}$$

In order that these two linear homogeneous equations have a solution other than the trivial one $A_1 = A_2 = 0$, the determinant of the coefficients of A_1 and A_2 must vanish, i.e.,

$$\begin{vmatrix} m_1\alpha^2 + s_1 + s & -s \\ -s & m_2\alpha^2 + s_2 + s \end{vmatrix} = 0. \tag{2.77}$$

Expanding the determinant gives

$$(m_1\alpha^2 + s_1 + s)(m_2\alpha^2 + s_2 + s) - s^2 = 0$$

$$\text{or,} \qquad \left(\alpha^2 + \frac{s_1 + s}{m_1}\right)\left(\alpha^2 + \frac{s_2 + s}{m_2}\right) - \frac{s^2}{m_1 m_2} = 0. \tag{2.78}$$

Equation (2.78) is called the *characteristic equation* or the *secular equation* of the system. Putting

$$\omega_{11}^2 = \frac{s_1 + s}{m_1}, \quad \omega_{22}^2 = \frac{s_2 + s}{m_2} \quad \text{and} \quad w_{12}^2 = \frac{s}{\sqrt{m_1 m_2}}, \tag{2.79}$$

we can write the characteristic equation as

$$(\alpha^2 + \omega_{11}^2)(\alpha^2 + \omega_{22}^2) - \omega_{12}^4 = 0$$

$$\text{or,} \qquad \alpha^4 + \alpha^2(\omega_{11}^2 + \omega_{22}^2) + \omega_{11}^2\omega_{22}^2 - \omega_{12}^4 = 0. \tag{2.80}$$

The roots of this quadratic equation in α^2 are

$$\alpha^2 = -\frac{1}{2}(\omega_{11}^2 + \omega_{22}^2) \pm \frac{1}{2}\left[(\omega_{11}^2 - \omega_{22}^2)^2 + 4\omega_{12}^4\right]^{1/2}. \tag{2.81}$$

The quantity under the radical in Eq. (2.81) is positive and the absolute value of the first term is greater than that of the second term. So, α^2 is real and negative, and we put

$$\alpha^2 = -\omega^2. \tag{2.82}$$

Therefore, Eq. (2.81) gives

$$\omega^2 = \frac{1}{2}(\omega_{11}^2 + \omega_{22}^2) \mp \frac{1}{2}\left[(\omega_{11}^2 - \omega_{22}^2)^2 + 4\omega_{12}^4\right]^{1/2}. \tag{2.83}$$

The solution of Eq. (2.83) yields two positive values of ω, say, ω_1 and ω_2, where

$$\omega_1 = \left[\frac{1}{2}(\omega_{11}^2 + \omega_{22}^2) - \frac{1}{2}\left\{(\omega_{11}^2 - \omega_{22}^2)^2 + 4\omega_{12}^4\right\}^{1/2}\right]^{1/2} \tag{2.84}$$

and
$$\omega_2 = \left[\frac{1}{2}(\omega_{11}^2 + \omega_{22}^2) + \frac{1}{2}\left\{(\omega_{11}^2 - \omega_{22}^2)^2 + 4\omega_{12}^4\right\}^{1/2}\right]^{1/2}. \tag{2.85}$$

The roots ω_1^2 and ω_2^2 of the characteristic equation are called the *eigenvalues* or the *characteristic values*, each being the square of a *natural* (or *normal*) frequency. Each natural frequency (ω_1 or ω_2) defines a *normal mode of vibration*. These vibrations are independent of each other, and any general solution of the system is a linear combination of the simple harmonic vibrations of the normal frequencies.

If, instead of two masses, we have n number of coupled masses, the characteristic equation would be an nth degree algebraic equation in ω^2. In this case, we get n number of normal modes of vibration with normal frequencies $\omega_1, \omega_2, \cdots, \omega_n$.

We consider a *special case* where $m_1 = m_2 = m$ and $s_1 = s_2 = s_0$. Then $\omega_{11}^2 = \omega_{22}^2$, and

$$\omega_1 = (\omega_{11}^2 - \omega_{12}^2)^{1/2} = \sqrt{\frac{s_0}{m}} \tag{2.86}$$

$$\omega_2 = (\omega_{11}^2 + \omega_{12}^2)^{1/2} = \sqrt{\frac{s_0 + 2s}{m}}. \tag{2.87}$$

Putting Eq. (2.86) in Eq. (2.75), we have $A_1 = A_2$. Thus, in the normal mode of vibration with the frequency ω_1, the two masses move in *the same direction with equal amplitudes.* In this mode, the coupling spring is not stressed, the frequency ω_1 being the natural frequency of each *decoupled* mass, obtained with $s = 0$. With Eq. (2.87) substituted in Eq. (2.75), we get $A_1 = -A_2$. Hence in the normal mode of frequency ω_2, the masses move in *opposite directions with equal amplitudes.* The general solution, obtained by superposing the normal modes, is

$$x_1 = A\cos(\omega_1 t + \theta_1) + B\cos(\omega_2 t + \theta_2), \tag{2.88}$$

$$x_2 = A\cos(\omega_1 t + \theta_1) - B\cos(\omega_2 t + \theta_2). \tag{2.89}$$

The coordinates $y_1 = (m/2)^{1/2}(x_1 + x_2)$ and $y_2 = (m/2)^{1/2}(x_1 - x_2)$ are called the *normal coordinates* of the system. Thus, the displacements of the masses (with a scale factor) under the normal modes of vibration give the normal coordinates.

Let the motion be initiated by displacing the mass 1 by a distance x_0 and then releasing it. Thus, the boundary conditions are

$$x_1 = x_0, \quad x_2 = 0, \quad \frac{dx_1}{dt} = 0, \quad \text{and} \quad \frac{dx_2}{dt} = 0 \text{ at } t = 0.$$

Putting these conditions in Eqs. (2.88) and (2.89) give

$$x_0 = A\cos\theta_1 + B\cos\theta_2, \qquad \omega_1 A\sin\theta_1 + \omega_2 B\sin\theta_2 = 0,$$
$$A\cos\theta_1 - B\cos\theta_2 = 0, \quad \text{and} \quad -\omega_1 A\sin\theta_1 + \omega_2 B\sin\theta_2 = 0.$$

These relationships are satisfied, if $A = B = \frac{x_0}{2}$, and $\theta_1 = \theta_2 = 0$. Hence Eqs. (2.88) and (2.89) reduce to

$$\begin{aligned} x_1 &= \frac{x_0}{2}(\cos\omega_1 t + \cos\omega_2 t) \\ &= x_0 \cos\left(\frac{\omega_1+\omega_2}{2}t\right)\cos\left(\frac{\omega_1-\omega_2}{2}t\right) \end{aligned} \tag{2.90}$$

$$\begin{aligned} x_2 &= \frac{x_0}{2}(\cos\omega_1 t - \cos\omega_2 t) \\ &= x_0 \sin\left(\frac{\omega_1+\omega_2}{2}t\right)\sin\left(\frac{\omega_2-\omega_1}{2}t\right). \end{aligned} \tag{2.91}$$

These equations show that the motion involves the sum and the difference of the *half* normal frequencies.

For *weak coupling*, $s \ll s_0$, and Eq. (2.87) gives

$$\omega_2 = \sqrt{\frac{s_0}{m}}\left(1+\frac{2s}{s_0}\right)^{1/2} \simeq \omega_1\left(1+\frac{s}{s_0}\right) = \omega_1 + 2\delta, \text{ (say)}, \tag{2.92}$$

where $2\delta(=\omega_1 s/s_0)$ is much smaller than ω_1. Hence $\omega_1 + \omega_2 \simeq 2\omega_1$ and $\omega_2 - \omega_1 \simeq 2\delta$. Equations (2.90) and (2.91) then reduce to

$$x_1 = x_0 \cos\omega_1 t \cos\delta t \tag{2.93}$$

$$x_2 = x_0 \sin\omega_1 t \sin\delta t. \tag{2.94}$$

These equations reveal that each mass oscillates with an angular frequency ω_1, the amplitude of oscillation changing slowly with an angular frequency δ. The amplitude of one mass is a maximum when the other

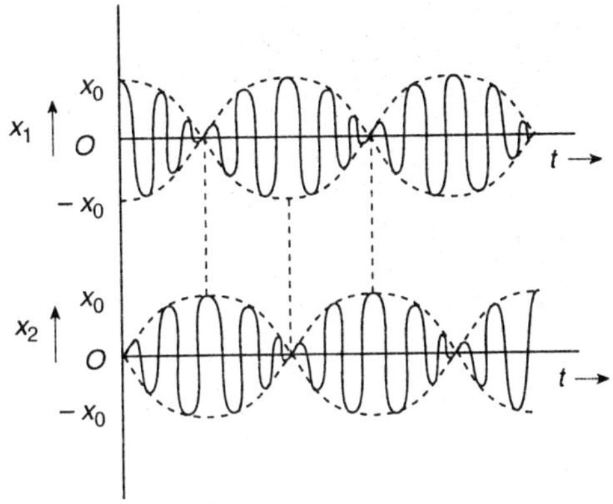

Fig. 2.11 Displacement versus time plots of the coupled masses

mass is momentarily at rest (Fig. 2.11). We get here the phenomenon of *beats*, the amplitude changing at a low frequency that is equal to the *difference* of the two closely spaced half normal frequencies.

For very weak coupling, we can neglect the energy stored in the coupling string, so that the whole process is a perpetual exchange of energy between mass 1 (which has all the energy originally) and mass 2. The two masses alternately act as the 'driver' and as the 'driven' vibrator. When the driver gives up its entire energy to the driven mass, the roles of the driver and the driven are interchanged.

2.13 Energy of Coupled Vibrations

The kinetic energy of the coupled system with $m_1 = m_2 = m$ is

$$T = \frac{1}{2}m\dot{x}_1^2 + \frac{1}{2}m\dot{x}_2^2. \tag{2.95}$$

The potential energy with $s_1 = s_2 = s_0$ is

$$\begin{aligned} V &= \frac{1}{2}s_0 x_1^2 + \frac{1}{2}s_0 x_2^2 + \frac{1}{2}s(x_2 - x_1)^2 \\ &= \frac{1}{2}(s_0 + s)x_1^2 + \frac{1}{2}(s_0 + s)x_2^2 - s x_1 x_2. \end{aligned} \tag{2.96}$$

We shall now express T and V in terms of the normal co-ordinates y_1 and y_2 introduced earlier:

$$y_1 = \left(\frac{m}{2}\right)^{1/2}(x_1 + x_2) \tag{2.97}$$

$$\text{and} \quad y_2 = \left(\frac{m}{2}\right)^{1/2}(x_1 - x_2). \tag{2.98}$$

Adding and subtracting Eqs. (2.97) and (2.98), we get

$$x_1 = \left(\frac{1}{2m}\right)^{1/2}(y_1 + y_2) \tag{2.99}$$

$$\text{and} \quad x_2 = \left(\frac{1}{2m}\right)^{1/2}(y_1 - y_2). \tag{2.100}$$

Substituting Eqs. (2.99) and (2.100) into Eqs. (2.95) and (2.96), we obtain, respectively,

$$T = \frac{1}{2}\dot{y}_1^2 + \frac{1}{2}\dot{y}_2^2 \tag{2.101}$$

$$\begin{aligned} \text{and} \quad V &= \frac{1}{2}\left(\frac{s_0}{m}\right)y_1^2 + \frac{1}{2}\left(\frac{s_0 + 2s}{m}\right)y_2^2 \\ &= \frac{1}{2}\omega_1^2 y_1^2 + \frac{1}{2}\omega_2^2 y_2^2 \end{aligned} \tag{2.102}$$

where ω_1 and ω_2 are the normal frequencies given by Eqs. (2.86) and (2.87), respectively. Note from Eq. (2.101) that T has no cross product term in $\dot{y}_1, \dot{y}_2$, also observe from Eq. (2.102) that V contains no cross product term in $y_1 y_2$. This is a useful property, which serves as a *check* for normal coordinates. In Eq. (2.102), the coefficients of y_1^2 and y_2^2 depend only on the rspective normal frequencies. Neither T nor V explicitly contains any mass or stiffness. This is possible due to the scale factor $(m/2)^{1/2}$ in Eqs. (2.97) and (2.98).

The total energy of the system is

$$\begin{aligned} W = T + V &= \left(\frac{1}{2}\dot{y}_1^2 + \frac{1}{2}\dot{y}_2^2\right) + \left(\frac{1}{2}\omega_1^2 y_1^2 + \frac{1}{2}\omega_2^2 y_2^2\right) \\ &= \left(\frac{1}{2}\dot{y}_1^2 + \frac{1}{2}\omega_1^2 y_1^2\right) + \left(\frac{1}{2}\dot{y}_2^2 + \frac{1}{2}\omega_2^2 y_2^2\right). \end{aligned} \tag{2.103}$$

Thus, W is the *sum of the energies the system would have when the modes are excited separately.*

2.14 A Number of Coupled Oscillators

Here we shall deal with the problem of an arbitrary number of similar oscillators coupled together. This has a bearing on a physical system like a crystal or a fluid where many atoms or molecules vibrate, being coupled to each other by cohesive forces.

We consider N identical particles, each of mass m, arranged in a line at equal distances a apart. The particles are coupled by $(N + 1)$ identical springs of spring constant s (Fig. 2.12). The particles execute longitudinal oscillations, their instantaneous displacements from the equilibrium

Fig. 2.12 N particles coupled by identical springs. The dashed lines indicate the equilibrium positions of the vibrating particles

positions being $x_1, x_2, \cdots, x_N$. The outer ends of the extreme springs are attached to rigid supports to give the boundary conditions that at $x = 0$ and $x = (N+1)a$, the displacement is zero.

The equation of motion of the nth particle is

$$m\frac{d^2x_n}{dt^2} = s(x_{n+1} - x_n) - s(x_n - x_{n-1})$$

$$\text{or,} \quad m\frac{d^2x_n}{dt^2} = s(x_{n+1} + x_{n-1} - 2x_n) \tag{2.104}$$

To determine the normal modes, we assume $x_n = A_n e^{j\omega t}$, $x_{n-1} = A_{n-1}e^{j\omega t}$, and $x_{n+1} = A_{n+1}e^{j\omega t}$ where A_n, A_{n-1}, and A_{n+1} are the amplitudes of oscillation, and ω is a normal frequency. Substituting for x_n, x_{n-1}, and x_{n+1} in Eq. (2.104) and dividing through by $e^{j\omega t}$, we obtain

$$(2s - m\omega^2)A_n - s(A_{n+1} + A_{n-1}) = 0. \tag{2.105}$$

Since $n = 1, 2, 3, \cdots, N$, we have N equations like Eq. (2.105), the nontrivial solutions of which give N possible normal frequencies ω. For two coupled oscillators, there are two equations giving two normal frequencies (see Sec. 2.12).

Equation (2.105) gives

$$\frac{A_{n+1} + A_{n-1}}{A_n} = 2 - \frac{m\omega^2}{s} = \frac{2\omega_0^2 - \omega^2}{\omega_0^2} \tag{2.106}$$

where $\omega_0^2 = s/m$; and $n = 1, 2, 3, \cdots, N$. Let us choose

$$A_n = C\sin n\theta \tag{2.107}$$

which satisfies the boundary condition that at $x = 0$, the displacement is zero, i.e., $A_0 = 0$. Here C and θ are constants. Then $A_{n-1} = C\sin(n-1)\theta$ and $A_{n+1} = C\sin(n+1)\theta$, and Eq. (2.106) gives

$$\frac{\sin(n+1)\theta + \sin(n-1)\theta}{\sin n\theta} = \frac{2\omega_0^2 - \omega^2}{\omega_0^2} \tag{2.108}$$

Since $\sin(n+1)\theta + \sin(n-1)\theta = 2\sin n\theta\cos\theta$, Eq. (2.108) simplifies to

$$\frac{2\omega_0^2 - \omega^2}{\omega_0^2} = 2\cos\theta. \tag{2.109}$$

In order that the boundary condition of zero displacement at $x = (N+1)a$ is satisfied, we must have $A_{N+1} = 0$, i.e., $\sin(N+1)\theta = 0$, or, $(N+1)\theta = p\pi$, or

$$\theta = \frac{p\pi}{N+1} \quad (p = 1, 2, 3, \cdots, N) \tag{2.110}$$

Using Eq. (2.110) into Eq. (2.107), we obtain

$$A_n = C \sin\left(\frac{pn\pi}{N+1}\right). \tag{2.111}$$

Substituting for θ from Eq. (2.110) into Eq. (2.109) gives the normal frequency ω. We put a subscript p to ω to obtain for the characteristic frequency ω_p for the pth mode

$$\omega_p^2 = 2\omega_0^2\left[1 - \cos\left(\frac{p\pi}{N+1}\right)\right] = 4\omega_0^2 \sin^2\left[\frac{p\pi}{2(N+1)}\right]$$

$$\text{or,} \quad \omega_p = 2\omega_0 \sin\left[\frac{p\pi}{2(N+1)}\right]. \tag{2.112}$$

There are N possible values of ω_p corresponding to $p = 1, 2, 3, \cdots, N$.

It is instructive to extract the results for a single oscillator and for two coupled oscillators from the above analysis. For *a single oscillator*, we have $N = 1$, $p = 1$, and $n = 1$. Then Eq. (2.112) yields

$$\omega_1 = 2\omega_0 \sin\frac{\pi}{4} = \sqrt{2}\omega_0 = \sqrt{\frac{2s}{m}}.$$

Also, Eq. (2.111) gives

$$A_1 = C \sin\frac{\pi}{4} = \frac{C}{\sqrt{2}}.$$

For *two coupled oscillators*, $N = 2$, $p = 1$ and 2, and $n = 1$ and 2. For the first mode ($p = 1$), Eq. (2.112) gives for the mode frequency

$$\omega_1 = 2\omega_0 \sin\frac{\pi}{6} = \omega_0 = \sqrt{\frac{s}{m}}.$$

The amplitudes of the two masses for this mode are

$$A_1 = C \sin\frac{\pi}{3} = \frac{\sqrt{3}}{2}C, \quad \text{and} \quad A_2 = C \sin\frac{2\pi}{3} = \frac{\sqrt{3}}{2}C.$$

Thus $A_1 = A_2$. This is the in-phase mode (see Sec. 2.12).

For the second mode ($p = 2$), the characteristic frequency is

$$\omega_2 = 2\omega_0 \sin\frac{\pi}{3} = \sqrt{3}\omega_0 = \sqrt{\frac{3s}{m}}.$$

The amplitudes of the two masses are

$$A_1 = C \sin\frac{2\pi}{3} = \frac{\sqrt{3}}{2}C \quad \text{and} \quad A_2 = C \sin\frac{4\pi}{3} = -\frac{\sqrt{3}}{2}C.$$

Here $A_1 = -A_2$. This is the out-of-phase mode (see Sec. 2.12).

2.15 Solved Problems

Problem 1. *In a one dimensional motion of a mass* 10 g, *it is acted on by a restoring force* 10 dyne/cm *and a resisting force* 2 dyne. s/cm. *Find* (i) *whether the motion is aperiodic or oscillatory;* (ii) *the resisting force per unit velocity which will make the motion critically damped;* (iii) *the mass for which the given forces will make the motion critically damped.*

[C.U. 1990, '92; cf. C.U. 1999, 2010]

Solution. Here $m = 10$g, $s = 10$ dyne/cm, $k = 2$ dyne. s/cm.

So, $2b = k/m = 0.2$ neper/s, and $\omega_0^2 = s/m = 1$ rad^2/s^2.

(i) Here $b^2/\omega_0^2 = 0.01 < 1$, showing that the motion is oscillatory.

(ii) For critical damping, $b = \omega_0$, i.e., $k^2/(4m^2) = s/m$. Hence $k = 2\sqrt{sm} = 2\sqrt{10 \times 10} = 20$ dyne. s/cm.

(iii) We have, $k^2/(4m^2) = s/m$, giving $m = \frac{k^2}{4s} = \frac{4}{4\times 10} = 0.1$g.

Problem 2. *If in the previous problem, the mass is given an initial impulse* 20 g.cm/s *while at rest, obtain an expression for the displacement x at time t.*

[C.U. 1990]

Solution. The general equation for the displacement in oscillatory motion is

$$x = e^{-bt}\left(A\cos\sqrt{\omega_0^2 - b^2}\,t + B\sin\sqrt{\omega_0^2 - b^2}\,t\right)$$

Here $b = k/(2m) = 0.1$ neper/s, and $\omega_0^2 = s/m = 1$ rad^2/s^2. Therefore, $(\omega_0^2 - b^2)^{1/2} = \sqrt{0.99}$ rad/s.

Since at $t = 0$, $x = 0$, we have $A = 0$. So,

$$x = Be^{-bt}\sin\sqrt{\omega_0^2 - b^2}\,t$$

and
$$\frac{dx}{dt} = -Bbe^{-bt}\sin\sqrt{\omega_0^2 - b^2}\,t + B\sqrt{\omega_0^2 - b^2}e^{-bt}\cos\sqrt{\omega_0^2 - b^2}\,t.$$

At $t = 0$, $\frac{dx}{dt} = \dot{x}_0$, say, the motion being excited by an impulsive force. So,

$$\dot{x}_0 = B\sqrt{\omega_0^2 - b^2}, \quad \text{or,} \quad B = \frac{\dot{x}_0}{\sqrt{\omega_0^2 - b^2}}.$$

The impulse of the force equals the change in momentum of the mass, so that $20 = m\dot{x}_0 = 10\dot{x}_0$ giving $\dot{x}_0 = 2$ cm/s.

Hence $B = \frac{2}{\sqrt{0.99}}$, and $x = \frac{2}{\sqrt{0.99}}e^{-0.1t}\sin\sqrt{0.99}\,t$ cm.

Problem 3. *A massless spring whose upper end is fixed to a rigid support carries a horizontal disc of mass* 100 g *at the lower end. It is observed that the* *system oscillates with a frequency of* 10 Hz *and the amplitude of the damped oscillations reduces to half its undamped value in one minute. Calculate* (i) *the resistive force constant,* (ii) *the relaxation time of the system,* (iii) *its quality factor and* (iv) *the force constant of the spring.* **[Burd. U. 1995]**

Solution. The angular frequency of oscillation is $\omega = \sqrt{\omega_0^2 - b^2} = 2\pi\times$ frequency $= 20\pi$. Let the undamped amplitude be A. The damped amplitude at time t is Ae^{-bt}. By the given condition

$$\frac{1}{2}A = Ae^{-60b}$$

whence $b = \frac{1}{60}\ln 2 = 0.01155$ neper/s.

(i) The resistive force constant is $k = 2mb = 2 \times 100 \times \frac{1}{60}\ln 2$

$$= \tfrac{10}{3}\ln 2 = 2.31 \text{ dyne.s/cm.}$$

(ii) The relaxation time is $\tau = \frac{1}{b} = \frac{60}{\ln 2} = 86.57$s.

(iii) The quality factor is $Q = \frac{\pi}{2\lambda}$, where λ is the log decrement.

We have $\lambda = \frac{bT}{2}$, where the time period T is given by

$$T = \frac{2\pi}{\omega} = \frac{2\pi}{\sqrt{\omega_0^2 - b^2}}.$$

So,

$$Q = \frac{\pi}{bT} = \frac{\sqrt{\omega_0^2 - b^2}}{2b} = \frac{20\pi \times 60}{2 \times \ln 2} = \frac{600\pi}{\ln 2} = 2.72 \times 10^3.$$

(iv) The force constant of the spring is

$$s = \omega_0^2 m = 100\omega_0^2 = 100(b^2 + 400\pi^2) = 3.95 \times 10^5 \text{ dyne/cm.}$$

Problem 4. *Show that for a forced vibration the total energy of the vibrating system is not constant. Prove that for such a case,*

$$\frac{\textit{average potential energy}}{\textit{average kinetic energy}} = \frac{\omega_0^2}{\omega^2}$$

where $\omega_0 = \sqrt{s/m}$. **[C.U. 1989, 2009]**

Solution. The displacement of a particle of mass m in forced vibration is $x = A\cos(\omega t - \phi)$. The kinetic energy of the particle is $\frac{1}{2}m\dot{x}^2 = \frac{1}{2}mA^2\omega^2\sin^2(\omega t - \phi)$. The potential energy is $\frac{1}{2}sx^2 = \frac{1}{2}sA^2\cos^2(\omega t - \phi)$. So, the total energy, i.e., the sum of the kinetic and the potential energies is

$$E = \frac{1}{2}mA^2\omega^2\sin^2(\omega t - \phi) + \frac{1}{2}sA^2\cos^2(\omega t - \phi).$$

Clearly, E varies with time and is not a constant. The average kinetic energy is

$$\bar{E}_k = \frac{1}{4} m A^2 \omega^2,$$

the average value of $\sin^2(\omega t - \phi)$ over a full cycle being 1/2.
The average potential energy is

$$\bar{E}_p = \frac{1}{4} s A^2,$$

the average value of $\cos^2(\omega t - \phi)$ over a full cycle being 1/2.
So,

$$\frac{\bar{E}_p}{\bar{E}_k} = \frac{s}{m\omega^2} = \frac{\omega_0^2}{\omega^2}$$

where $\omega_0^2 = s/m$, ω_0 being the natural angular frequency.

Problem 5. *An object of mass* 2 kg *hangs from a spring of negligible mass. The spring is extended by* 2.5 cm *when the object is attached. The top end of the spring is oscillated up and down in SHM with an amplitude of* 2 mm. *The Q of the system is* 20. *If* $g = 10\,\mathrm{m/s^2}$, (i) *what is the angular frequency* ω_0 *of the free undamped oscillations?* (ii) *what is the amplitude of forced oscillations at* $\omega = \omega_0$*?* (iii) *what is the mean power input to maintain the forced oscillations at an angular frequency* ω *one per cent greater than* ω_0*?*

[Burd. U. 1994]

Solution. If s is the spring constant, at equilibrium of the 2 kg suspended mass, we have $s \times 2.5 \times 10^{-2} = 2\,g = 2 \times 10$, whence $s = 800$ N/m.

(i) $\omega_0 = \sqrt{s/m} = \sqrt{800/2} = 20$ rad/s.

(ii) Q being large, b is small so that the displacement amplitude attains its maximum value practically at $\omega = \omega_0$. The maximum displacement amplitude is

$$A_m = \frac{F/m}{2b\sqrt{\omega_0^2 - b^2}} \simeq \frac{F}{2bm\omega_0} = \frac{FQ}{m\omega_0^2}.$$

The amplitude of the driving force is $F = s \times 2 \times 10^{-3} = 800 \times 2 \times 10^{-3} =$ 1.6 N. Therefore,

$$A_m = \frac{1.6 \times 20}{2 \times 20^2} = 0.04\,\mathrm{m} = 4\ \mathrm{cm}.$$

(iii) The average input power is

$$P_{av} = \frac{F^2 k}{2|Z_m|^2},$$

where $k = 2bm = \frac{\omega_0}{Q} m = \frac{20}{20} \times 2 = 2$ kg/s.

$$|Z_m|^2 = k^2 + \left(\omega m - \frac{s}{\omega}\right)^2 = 4 + \left(20.2 \times 2 - \frac{800}{20.2}\right)^2$$

$$= 4 + (40.4 - 39.6)^2 = 4.64\ \text{kg}^2/\text{s}^2.$$

So,

$$P_{av} = \frac{1.6^2 \times 2}{2 \times 4.64} = 0.55\ \text{W}.$$

Problem 6. *If ω_1 and ω_2 are the half-power frequencies and ω_0 is the resonant frequency of a forced system, show that $\omega_0^2 = \omega_1\omega_2$.*

Solution. The half-power frequencies satisfy Eq. (2.62), viz.,

$$\omega^2 \mp 2b\omega - \omega_0^2 = 0.$$

ω_1 and ω_2 are the positive roots of this equation:

$$\omega_1 = \sqrt{b^2 + \omega_0^2} - b$$

$$\text{and} \quad \omega_2 = \sqrt{b^2 + \omega_0^2} + b.$$

So,

$$\omega_1\omega_2 = (b^2 + \omega_0^2) - b^2 = \omega_0^2.$$

Problem 7. *The equations of motion of two coupled oscillators are*

$$\ddot{q}_1 + \sqrt{2}q_2 + 4q_1 = 0 \quad \textit{and} \quad \ddot{q}_2 + \sqrt{2}q_1 + 5q_2 = 0.$$

Find the normal frequencies and the ratios of the amplitudes of the normal modes. **[Burd. U. 1996, 98, 2000]**

Solution. Let $q_1 = A_1e^{\alpha t}$ and $q_2 = A_2e^{\alpha t}$, where A_1 and A_2 are complex in general. Substituting for q_1 and q_2 in the given differential equations and dividing through by $e^{\alpha t}$, we obtain

$$(\alpha^2 + 4)A_1 + \sqrt{2}A_2 = 0 \tag{i}$$

$$\text{and} \quad \sqrt{2}A_1 + (\alpha^2 + 5)A_2 = 0. \tag{ii}$$

For non zero solutions for A_1 and A_2, we have

$$\begin{vmatrix} \alpha^2 + 4 & \sqrt{2} \\ \sqrt{2} & \alpha^2 + 5 \end{vmatrix} = 0$$

$$\text{or,} \quad (\alpha^2 + 4)(\alpha^2 + 5) - 2 = 0$$

$$\text{or,} \quad (\alpha^2)^2 + 9\alpha^2 + 18 = 0$$

$$\text{or,} \quad (\alpha^2 + 3)(\alpha^2 + 6) = 0.$$

The roots are $\alpha^2 = -3$ and $\alpha^2 = -6$. So the normal frequencies are $\omega_1 = \sqrt{3}$ rad/s and $\omega_2 = \sqrt{6}$ rad/s.

Putting $\alpha^2 = -3$ in (i) gives $A_1 + \sqrt{2}A_2 = 0$. So the ratio of the amplitudes for this normal mode is $A_1/A_2 = -\sqrt{2}$. Putting $\alpha^2 = -6$ in (i) yields $-2A_1 + \sqrt{2}A_2 = 0$. For this normal mode, $A_1/A_2 = 1/\sqrt{2}$.

Problem 8. *Two particles of the same mass m are attached to a weightless spring of length l with fixed ends stretched under a tension T. Find the normal frequencies of vibration of the masses if they are attached at distances $l/3$ and $2l/3$ from one end. Assume that the tension in the string remains constant during the vibration.*

Solution. To ignore the effect of gravity, we assume that the string is initially horizontal and that the coupled masses move on a smooth horizontal surface. At any instant of time t, let x_1 and x_2 be the displacements of the masses

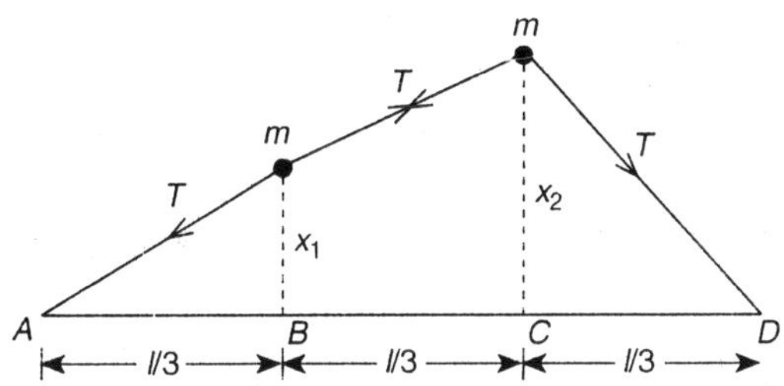

Fig. 2.13 Figure for Problem 8

(Fig. 2.13). As x_1 and x_2 are small, the tension T in the string does not change as the particles oscillate.

Resolving the forces on a mass in the direction of its displacement, we can write the equations of motion as

$$m\frac{d^2x_1}{dt^2} = -T\frac{x_1}{l/3} + T\frac{x_2 - x_1}{l/3}$$

$$\text{and} \quad m\frac{d^2x_2}{dt^2} = -T\frac{x_2}{l/3} - T\frac{x_2 - x_1}{l/3}.$$

Putting $\omega_0^2 = \frac{3T}{lm}$, these equations take the form

$$\frac{d^2x_1}{dt^2} + 2\omega_0^2 x_1 - \omega_0^2 x_2 = 0$$

$$\text{and} \quad \frac{d^2x_2}{dt^2} - \omega_0^2 x_1 + 2\omega_0^2 x_2 = 0.$$

Writing $x_1 = A_1 e^{j\omega t}$ and $x_2 = A_2 e^{j\omega t}$, where A_1 and A_2 are generally complex, we get from the above equations:

$$(2\omega_0^2 - \omega^2)A_1 - \omega_0^2 A_2 = 0 \qquad \text{(i)}$$

$$\text{and} \quad -\omega_0^2 A_1 + (2\omega_0^2 - \omega^2)A_2 = 0. \qquad \text{(ii)}$$

For non-zero A_1 and A_2, we have

$$(2\omega_0^2 - \omega^2)^2 - \omega_0^4 = 0$$

$$\text{or,} \quad (2\omega_0^2 - \omega^2 + \omega_0^2)(2\omega_0^2 - \omega^2 - \omega_0^2) = 0$$

So, $\omega = \omega_0 = \sqrt{\frac{3T}{lm}}$ or, $\omega = \sqrt{3}\omega_0 = 3\sqrt{\frac{T}{lm}}$.

These are the normal frequencies.

Problem 9. *In problem No. 8, show that the ratio of the amplitudes of the vibrating coupled masses is 1 for the normal frequency $\sqrt{3T/lm}$ and -1 for the normal frequency $3\sqrt{T/lm}$. Show the displacements of the masses for the normal modes of vibration. How can you excite these modes?*

Solution. Substituting $\omega = \sqrt{3T/lm}$ in Eq. (i) of the above problem 8, we obtain $A_1 = A_2$, or $A_1/A_2 = 1$. (proved). Following the same procedure for $\omega = 3\sqrt{T/lm}$, we get $A_1 + A_2 = 0$, or $A_1/A_2 = -1$ (proved).

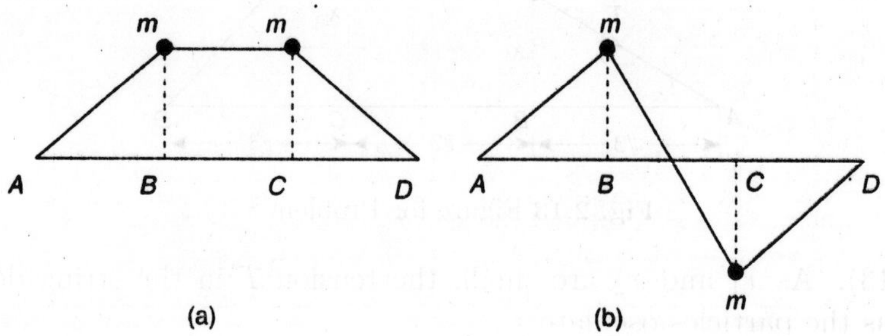

Fig. 2.14 Displacements of the masses for the normal modes of vibration

In the normal mode of vibration with frequency $\sqrt{3T/lm}$, the masses move in the same direction, as shown in Fig. 2.14 (a). For the other mode with frequency $3\sqrt{T/lm}$, the masses move in the opposite direction [Fig. 2.14 (b)].

The normal mode with the lower frequency can be excited by initially displacing the masses equally in the same direction and then releasing them. The normal mode with the higher frequency can be excited if the masses are displaced equally in the opposite direction at time $t = 0$ and then released.

Problem 10. *Show that the steady-state response in forced vibration demands that the damping force be proportional to the velocity and not other functions of the velocity.*

Solution. The equation of forced vibration (in complex form) is

$$m\frac{d^2S}{dt^2} + k\frac{dS}{dt} + sS = Fe^{j\omega t}.$$

The steady-state response is given by $S = S_0 e^{j\omega t}$. Substitution gives

$$(-m\omega^2 + j\omega k + s)S_0 e^{j\omega t} = Fe^{j\omega t},$$

so that $e^{j\omega t}$ cancels out from the two sides, making S_0 independent of time. If the damping force is not $k\frac{dS}{dt}$ but involves some other functions of it, the time dependences would not disappear from the two sides of the above equation. Hence the steady-state solution is not achieved.

Problem 11. *Two particles of masses m_1 and m_2 are connected by a massless spring of spring constant s. If the system is free to oscillate along the length of the spring, show that the frequency of oscillation is $\frac{1}{2\pi}\sqrt{\frac{s}{\mu}}$, where $\mu = \frac{m_1 m_2}{m_1+m_2}$. Also find the ratio of the amplitudes of the oscillating particles.*

Solution. Let x_1 and x_2 be the instantaneous displacements of the particles from their equilibrium positions (Fig. 2.15). For m_1, the restoring force is $-s(x_1-x_2)$ whereas for m_2 the same is $-s(x_2-x_1)$. The equations of motion of the masses are

$$m_1\frac{d^2x_1}{dt^2} = -s(x_1 - x_2) \tag{i}$$

and $$m_2\frac{d^2x_2}{dt^2} = -s(x_2 - x_1). \tag{ii}$$

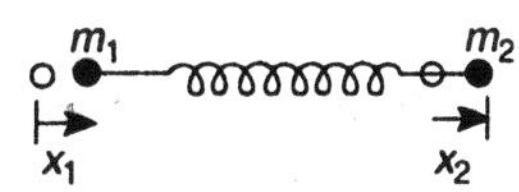

Fig. 2.15 Two masses coupled by a spring. The open circles denote the equilibrium positions of the masses

Assuming that $x_1 = A_1 e^{j\omega t}$ and $x_2 = A_2 e^{j\omega t}$, where A_1 and A_2 are in general complex, we obtain from Eqs. (i) and (ii)

$$(s - m_1\omega^2)A_1 - sA_2 = 0 \tag{iii}$$

and $$-sA_1 + (s - m_2\omega^2)A_2 = 0. \tag{iv}$$

For nonzero A_1 and A_2, we have from Eqs. (iii) and (iv)

$$(s - m_1\omega^2)(s - m_2\omega^2) - s^2 = 0$$

or, $$m_1 m_2\omega^2 = s(m_1 + m_2)$$

or, $$\omega = \sqrt{\frac{s}{\mu}}, \tag{v}$$

where $\mu = m_1 m_2/(m_1 + m_2)$. The quantity μ is the *reduced mass* of the system because it is smaller than both m_1 and m_2. The frequency of oscillation is

$$f = \frac{\omega}{2\pi} = \frac{1}{2\pi}\sqrt{\frac{s}{\mu}}, \text{ proved.}$$

$$\frac{A_1}{A_2} = -\frac{m_2}{m_1},$$

which gives the ratio of the amplitudes of the oscillating particles.

Problem 12. *Two identical particles, each of mass m and lying on a smooth horizontal platform, are connected to two identical massless springs of spring constant k (Fig. 2.16). Determine the characteristic frequencies for the normal modes of oscillation. Also find the ratio of the amplitudes of the masses for each mode.*

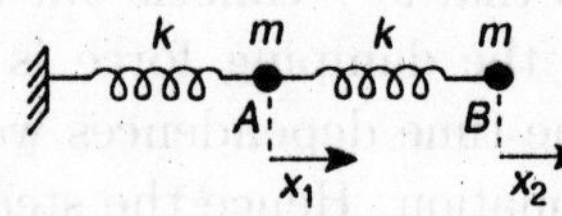

Fig. 2.16 Figure for problem 12

Solution. Let x_1 and x_2 be the instantaneous displacements of the masses. The restoring force on mass A is $-kx_1 - k(x_1 - x_2)$ and that on mass B is $-k(x_2 - x_1)$. So the equations of motion of the masses are

$$m\frac{d^2x_1}{dt^2} + k(2x_1 - x_2) = 0 \tag{i}$$

$$\text{and} \quad m\frac{d^2x_2}{dt^2} + k(x_2 - x_1) = 0 \tag{ii}$$

To find the normal frequencies we assume $x_1 = A_1e^{j\omega t}$ and $x_2 = A_2e^{j\omega t}$. Substituting for x_1 and x_2 in Eqs. (i) and (ii), we obtain

$$(2k - m\omega^2)A_1 - kA_2 = 0 \tag{iii}$$

$$-kA_1 + (k - m\omega^2)A_2 = 0 \tag{iv}$$

For nontrivial solutions of Eqs. (iii) and (iv), we have

$$\begin{vmatrix} 2k - m\omega^2 & -k \\ -k & k - m\omega^2 \end{vmatrix} = 0$$

$$\text{or,} \quad \omega^4 - \frac{3k}{m}\omega^2 + \frac{k^2}{m^2} = 0,$$

of which the roots are

$$\omega^2 = (3 \pm \sqrt{5})\frac{k}{2m}.$$

So the characteristic frequencies are

$$\omega_1 = \left[\left(\frac{3-\sqrt{5}}{2}\right)\frac{k}{m}\right]^{1/2} \quad \text{and} \quad \omega_2 = \left[\left(\frac{3+\sqrt{5}}{2}\right)\frac{k}{m}\right]^{1/2}.$$

Substituting for $\omega(=\omega_1)$ in Eq. (iii) we have

$$\frac{A_1}{A_2} = \frac{k}{2k - m\omega_1^2} = \frac{2}{\sqrt{5}+1} = \frac{\sqrt{5}-1}{2}$$

which gives the ratio of the amplitudes of the two masses for the normal frequency ω_1. Similarly, substituting for $\omega(=\omega_2)$ in Eq. (iii) we have

$$\frac{A_1}{A_2} = -\frac{2}{\sqrt{5}-1} = -\frac{\sqrt{5}+1}{2}$$

which gives the ratio of the amplitudes for the frequency ω_2.

Problem 13. *Find the normal coordinates of the system considered in Fig. 2.16 (Problem 12).*

Solution. The general expressions for the displacements of the masses, obtained by the superposition of the normal modes, are

$$x_1 = A\cos(\omega_1 t + \theta_1) + B\cos(\omega_2 t + \theta_2)$$

$$\text{and} \quad x_2 = A'\cos(\omega_1 t + \theta_1) + B'\cos(\omega_2 t + \theta_2)$$

where

$$\omega_1 = \left[\left(\frac{3-\sqrt{5}}{2}\right)\frac{k}{m}\right]^{1/2} \quad \text{and} \quad \omega_2 = \left[\left(\frac{3+\sqrt{5}}{2}\right)\frac{k}{m}\right]^{1/2}.$$

Also,

$$\frac{A}{A'} = \frac{\sqrt{5}-1}{2} = \alpha \ (\text{say})$$

$$\text{and} \quad \frac{B}{B'} = -\frac{\sqrt{5}+1}{2} = -\frac{(\sqrt{5}+1)(\sqrt{5}-1)}{2(\sqrt{5}-1)} = -\frac{2}{\sqrt{5}-1} = -\frac{1}{\alpha}.$$

(see solved problem 12)

So,

$$x_1 = A\cos(\omega_1 t + \theta_1) + B\cos(\omega_2 t + \theta_2)$$

$$\text{and} \quad x_2 = \frac{A}{\alpha}\cos(\omega_1 t + \theta_1) - \alpha B\cos(\omega_2 t + \theta_2)$$

With unity scale factors the normal coordinates are $y_1 = A\cos(\omega_1 t + \theta_1)$ and $y_2 = B\cos(\omega_2 t + \theta_2)$, so that $x_1 = y_1 + y_2$ and $x_2 = \frac{y_1}{\alpha} - \alpha y_2$. Solving for y_1 and y_2, we get

$$y_1 = \frac{\alpha^2}{\alpha^2+1}x_1 + \frac{\alpha}{\alpha^2+1}x_2 = \frac{5-\sqrt{5}}{10}x_1 + \frac{\sqrt{5}}{5}x_2$$

$$\text{and} \quad y_2 = \frac{x_1}{\alpha^2+1} - \frac{\alpha}{\alpha^2+1}x_2 = \frac{5+\sqrt{5}}{10}x_1 - \frac{\sqrt{5}}{5}x_2.$$

Problem 14. *Consider a symmetrical linear triatomic molecule such as carbon dioxide, which has the chemical structure* O-C-O. *Find the normal frequencies of the longitudinal vibration of the molecule taking into account only nearest neighbour interactions.*

Solution. The system is equivalent to a set of three particles of masses m_a, m_b and m_a, respectively connected by elastic springs, each of spring constant s (Fig. 2.17).

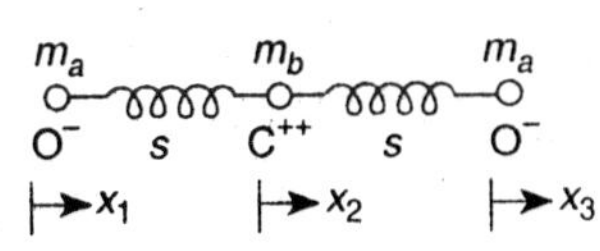

Fig. 2.17 A Linear symmetrical triatomic molecule

Let x_1 and x_2, and x_3 be the instantaneous displacement of the masses m_a, m_b, and m_a, respectively. The equations of motion are

$$m_a\frac{d^2x_1}{dt^2}+s(x_1-x_2)=0 \tag{i}$$

$$m_b\frac{d^2x_2}{dt^2}+s(x_2-x_1)+s(x_2-x_3)=0 \tag{ii}$$

and $$m_a\frac{d^2x_3}{dt^2}+s(x_3-x_2)=0. \tag{iii}$$

Let the solutions be

$$x_1=A_1e^{j\omega t},\quad x_2=A_2e^{j\omega t},\quad \text{and}\quad x_3=A_3e^{j\omega t}.$$

Substituting these into Eqs. (i), (ii) and (iii), we obtain

$$(-m_a\omega^2+s)A_1-sA_2=0 \tag{iv}$$

$$-sA_1+(-m_b\omega^2+2s)A_2-sA_3=0 \tag{v}$$

and $$-sA_2+(-m_a\omega^2+s)A_3=0 \tag{vi}$$

For nonzero solutions

$$\begin{vmatrix} -m_a\omega^2+s & -s & 0 \\ -s & -m_b\omega^2+2s & -s \\ 0 & -s & -m_a\omega^2+s \end{vmatrix}=0 \tag{vii}$$

or, $$\omega^2(-m_a\omega^2+s)(-m_am_b\omega^2+sm_b+2sm_a)=0,$$

from which the normal frequencies are found to be

$$\omega_1=0,\quad \omega_2=\sqrt{\frac{s}{m_a}},\quad \omega_3=\sqrt{\frac{s}{m_a}+\frac{2s}{m_b}}.$$

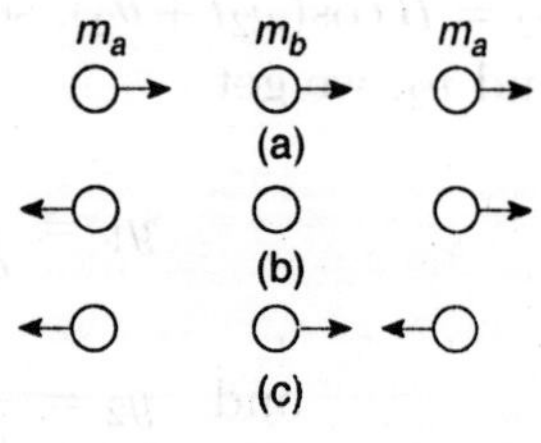

Fig. 2.18 Longitudinal vibrational modes of the triatomic molecule

Putting $\omega=\omega_1=0$ in Eqs. (iv), (v), and (vi), we have $A_1=A_2=A_3$. So this mode is no oscillation at all, but is a translation of the whole system (Fig. 2.18a).

Putting $\omega=\omega_2=\sqrt{s/m_a}$ in Eqs. (iv) through (vi), we get $A_2=0$ and $A_1=-A_3$.

So, the mass m_b at the middle remains at rest while the two end masses m_a, m_a vibrate in opposite directions with equal amplitude (Fig. 2.18b).

Substituting $\omega=\omega_3=\sqrt{\frac{s}{m_a}+\frac{2s}{m_b}}$ in Eqs. (iv) through (vi) we obtain

$$A_1=A_3\quad \text{and}\quad A_2=-2A_1\left(\frac{m_a}{m_b}\right)=-2A_3\left(\frac{m_a}{m_b}\right).$$

Thus, in this mode, the two end masses m_a, m_a vibrate in phase with equal amplitude but the middle mass m_b vibrates with a different amplitude in the opposite direction (Fig. 2.18c).

Problem 15. *For the symmetrical linear triatomic molecule of Problem No. 14 determine the normal frequencies of transverse vibration.*

Solution. For transverse motion of the atoms, the longitudinal separation l between the atoms is not altered. Let y_1, y_2, and y_3 be the instantaneous transverse displacements of the masses m_a, m_b, and m_a respectively, from their equilibrium positions. The equations of motion are

$$m_a \frac{d^2y_1}{dt^2} + s_t(y_1 - y_2) = 0 \tag{i}$$

$$m_b \frac{d^2y_2}{dt^2} - s_t(y_1 - y_2) - s(y_3 - y_2) = 0 \tag{ii}$$

and $$m_a \frac{d^2y_3}{dt^2} + s_t(y_3 - y_2) = 0 \tag{iii}$$

where s_t is the spring constant.

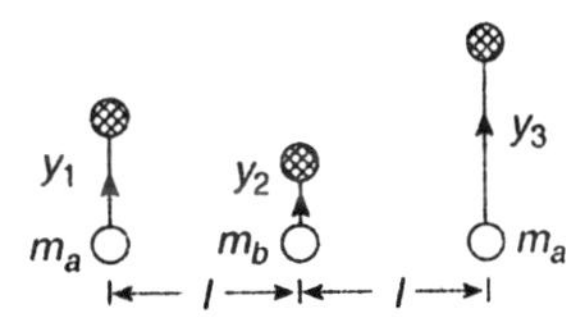

Fig. 2.19 Transverse displacements of the atoms. The open circles are the equilibrium positions and the solid circles are the displaced positions of the atoms

Let the solutions be $y_1 = A_1e^{j\omega t}$, $y_2 = A_2e^{j\omega t}$, and $y_3 = A_3e^{j\omega t}$. Substituting these into Eqs. (i), (ii), and (iii) we get

$$(s_t - m_a\omega^2)A_1 - s_tA_2 = 0 \tag{iv}$$

$$-s_tA_1 + (2s_t - m_b\omega^2)A_2 - s_tA_3 = 0 \tag{v}$$

$$-s_tA_2 + (s_t - m_a\omega^2)A_3 = 0 \tag{vi}$$

These equations are the same as the corresponding equations (iv), (v), and (vi) of Problem No. 14. Thus we get similar solutions for the normal frequencies with s replaced by s_t. These normal frequencies are

$$\omega_1 = 0, \quad \omega_2 = \sqrt{\frac{s_t}{m_a}}, \quad \text{and } \omega_3 = \sqrt{\frac{s_t}{m_a} + \frac{2s_t}{m_b}}.$$

For $\omega = \omega_1 = 0$, we have $A_1 = A_2 = A_3$. This mode corresponds to a pure translation of the whole system [Fig. 2.20(a)].

For $\omega = \omega_2 = \sqrt{s_t/m_a}$, we get $A_2 = 0$ and $A_1 = -A_3$.

Here the centre mass remains at rest and the two end masses vibrate in opposite directions with equal amplitudes [Fig. 2.20(b)].

For $\omega = \omega_3 = \sqrt{\frac{s_t}{m_a} + \frac{2s_t}{m_b}}$, we have

$$A_1 = A_3$$

$$\text{and} \quad \frac{A_2}{A_1} = \frac{A_2}{A_3} = -\frac{2m_a}{m_b}.$$

In this mode, the end masses vibrate in unison with the same amplitude, but the centre mass vibrates in the opposite direction with a different amplitude [Fig. 2.20(c)].

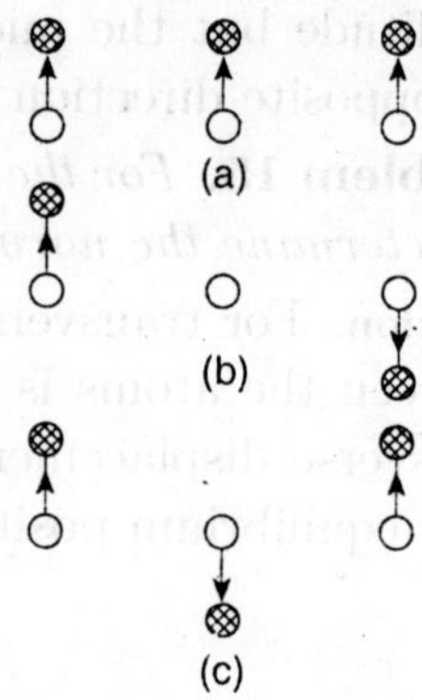

Fig. 2.20 Transverse vibrational modes of the triatomic molecule

Problem 16. *The displacement of a particle of mass m executing underdamped oscillatory motion is given by $x(t) = \rho e^{-bt}\cos(n't - \alpha)$ where $n' = \sqrt{n^2 - b^2}$, n is the undamped angular frequency of the particle, and b is the decay constant.* (i) *Calculate ρ and α when the motion starts with the initial velocity v_0 only,* (ii) *Calculate the energy of vibration and show that for small damping it approximates to $E(t) = \frac{1}{2}mv_0^2 e^{-2bt}$.* **[C.U. 2002, '09]**

Solution. (i) Given, $x = \rho e^{-bt}\cos(n't - \alpha)$.

At $t = 0$, $x = 0$. So, $0 = \rho e^0 \cos(0 - \alpha)$, i.e., $\rho\cos\alpha = 0$.

Since $\rho \neq 0$ ($\rho = 0$ will imply that $x = 0$ for all values of t), $\cos\alpha = 0$, or $\alpha = \pi/2$. Therefore,

$$x = \rho e^{-bt}\sin n't \quad \text{and} \quad \frac{dx}{dt} = \rho e^{-bt}\ [n'\cos n't - b\sin n't]$$

Given : at $t = 0$, $\frac{dx}{dt} = v_0$. Hence $v_0 = \rho e^0[n' - 0]$, i.e., $\rho = v_0/n'$.

Thus

$$x = \frac{v_0}{n'}e^{-bt}\sin n't.$$

(ii) The energy is

$$E(t) = \frac{1}{2}m\left(\frac{dx}{dt}\right)^2 + \frac{1}{2}mn^2x^2 = \frac{1}{2}m\left(\frac{v_0}{n'}\right)^2 e^{-2bt}\left[n'\cos n't - b\sin n't\right]^2$$

$$+\frac{1}{2}mn^2\left(\frac{v_0}{n'}\right)^2 e^{-2bt}\sin^2 n't$$

$$= \frac{1}{2}mv_0^2e^{-2bt}\left[\cos^2 n't - 2\frac{b}{n'}\sin n't\cos n't + \left(\frac{b}{n'}\right)^2\sin^2 n't\right.$$

$$\left. + \left(\frac{n}{n'}\right)^2\sin^2 n't\right]$$

For small damping $n' \simeq n$, and the second and the third terms within the parentheses can be neglected. Hence

$$E(t) \simeq \frac{1}{2} m v_0^2 e^{-2bt}, \text{ proved.}$$

Observe that the energy of vibration decays with time.

Problem 17. *A particle of mass* 1 kg *moving along the x-axis is acted on by a restoring force* $4x$ newton *towards the origin, and a damping force* $4\dot{x}$ newton. *The particle is initially at rest at* $x = 1$ m. *Find the instant of time when the velocity of the particle is a maximum. What is the maximum velocity?*

Solution. The equation of motion of the particle is

$$\ddot{x} + 4\dot{x} + 4x = 0$$

Here $b = \omega_0 = 2$, and the motion is critically damped (see Sec. 2.2). The solution of the equation is

$$x = (A_1 + A_2 t)e^{-2t}$$

where A_1 and A_2 are constants. At $t = 0$, $x = 1$ m and $\dot{x} = 0$. Hence $A_1 = 1$ and $A_2 = 2$; and so

$$x = (1 + 2t)e^{-2t} \text{ m.}$$

The velocity of the particle is

$$v = \dot{x} = -4te^{-2t} \text{ m/s.}$$

When v is a maximum, $\dot{v} = -4e^{-2t} + 8te^{-2t} = 0$, or $t = \frac{1}{2}$s. The maximum velocity is

$$v_{\max} = -4 \times \frac{1}{2} e^{-2\times\frac{1}{2}} = -\frac{2}{e} \text{ m/s.}$$

The negative sign indicates that the velocity is directed towards the origin.

Problem 18. *In Prob.* 17, *what is the energy of the particle at* $t = 0$ *and at* $t = \frac{1}{2}$s*? Show that the loss of energy is equal to the work done against the damping force.*

Solution. As the velocity is zero at $t = 0$, the initial energy of the particle is totally the potential energy. Since the spring constant is $s = 4$N/m, the potential energy at $t = 0$ is $E_i = \frac{1}{2}sx_i^2 = \frac{1}{2} \times 4 \times 1^2 = 2$ J.

At $t = \frac{1}{2}$s, $$x = \left(1 + 2 \times \frac{1}{2}\right) e^{-2\times\frac{1}{2}} = \frac{2}{e} \text{ m}$$

The potential energy is $E_p = \frac{1}{2}sx^2 = 2 \times \frac{4}{e^2} = \frac{8}{e^2}$ J.

The velocity at $t = \frac{1}{2}$s is $v = -\frac{2}{e}$ m/s. The kinetic energy is

$$E_k = \frac{1}{2} \times 1 \times v^2 = \frac{2}{e^2}\text{ J}.$$

The total energy at $t = \frac{1}{2}$s is

$$E = E_p + E_k = \frac{8}{e^2} + \frac{2}{e^2} = \frac{10}{e^2}\text{ J}.$$

The loss of energy is

$$E_d = E_i - E = \left(2 - \frac{10}{e^2}\right)\text{ J}. \qquad \text{(i)}$$

The work done against the damping force is

$$W = 4\int_{x_i}^{x} \frac{dx}{dt}dx = 4\int_0^{1/2}\left(\frac{dx}{dt}\right)^2 dt = 64\int_0^{1/2} t^2 e^{-4t} dt$$

$$= (16t^2 + 8t + 2)e^{-4t}\Big|_{1/2}^{0} = \left(2 - \frac{10}{e^2}\right)\text{ J}. \qquad \text{(ii)}$$

From (i) and (ii) it is proved that $E_d = W$.

Problem 19. *Two identical pendulums, each of bob mass m and string length l, are coupled by a massless spring of stiffness factor s, the unstretched length of the spring being equal to the distance between the bobs in their equilibrium positions (Fig. 2.21). For small oscillations of the system along the length of the spring,* (i) *find the normal frequencies and* (ii) *show diagramatically the normal modes of vibration. What are the normal coordinates?*

Solution. Let x_1 and x_2 be the instantaneous displacements of the masses A and B at time t. The equations of motion of the two masses can be written as

$$m\frac{d^2x_1}{dt^2} = -\frac{mg}{l}x_1 - s(x_1 - x_2) \qquad \text{(i)}$$

and $$m\frac{d^2x_2}{dt^2} = -\frac{mg}{l}x_2 - s(x_2 - x_1) \qquad \text{(ii)}$$

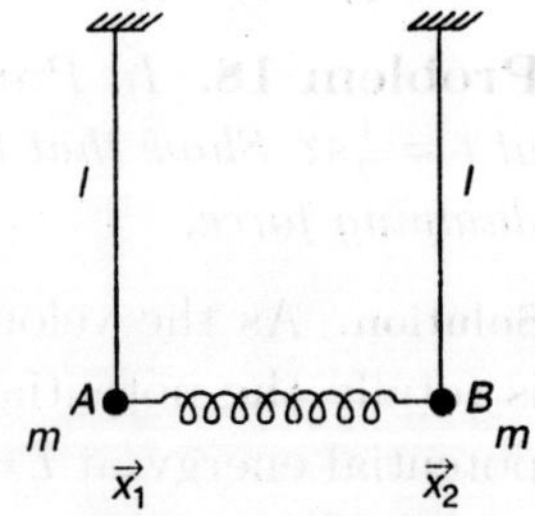

Fig. 2.21 Figure for Problem 19

Putting $\omega_0^2 = \frac{g}{l}$ and $\frac{s}{m} = \omega_c^2$, and assuming that $x_1 = A_1e^{j\omega t}$ and $x_2 = A_2e^{j\omega t}$, where A_1 and A_2 are in general complex, we get from Eqs. (i) and (ii)

$$\left(\omega_0^2 + \omega_c^2 - \omega^2\right)A_1 - \omega_c^2 A_2 = 0 \qquad \text{(iii)}$$

and $$-\omega_c^2 A_1 + \left(\omega_0^2 + \omega_c^2 - \omega^2\right)A_2 = 0 \qquad \text{(iv)}$$

For nonzero A_1 and A_2, Eqs. (iii) and (iv) give

$$\begin{vmatrix} \omega_0^2 + \omega_c^2 - \omega^2 & -\omega_c^2 \\ -\omega_c^2 & \omega_0^2 + \omega_c^2 - \omega^2 \end{vmatrix} = 0$$

$$\text{or,} \quad \left(\omega_0^2 + \omega_c^2 - \omega^2\right)^2 - \omega_c^4 = 0$$

$$\text{or,} \quad \omega_0^2 + \omega_c^2 - \omega^2 = \pm\omega_c^2$$

yielding two normal frequencies $\omega = \omega_1 = \omega_0$ and $\omega = \omega_2 = \sqrt{\omega_0^2 + 2\omega_c^2}$.

(i) Thus the normal frequencies are

$$\omega_1 = \sqrt{\frac{g}{l}} \quad \text{and} \quad \omega_2 = \sqrt{\frac{g}{l} + \frac{2s}{m}}$$

Substituting $\omega = \omega_1 = \omega_0$ in Eq. (iii) we get $A_1 = A_2$, showing that the masses move in the same direction with the same amplitude and are in phase. This is the in-phase mode.

Putting $\omega = \omega_2 = \sqrt{\omega_0^2 + 2\omega_c^2}$ in Eq. (iii) we obtain $A_1 = -A_2$. Here the masses move in the opposite direction with the same amplitude. This is the out-of-phase mode.

In the in-phase mode, the coupling spring is not stretched. This is not true for the out-of-phase mode.

(ii) The displacements of the masses in the two normal modes of vibration are shown in Fig. 2.22 (a) and (b).

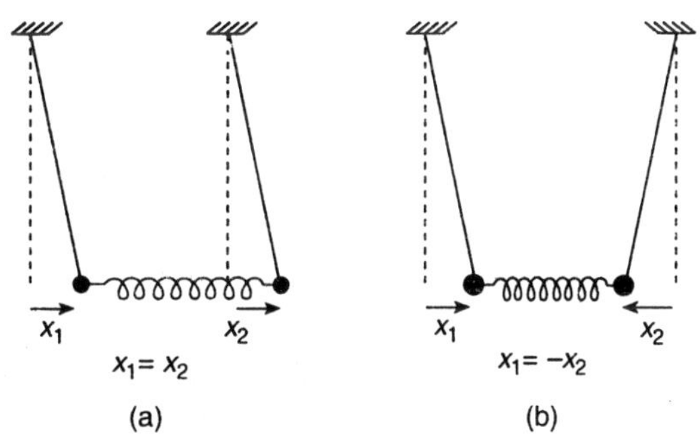

Fig. 2.22 Displacements of the masses for (a) the in-phase mode and (b) the out-of-phase mode

The general motion is obtained by superimposing the normal modes. Thus, if x_1 and x_2 are the displacements of the masses in the general motion, we have

$$x_1 = A\cos(\omega_1 t + \theta_1) + B\cos(\omega_2 t + \theta_2)$$

$$\text{and} \quad x_2 = A\cos(\omega_1 t + \theta_1) - B\cos(\omega_2 t + \theta_2)$$

The normal coordinates are $y_1 = A\cos(\omega_1 t + \theta_1)$ and $y_2 = B\cos(\omega_2 t + \theta_2)$. Hence, $x_1 = y_1 + y_2$ and $x_2 = y_1 - y_2$. Solving for y_1 and y_2, we obtain

$$y_1 = \frac{1}{2}(x_1 + x_2) \quad \text{and} \quad y_2 = \frac{1}{2}(x_1 - x_2).$$

Problem 20. *The steady-state displacement of a particle acted on by a force $F\cos\omega t$ is $x = a\cos\omega t + b\sin\omega t$. Calculate the average power. Hence interpret the amplitudes a and b physically.*

Solution. The instantaneous power of the driver is

$$P = (F\cos\omega t)\frac{dx}{dt} = (F\cos\omega t)(-a\omega\sin\omega t + b\omega\cos\omega t)$$

$$= F\omega\left(b\cos^2\omega t - \frac{1}{2}a\sin 2\omega t\right)$$

The average power over a complete cycle of time period $T\left(=\frac{2\pi}{\omega}\right)$ is

$$P_{av} = \frac{1}{T}\int_0^T P\,dt = \frac{1}{2}F\omega b$$

since the average values of $\cos^2\omega t$ and $\sin 2\omega t$ over a cycle are $\frac{1}{2}$ and zero, respectively.

The average power P_{av} is found to be proportional to b which is the amplitude of that component of the steady-state displacement x which is 90° out of phase with the driving force $F\cos\omega t$. Hence b is referred to as the *absorptive amplitude*. The component $a\cos\omega t$ of the displacement x which is in phase with the driving force contributes nothing to P_{av}. Hence a is called the *dispersive* or the *elastic amplitude*.

Problem 21. *The steady-state velocity of a particle in forced vibration is $v = v_0\sin\omega t$. Find the total dissipated energy W in the time interval 0 to t. Sketch W against t.*

Solution. The resistive force is $k\frac{dx}{dt} = kv = kv_0\sin\omega t$. The instantaneous rate of dissipation of energy is the product of the resistive force and the velocity. Thus

$$\frac{dW}{dt} = kv_0^2\sin^2\omega t = \frac{1}{2}kv_0^2(1 - \cos 2\omega t)$$

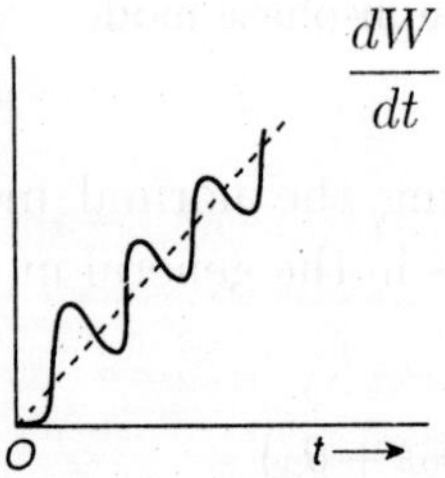

Fig. 2.23 Variation of W against t

The total dissipated energy in the time interval 0 to t is

$$W = \frac{1}{2}kv_0^2\int_0^t (1 - \cos 2\omega t)dt$$

$$= \frac{1}{2}kv_0^2\left(t - \frac{1}{2\omega}\sin 2\omega t\right) \qquad \text{(i)}$$

W is sketched against t in Fig. 2.23. The first term within parentheses on the right-hand side of Eq. (i) represents a linear increase of W with t, as shown by the broken line in Fig. 2.23. Superimposed on it is the oscillatory variation represented by the second term.

So, the net variation of W is given by the solid curve in Fig. 2.23.

Problem 22. *The equation of motion of a particle in damped motion is given (with usual symbols) by*

$$m\frac{d^2x}{dt^2} + k\frac{dx}{dt} + sx = 0.$$

Hence show that the instantaneous power spent in overcoming the resistive force is equal to the time rate of decay of the energy of the particle.

Solution. The resistive force is $k\frac{dx}{dt}$. The instantaneous power spent in overcoming it is the product of this force and the velocity, i.e., $k\left(\frac{dx}{dt}\right)^2$. From the given equation of motion, we get

$$\begin{aligned} k\left(\frac{dx}{dt}\right)^2 &= -\left(m\frac{dx}{dt}\frac{d^2x}{dt^2} + sx\frac{dx}{dt}\right) \\ &= -\frac{d}{dt}\left[\frac{1}{2}m\left(\frac{dx}{dt}\right)^2 + \frac{1}{2}sx^2\right] = -\frac{dE}{dt}, \qquad \text{(i)} \end{aligned}$$

where

$$E = \frac{1}{2}m\left(\frac{dx}{dt}\right)^2 + \frac{1}{2}sx^2. \qquad \text{(ii)}$$

The first term on the right-hand side of Eq. (ii) is the kinetic energy, and the second term is the potential energy of the particle. Thus E is the total energy of the particle. Equation (i) shows that the instantaneous power spent in overcoming the resistive force is equal to the time rate of decay of E.

REVIEW QUESTIONS

1. A particle executes a one-dimensional motion along $X'OX$ under (i) a restoring force proportional to its displacement from the equilibrium position 'O' and (ii) a damping force proportional to its velocity. Set up the equation of motion of the particle, and solve it. Show that the logarithm of the ratio of successive amplitudes on either side of the equilibrium position is a constant. What is the solution for undamped motion? **[cf. C.U. 1997]**

2. Write the differential equation for the damped natural oscillations of a system from the energy principle and solve it. Give the conditions for over damped, critically damped, and under damped motions. Write the time-displacement relationship in each case.

3. Define the following terms in connection with the damped simple harmonic motion of a particle (i) decay constant, (ii) modulus of decay, and (iii) logarithmic decrement.

4. Show that, for small damping, the average fractional loss of energy of a particle executing a damped simple harmonic motion in a cycle is four times the logarithmic decrement. **[cf. C.U. 2007]**

5. The differential equation for a one-dimensional damped harmonic oscillator is

$$m\frac{d^2x}{dt^2} + k\frac{dx}{dt} + sx = 0.$$

Explain the significance of each term in the equation. Solve the equation in the case of critical damping. **[C.U. 2004]**

6. Set up the differential equation of motion of a simple harmonic oscillator subjected to a damping force and an external simple harmonic force. Obtain expressions for the amplitude and the phase angle of the displacement in the steady state. **[cf. C.U. 1996]**

7. Give the transient solution and the steady-state solution of a damped simple harmonic oscillator acted on by an external simple harmonic force.

8. What are the electrical analogues of the following mechanical quantities: mass, compliance, velocity, and force?

Define mechanical impedance, inertial reactance, and reactance of compliance.

9. (a) Distinguish between amplitude resonance and velocity resonance. **[C.U. 1996, 2008]**

(b) Show that at velocity resonance, the maximum velocity is inversely proportional to damping and the velocity of the oscillator is in phase with the driving force. **[C.U. 1996]**

10. For the forced vibration of a mechanical damped harmonic oscillator driven by a sinusoidal force of angular frequency ω, obtain the value of ω for amplitude resonance. What is the amplitude of the oscillator at resonance? What are the time averaged kinetic and potential energies of the oscillator? Determine the value of ω for which they are equal. **[Burd. U. 1992]**

11. For a damped mechanical harmonic oscillator driven by a sinusoidal force of amplitude F and angular frequency ω, find an expression for the average power supplied by the driving force in one cycle. Define the Q factor of the driven oscillator. **[Burd. U. 1994]**

12. What is resonance? Draw the plots of the displacement amplitude and the velocity amplitude against the frequency of the sinusoidal force driving a mechanical oscillator for different values of damping. What is the power factor at velocity resonance? What is sharpness of resonance? Give a measure for it. **[cf. C.U. 1992]**

13. (a) Derive an expression for the average power supplied by the sinusoidal force driving a mechanical oscillator over a complete cycle in the steady-state. What is power factor and what is its value at resonance?

(b) Show that in the steady-state, the time-averaged input power equals the time-averaged power dissipated through friction.

[C.U. 1996, 2006]

14. (a) What are half-power frequencies? How are they related to sharpness of resonance?

(b) Show that the resonant frequency is the geometric mean between the half-power frequencies.

(c) When is the motion of a harmonic oscillator executing forced oscillations stiffness-controlled, mass-controlled, and resistance-controlled?

(d) Explain whether resonance should be sharp or flat (i) in detecting a particular frequency with the help of a Helmholtz resonator, and (ii) in the design of loudspeakers and microphones.

(e) Why are soldiers asked to break steps while crossing a bridge?

15. Show how the phase angle between the displacement and the exciting force and that between the velocity and the exciting force vary with the frequency of the exciting force in the case of forced vibrations. What are the phase angles at resonance?

16. Using Eq. (2.30) for the steady state displacement of a particle serving as a linear harmonic oscillator under the action of an external periodic force $F\cos\omega t$, answer the following:

(i) Find the instantaneous velocity of the particle and show that the velocity and the displacement differ in phase by $\pi/2$. (ii) Calculate the average power. (iii) Obtain the forcing frequency ω for which the average power is a maximum. Calculate this maximum average power. What is this situation called? (iv) Determine the mistuning factor $\Delta_{1/2} = \left|\frac{\omega_0}{\omega} - \frac{\omega}{\omega_0}\right|$ for which the average power is half that at resonance. **[cf. C.U. 1998]**

17. Derive the conditions for amplitude resonance and velocity resonance. Show that the displacement amplitude at velocity resonance is QF/s, where the symbols are standard. Prove that, for very small damping, this is also the maximum displacement amplitude at amplitude resonance.

18. Write the equations of motion of two masses coupled together and connected to two rigid supports by three springs of the same stiffness constant. Derive the characteristic equation and the frequencies of the normal modes of vibration. Write the general equations for the displacement of the masses in terms of the normal modes.

19. What are the normal co-ordinates of a coupled system? Show that the energy of the coupled vibrations of two equal masses is the sum of the energies the system would have if the normal modes were separately excited.

20. What is the difference between forced vibration and coupled vibration? **[C.U. 2008]**

PROBLEMS

1. A particle of mass 2g capable of executing a damped SHM, is acted on by an impulsive force while at rest. Its successive maximum displacements on the opposite sides immediately after it starts to move are 13 cm and 10 cm, respectively. Find the logarithmic decrement and the mechanical resistance if the time period is 10s. What would be the amplitude of oscillation in the absence of damping?
   [**Ans.** 0.262, 0.21 g/s, 14.82 cm]

2. A particle of mass 1g is displaced from its position of rest. It is acted on by a restoring force 5 dyne/cm and a damping force 1 dyne. s/cm. Examine whether the motion is oscillatory or not, if the particle is released. If the initial displacement of the particle is 5 cm, what is its displacement after 5s?
   [**Ans.** oscillatory, −0.134 cm]

3. The motion of a particle of mass 0.1 kg subjected to a restoring force 0.1 N/m is critically damped. While at rest, its motion is started by an initial velocity of 0.5 m/s. Find the maximum displacement of the particle.
   [**Ans.** 0.184 m]

4. A particle of mass 0.01 kg is subjected to a restoring force 0.5 N/m, a damping force 10^{-4} kg/s, and an external sinusoidal force of constant amplitude. For what frequency of the driving force is velocity resonance achieved? Calculate the Q-factor and the bandwidth of resonance.

   [**Ans.** 7.07 rad/s, 707, 10^{-2} rad/s]

5. Show that in the steady-state the frequency of vibration of a damped oscillator subjected to an external simple harmonic force must be the same as that of the periodic force.

   [HINTS. If possible, let the frequency of oscillation be ω' where ω' is different from ω, the frequency of the simple harmonic force. The equation of motion is $m\ddot{x} + k\dot{x} + sx = Fe^{j\omega t}$. The solution is $x = x_0 e^{j\omega' t}$, whence $(-m\omega'^2 + jk\omega' + s)x_0 e^{j\omega' t} = Fe^{j\omega t}$, giving $x_0 = Fe^{j(\omega-\omega')t}/(-m\omega'^2 + jk\omega' + s)$. Since x_0 must be independent of t, we must have $\omega' = \omega$.]

6. A particle of mass 0.1 kg is suspended from a spring of spring constant 200 N/m. A resistive force $-kv$ acts on the particle, v being the velocity of the particle in m/s and $k = 1$ N.s/m. A harmonic driving force $2.5 \cos 50t$ N is applied to the particle. Calculate the amplitude of oscillation of the particle, and the phase difference between the driving force and the particle displacement in the steady state.

   [**Ans.** 3.54 cm, $-45°$]

7. A particle of mass 0.1 kg is connected to the lower end of a massless spring suspended vertically from a rigid support. The spring has the force constant 10 N/m. A resistive force $-kv$ acts on the oscillating particle where k is a constant and v is the particle velocity. The time constant of the amplitude decay of the damped oscillatory motion of the particle is 25 s. Determine (i) the value of k, (ii) the Q-value of the oscillator, and (iii) the percentage change in frequency of oscillation due to damping. **[cf. Burd. U. 1997]**

   [**Ans.** 8×10^{-3} kg/s, 125, 0.02]

8. A particle of mass 0.005 kg performs a one-dimensional damped oscillatory motion, the resisting force per unit velocity being 10^{-3} N.s/m. The amplitude of oscillation is found to decay to 20% of its initial value in 5 full cycles. Find the time periods of the damped and the undamped oscillations of the particle.

   [**Ans.** 0.3219 s, 0.3219 s]

9. A vibrator of mass 1 gm. is acted upon by a restoring force of 10^7 dyne/cm, a retarding force of 4×10^3 dyne.sec/cm., and a driving force of $10^7 \cos \omega t$ dyne. Find the maximum possible amplitude.

   [**Ans.** 1.02 cm] **[cf. C.U. 2001, '05]**

10. A particle of mass 1 g whose undamped frequency is 4 rad/s, is subjected to a damping force of 10 dyne.s/cm. Show that the motion of the particle is overdamped. If the particle is initially at 0.5 cm. from its equilibrium position, what minimum initial velocity towards the equilibrium position must be given to the particle so that its displacement can be negative?

[**Ans.** 4 cm/s]

11. The displacement of a particle of mass m under critical damping is $x = vte^{-bt}$ where v is the velocity imparted to the particle at time $t = 0$, and $b = k/(2m) = \sqrt{s/m}$; k is the resisting force per unit velocity and s is the stiffness factor. (i) Show that the particle attains its maximum displacement at $t = 1/b$ and that the total energy of the particle at that instant of time is $\frac{1}{2}mv^2 \exp(-2)$. (ii) What is the total energy of the particle at $t = 2/b$?

[**Ans.** $\frac{5}{2}mv^2 \exp(-4)$]

12. The displacement of a particle of mass m driven by a force $F \cos \omega t$ is

$$x = \frac{(F/m) \cos \omega t}{\left[(\omega^2 - \omega_0^2)^2 + 4b^2\omega^2\right]^{1/2}}$$

where ω_0 is the undamped natural frequency of the forced system and b is the decay constant. Find ω for which the acceleration of the forced system is a maximum. What is the electrical analogue of the maximum acceleration?

[**Ans.** $\omega_0/\sqrt{1 - 2b^2/\omega_0^2} = \omega_0/\sqrt{1 - \frac{1}{2Q^2}}$. The electrical analogue is the maximum time rate of change of current in a series LCR circuit driven by a sinusoidal voltage.]

[HINTS. Acceleration amplitude, $a_0 = \frac{(F/m)\omega^2}{\left[(\omega^2-\omega_0^2)^2+4b^2\omega^2\right]^{1/2}}$. For a maximum a_0, put $da_0/d\omega = 0$.]

13. The equation of motion of a particle of mass m is $\ddot{x} + 2b\dot{x} + b^2x = 0$ where $2b = k/m$, k being the damping force per unit velocity. Identify the nature of damping. If the particle starts with an initial velocity v_0, write x as a function of time t. Show that the initial kinetic energy of the particle is expended in doing work against the damping force.

[HINTS. The work done against the damping force is $k \int_0^\infty \dot{x}^2 dt$. Show that it is equal to the initial kinetic energy $(1/2)mv_0^2$.]

[**Ans.** critical damping; $x = v_0te^{-bt}$]

Chapter 3

Fourier Analysis

3.1 Fourier's Theorem

A function of time t, i.e., $f(t)$ is said to be *periodic* if it repeats itself after a regular interval of time, say T. Thus, if $f(t)$ is periodic we have $f(t) = f(t \pm nT)$, where n is a positive integer. The time interval T is called the *period* of the function. The functions $\sin \omega t$ and $\cos \omega t$ are periodic, the period being $T = 2\pi/\omega$. We may have functions of t which are periodic but not sinusoidal. A periodic saw-tooth waveform used in the time base of a cathode-ray oscilloscope, a square wave generated by an electronic function generator are examples of such periodic functions.

Fourier's theorem states as follows: *Any single-valued periodic function can be represented by a summation of simple harmonic terms having frequencies which are integral multiples,* i.e., *harmonics of the frequency of the periodic function.* The series of the simple harmonic terms expressing a periodic function is referred to as the *Fourier series.* If $f(t)$ is periodic of period T, the Fourier series is

$$\begin{aligned} f(t) &= a_0 + a_1 \cos \omega t + a_2 \cos 2\omega t + \cdots + a_n \cos n\omega t + \cdots \\ &\qquad + b_1 \sin \omega t + b_2 \sin 2\omega t + \cdots + b_n \sin n\omega t + \cdots \\ &= a_0 + \sum_{n=1}^{\infty} (a_n \cos n\omega t + b_n \sin n\omega t). \end{aligned} \tag{3.1}$$

Here ω is the *fundamental frequency*: $\omega = \frac{2\pi}{T}$. The integral multiples of ω, i.e., $2\omega, 3\omega, 4\omega, 5\omega$ etc. are termed the *harmonics* of the fundamental frequency ω. The frequency 2ω is the second harmonic, 3ω is the third harmonic, and so on, $n\omega$ being the nth harmonic of ω. The coefficients $a_0, a_1, a_2, \cdots, b_1, b_2, b_3, \cdots$ are known as the *Fourier coefficients.*

Fourier's theorem has applications in many branches of physics. In acoustics, it is used in the analysis of musical notes. In general, a musical note is a periodic function of time. By Fourier's theorem, it consists of a number of harmonically related sinusoidal functions which represent pure tones.

The sufficient conditions to be satisfied by a periodic function $f(t)$ so that it can be expressed as a convergent Fourier series are called *Dirichlet's conditions*. These conditions are:

(i) $f(t)$ is a finite single-valued function.

(ii) $f(t)$ can have a finite number of discontinuities in the interval T.

(iii) $f(t)$ can have a finite number of maxima and minima in the interval T.

(iv) $\int_{t_0}^{t_0+T} |f(t)|dt$ exists.

A function satisfying these conditions in a given interval is said to be *piece-wise regular* in that interval.

Fourier's theorem does not apply at the discontinuities of $f(t)$. At a discontinuity of $f(t)$, the value of the Fourier series is the mean of the values of $f(t)$ on both sides of the discontinuity. This point is illustrated in the Fourier series of square wave pulses and saw-tooth waveform, discussed in Sec. 3.5.

An alternative form of expressing the Fourier series given by Eq.(3.1) is obtained by noting that

$$a_n \cos n\omega t + b_n \sin n\omega t = A_n \cos(n\omega t - \alpha_n), \tag{3.2}$$

where $A_n = \sqrt{a_n^2 + b_n^2}$ and $\tan \alpha_n = b_n/a_n$.
Hence the Fourier series for $f(t)$, i.e., Eq. (3.1) can be rewritten as

$$f(t) = A_0 + \sum_{n=1}^{\infty} A_n \cos(n\omega t - \alpha_n), \tag{3.3}$$

where $A_0 = a_0$. The quantity A_n is the amplitude and α_n is the phase of the nth harmonic. The values of A_n and α_n for different values of n show the harmonic content of $f(t)$. The variation of A_n with n (or $n\omega$) gives the *amplitude spectrum*, and the variation of α_n with n (or $n\omega$) gives the *phase spectrum* of $f(t)$. Since the amplitude and the phase angle values appear only at $n = 1, 2, 3$, etc., these spectra are termed *line spectra*.

Usually the amplitudes of the higher harmonic terms fall off with increasing n, so that the infinite Fourier series can be truncated after a finite number of terms depending on the accuracy with which $f(t)$ is to be represented. The evaluation of the Fourier coefficients and a decision regarding the termination point of the series form the subject matter of *Fourier analysis*.

3.2 Evaluation of Fourier Coefficients

To determine the Fourier coefficients, we use the following results, m and n being integers.

$$\int_{t_0}^{t_0+T} \sin m\omega t\, dt = 0, \quad \text{for all integral values of } m. \tag{3.4}$$

$$\int_{t_0}^{t_0+T} \cos n\omega t\, dt = 0, \quad \text{for all integral values of } n. \tag{3.5}$$

$$\int_{t_0}^{t_0+T} \cos m\omega t \sin n\omega t\, dt = 0, \quad \text{for all integral values of } m \text{ and } n. \tag{3.6}$$

$$\int_{t_0}^{t_0+T} \sin m\omega t \sin n\omega t\, dt = 0, \quad \text{for } m \neq n \tag{3.7}$$

$$= \frac{T}{2}, \quad \text{for } m = n \tag{3.8}$$

and
$$\int_{t_0}^{t_0+T} \cos m\omega t \cos n\omega t\, dt = 0, \quad \text{for } m \neq n \tag{3.9}$$

$$= \frac{T}{2}, \quad \text{for } m = n. \tag{3.10}$$

Integrating Eq. (3.1), we get

$$\int_0^T f(t)dt = a_0 \int_0^T dt + \sum_{n=1}^{\infty} \int_0^T (a_n \cos n\omega t + b_n \sin n\omega t)dt = a_0 T,$$

which follows from Eqs. (3.5) and (3.4) with $t_0 = 0$. Therefore,

$$a_0 = \frac{1}{T} \int_0^T f(t)dt. \tag{3.11}$$

This shows that a_0 is *the average value* of $f(t)$ over the period T. Thus, $a_0 = \overline{f(t)}$, the bar denoting the mean value.

To find the coefficient a_k, Eq. (3.1) is multiplied by $\cos k\omega t$ and then integrated from 0 to T. Hence

$$\int_0^T f(t) \cos k\omega t\, dt = \int_0^T a_0 \cos k\omega t\, dt + \sum_{n=1}^{\infty} \int_0^T (a_n \cos n\omega t \cos k\omega t + b_n \sin n\omega t \cos k\omega t)dt$$

$$= 0 + a_k \frac{T}{2} + 0,$$

where Eqs. (3.5), (3.9), (3.10), and (3.6) have been used with $t_0 = 0$. So,

$$a_k = \frac{2}{T} \int_0^T f(t) \cos k\omega t\, dt. \tag{3.12}$$

Similarly, to determine the coefficient b_k, we multiply. Eq. (3.1) by $\sin k\omega t$ and integrate from 0 to T. Proceeding along the same lines followed in the evaluation of a_k, we obtain

$$b_k = \frac{2}{T}\int_0^T f(t)\sin k\omega t\,dt. \tag{3.13}$$

With known $f(t)$, the Fourier coefficients can be found from Eqs. (3.11), (3.12), and (3.13). The evaluation of the Fourier coefficients is generally a tedious task. Fortunately, some simplifications can be made if the periodic function has certain types of symmetries. We shall discuss this point in the following section. For this simplification, symmetry must be kept in mind in selecting the $t = 0$ point on the time axis.

3.3 Symmetry and Fourier Coefficients

(a) Even-function symmetry: The function $f(t)$ is an even function when

$$f(t) = f(-t). \tag{3.14}$$

The interval of integration is taken from $-\frac{T}{2}$ to $\frac{T}{2}$ by putting $t_0 = -\frac{T}{2}$. Then Eqs. (3.11) to (3.13) reduce to

$$a_0 = \frac{1}{T}\int_{-T/2}^{T/2} f(t)dt \tag{3.15}$$

$$a_k = \frac{2}{T}\int_{-T/2}^{T/2} f(t)\cos k\omega t\,dt \tag{3.16}$$

$$\text{and} \quad b_k = \frac{2}{T}\int_{-T/2}^{T/2} f(t)\sin k\omega t\,dt. \tag{3.17}$$

When $f(t)$ is an even periodic function, these equations give

$$\begin{aligned} a_0 &= \frac{1}{T}\left[\int_{-T/2}^{0} f(t)dt + \int_0^{T/2} f(t)dt\right] \\ &= \frac{1}{T}\left[\int_0^{T/2} f(-z)dz + \int_0^{T/2} f(t)dt\right], \end{aligned}$$

where the variable t is changed to $-z$ in the first integral. Since $f(-z) = f(z)$, we get

$$a_0 = \frac{2}{T}\int_0^{T/2} f(t)dt. \tag{3.18}$$

Again,

$$
\begin{aligned}
a_k &= \frac{2}{T}\left[\int_{-T/2}^{0} f(t)\cos k\omega t\,dt + \int_0^{T/2} f(t)\cos k\omega t\,dt\right] \\
&= \frac{2}{T}\left[\int_0^{T/2} f(-z)\cos k\omega(-z)dz + \int_0^{T/2} f(t)\cos k\omega t\,dt\right] \\
&= \frac{2}{T}\left[\int_0^{T/2} f(z)\cos k\omega z\,dz + \int_0^{T/2} f(t)\cos k\omega t\,dt\right].
\end{aligned}
$$

In the first integral we have used the fact that $f(t)$ is an even function and that $\cos(-k\omega z) = \cos k\omega z$. So,

$$a_k = \frac{4}{T}\int_0^{T/2} f(t)\cos k\omega t\,dt. \tag{3.19}$$

Furthermore,

$$
\begin{aligned}
b_k &= \frac{2}{T}\left[\int_{-T/2}^{0} f(t)\sin k\omega t\,dt + \int_0^{T/2} f(t)\sin k\omega t\,dt\right] \\
&= \frac{2}{T}\left[\int_0^{T/2} f(-z)\sin(-k\omega z)dz + \int_0^{T/2} f(t)\sin k\omega t\,dt\right] \\
&= \frac{2}{T}\left[-\int_0^{T/2} f(z)\sin k\omega z\,dz + \int_0^{T/2} f(t)\sin k\omega t\,dt\right].
\end{aligned}
$$

Here we have used the fact that $f(z)$ is an even function and that $\sin(-k\omega z) = -\sin k\omega z$. Hence

$$b_k = 0. \tag{3.20}$$

Equations (3.18), (3.19), and (3.20) show that *the Fourier series expansion of an even periodic function consists of only the cosine terms and a constant, all the sine terms being absent.*

(b) Odd-function symmetry: The function $f(t)$ is an odd function when

$$f(t) = -f(-t). \tag{3.21}$$

Proceeding along the lines similar to those in deriving Eqs. (3.18) to (3.20), we find that

$$a_0 = 0 \tag{3.22}$$

$$a_k = 0 \tag{3.23}$$

and
$$b_k = \frac{4}{T}\int_0^{T/2} f(t)\sin k\omega t\,dt. \tag{3.24}$$

So, *the Fourier series expansion of an odd-periodic function consists of only the sine terms, the constant and the cosine terms being zero.*

(c) Half-wave symmetry: A periodic function $f(t)$ is said to have half-wave symmetry when

$$f(t) = -f\left(t - \frac{T}{2}\right). \tag{3.25}$$

Thus, if the periodic function can be obtained by inverting $f(t)$ from 0 to $T/2$ and shifting it by $T/2$, then the function will have half-wave symmetry. Two examples of half-wave symmetry are given Fig. 3.1.

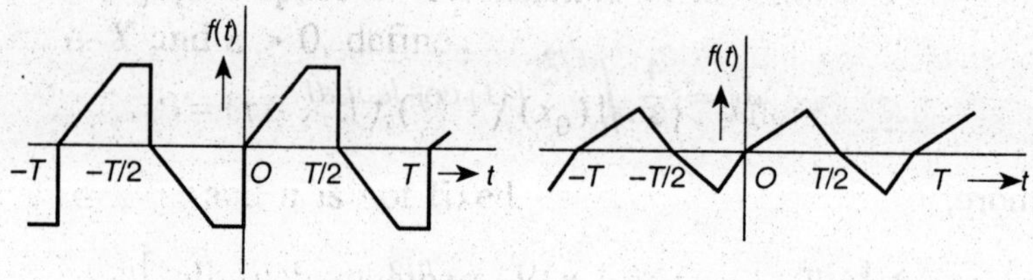

Fig. 3.1 Waveforms having half-wave symmetry

It is readily found that the average value of a periodic function possessing half-wave symmetry is zero. That is, $a_0 = 0$. We obtain

$$a_k = \frac{2}{T}\int_{-T/2}^{T/2} f(t)\cos k\omega t\,dt$$

$$= \frac{2}{T}\left[\int_{-T/2}^{0} f(t)\cos k\omega t\,dt + \int_{0}^{T/2} f(t)\cos k\omega t\,dt\right]. \tag{3.26}$$

The variable of integration in the first integral on the right-hand side of Eq. (3.26) is changed to z, where $z = t + T/2$. Then, $z = \frac{T}{2}$ at $t = 0$; and $z = 0$ at $t = -T/2$. Furthermore,

$$f(t) = f\left(z - \frac{T}{2}\right) = -f(z),$$

using Eq. (3.25), and $dt = dz$. Therefore,

$$\int_{-T/2}^{0} f(t)\cos k\omega t\,dt = -\int_{0}^{T/2} f(z)\cos k\omega\left(z - \frac{T}{2}\right)dz.$$

We have $\cos k\omega\left(z - \frac{T}{2}\right) = \cos(k\omega z - k\pi) = \cos k\pi\cos k\omega z$. So,

$$\int_{-T/2}^{0} f(t)\cos k\omega t\,dt = -\int_{0}^{T/2} f(z)\cos k\pi\cos k\omega z\,dz.$$

Hence Eq. (3.26) yields

$$a_k = \frac{2}{T}(1 - \cos k\pi)\int_0^{T/2} f(t)\cos k\omega t\, dt. \tag{3.27}$$

If k is even, $\cos k\pi = 1$, so that $a_k = 0$.
If k is odd, $\cos k\pi = -1$, and

$$a_k = \frac{4}{T}\int_0^{T/2} f(t)\cos k\omega t\, dt. \tag{3.28}$$

In a similar way, we can show that, when k is even $b_k = 0$, and when k is odd

$$b_k = \frac{4}{T}\int_0^{T/2} f(t)\sin k\omega t\, dt. \tag{3.29}$$

We thus conclude that *the Fourier series expansion of a periodic function with half-wave symmetry contains only odd harmonics, the constant term being zero.*

3.4 Exponential Form of Fourier Series

The Fourier series for a periodic function $f(t)$ is given by Eq. (3.1). Noting that

$$\cos n\omega t = \frac{e^{jn\omega t} + e^{-jn\omega t}}{2} \tag{3.30}$$

$$\text{and} \quad \sin n\omega t = \frac{e^{jn\omega t} - e^{-jn\omega t}}{2j}, \tag{3.31}$$

where $j = \sqrt{-1}$, we can rewrite Eq. (3.1) as

$$f(t) = a_0 + \sum_{n=1}^{\infty}\left[\frac{1}{2}a_n\left(e^{jn\omega t} + e^{-jn\omega t}\right) + \frac{1}{2j}b_n\left(e^{jn\omega t} - e^{-jn\omega t}\right)\right]. \tag{3.32}$$

Since $\frac{1}{j} = -j$, Eq. (3.32) reduces to

$$f(t) = a_0 + \sum_{n=1}^{\infty}\left[\left(\frac{a_n - jb_n}{2}\right)e^{jn\omega t} + \left(\frac{a_n + jb_n}{2}\right)e^{-jn\omega t}\right]. \tag{3.33}$$

We put

$$C_0 = a_0, \quad C_n = \frac{a_n - jb_n}{2}, \quad \text{and } C_{-n} = \frac{a_n + jb_n}{2}. \tag{3.34}$$

Then Eq. (3.33) gives

$$f(t) = C_0 + \sum_{n=1}^{\infty}\left(C_n e^{jn\omega t} + C_{-n}e^{-jn\omega t}\right) = \sum_{n=-\infty}^{\infty} C_n e^{jn\omega t}. \tag{3.35}$$

Equation (3.35) is referred to as the *exponential form* of the Fourier series. In terms of a_n and b_n, the coefficient C_n can be obtained as follows: We have

$$C_n = \frac{1}{T}\int_0^T f(t)\cos n\omega t\, dt - \frac{j}{T}\int_0^T f(t)\sin n\omega t\, dt, \tag{3.36}$$

since $C_n = \frac{a_n - jb_n}{2}$, and a_n and b_n are given by Eqs. (3.12) and (3.13), respectively, with k replaced by n. Equation (3.36) gives

$$\begin{aligned} C_n &= \frac{1}{T}\int_0^T f(t)\,(\cos n\omega t - j\sin n\omega t)\, dt \\ &= \frac{1}{T}\int_0^T f(t)e^{-jn\omega t} dt. \end{aligned} \tag{3.37}$$

This equation for C_n holds irrespective of whether n is positive, negative or zero. Since C_n is complex, we can write $C_n = |C_n|e^{j\theta_n}$. Therefore, $C_{-n} = C_n^* = |C_n|e^{-j\theta_n}$ and $|C_n| = \frac{1}{2}\sqrt{a_n^2 + b_n^2} = \frac{A_n}{2}$. The angle θ_n satisfies the relationship $\tan\theta_n = -b_n/a_n$. These expressions for $|C_n|$ and θ_n are valid for all values of n except $n = 0$, when $C_0 = a_0$, which is real and is the mean value of $f(t)$.

As $C_n = A_n/2$, a scale factor of 1/2 appears in the amplitude spectrum for the exponential form of the Fourier series compared to the ordinary form for all the lines except for the one with $n = 0$. For the exponential form, the spectral lines are drawn for both positive and negative values of n.

3.5 Fourier Series of Some Typical Waveforms

(a) Square wave pulses: The square wave pulses, shown in Fig. 3.2, have the period T and amplitude V. We have

$$\begin{aligned} f(t) &= V \quad \text{for } 0 < t < \frac{T}{2} \\ &= -V \quad \text{for } \frac{T}{2} < t < T. \end{aligned}$$

Hence

$$\begin{aligned} a_0 &= \frac{1}{T}\int_0^T f(t)dt = \frac{1}{T}\left[\int_0^{T/2} V\, dt + \int_{T/2}^T (-V)dt\right] \\ &= \frac{V}{T}\left[\frac{T}{2} - \left(T - \frac{T}{2}\right)\right] = 0. \end{aligned}$$

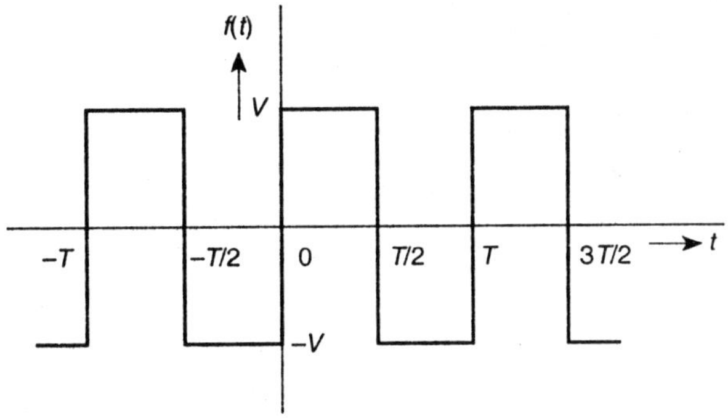

Fig. 3.2 A train of square wave pulses

This is the expected result, as it is evident from Fig. 3.2 that the average value of $f(t)$ is zero.

$$\begin{aligned}
a_k &= \frac{2}{T}\int_0^T f(t)\cos k\omega t\,dt \\
&= \frac{2}{T}\left[\int_0^{T/2} V\cos k\frac{2\pi}{T}t\,dt + \int_{T/2}^{T}(-V)\cos k\frac{2\pi}{T}t\,dt\right] \\
&= \frac{2V}{T}\times 0 = 0.
\end{aligned}$$

This result is also expected: since $f(t)$ is an odd function, all the cosine terms must be absent.

$$\begin{aligned}
b_k &= \frac{2}{T}\int_0^T f(t)\sin k\omega t\,dt \\
&= \frac{2V}{T}\left[\int_0^{T/2}\sin k\frac{2\pi}{T}t\,dt - \int_{T/2}^{T}\sin k\frac{2\pi}{T}t\,dt\right] \\
&= \frac{2V}{T}\frac{T}{2\pi k}\left[\cos k\frac{2\pi}{T}t\Big|_{T/2}^{0} + \cos k\frac{2\pi}{T}t\Big|_{T/2}^{T}\right] \\
&= \frac{V}{\pi k}(1-\cos k\pi+\cos 2k\pi-\cos k\pi) \\
&= \frac{2V}{k\pi}(1-\cos k\pi).
\end{aligned}$$

If k is even, $\cos k\pi = 1$ and so $b_k = 0$.
If k is odd, $\cos k\pi = -1$ and so $b_k = \frac{4V}{k\pi}$.

Therefore, all the even terms of the sine series are absent, and we have for the Fourier series

$$f(t) = \sum_{n=1,3,5,\cdots}^{\infty} b_n \sin n\omega t = \frac{4V}{\pi}\sum_{n=1,3,5,\cdots}^{\infty}\frac{1}{n}\sin n\omega t. \tag{3.38}$$

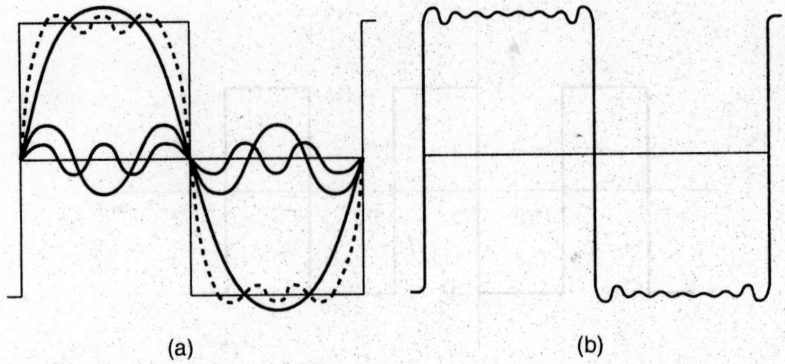

Fig. 3.3 Superposition of the first (a) three terms, (b) fifteen terms in the Fourier series of square wave pulses

In Fig. 3.3(a), the first three terms of the Fourier series have been added. The solid curves give the individual terms, and the resultant curve obtained by their superposition, is shown by the dashed curve. Clearly, the resultant curve roughly represents the square wave pulses. The resemblance to the square wave pulses can be improved by including more and more terms in the Fourier series. If fifteen terms are retained, the resultant curve is almost a square wave [Fig. 3.3(b)]. The superposition of the sinusoidal components of the Fourier series to reproduce the periodic waveform is known as *Fourier synthesis.*

Observe that at the discontinuities ($t = 0$ and $t = T/2$), the Fourier series attains the *mean value* at the discontinuity. In this case, it is zero. This is a *general property* of the Fourier series.

The amplitude and phase spectra of the train of square wave pulses are shown in Figs. 3.4(a) and (b), respectively. Note that the amplitudes fall off for higher harmonics.

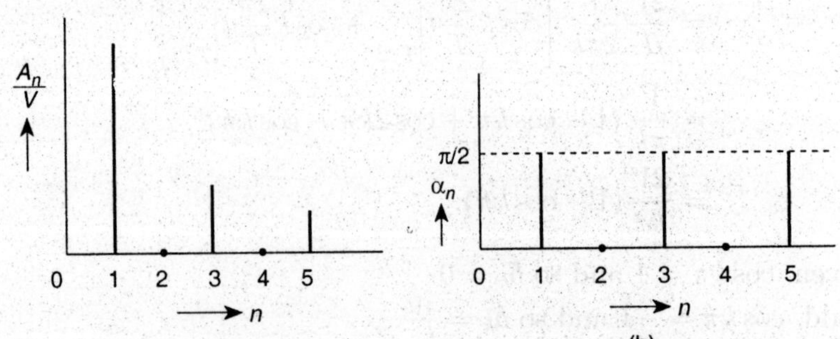

Fig. 3.4 Amplitude and phase spectra for square wave pulses

(b) Triangular waveform: Fig. 3.5 shows a triangular waveform of period T and amplitude V. The function $f(t)$ is an odd periodic function here. So, only the sine terms are present in the Fourier series. We have

$$\begin{aligned} f(t) &= \frac{4t}{T}V, && \text{for } 0 < t < \frac{T}{4} \\ &= \left(-\frac{4t}{T}+2\right)V, && \text{for } \frac{T}{4} < t < \frac{T}{2}. \end{aligned}$$

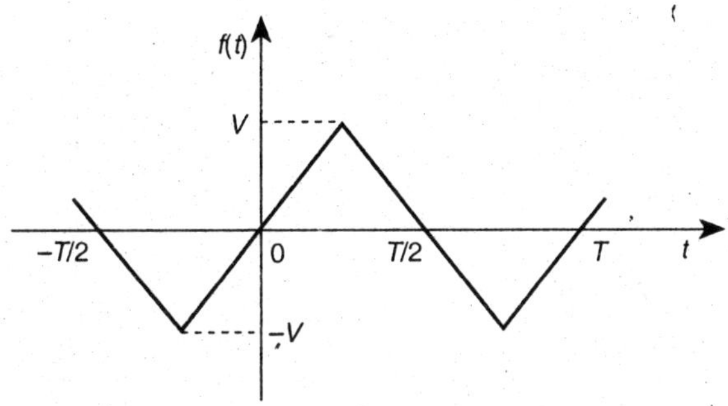

Fig. 3.5 Triangular waveform

Therefore,

$$\begin{aligned} b_k &= \frac{4}{T}\int_0^{T/2} f(t)\sin k\omega t\,dt \\ &= \frac{4V}{T}\left[\frac{4}{T}\int_0^{T/4} t\sin k\frac{2\pi}{T}t\,dt + \int_{T/4}^{T/2}\left(-\frac{4t}{T}+2\right)\sin k\frac{2\pi}{T}t\,dt\right] \\ &= \frac{8V}{k^2\pi^2}\sin\frac{k\pi}{2}. \end{aligned}$$

Thus the Fourier series representation of $f(t)$ is

$$\begin{aligned} f(t) &= \sum_{n=1}^{\infty} b_n \sin n\omega t = \frac{8V}{\pi^2}\sum_{n=1}^{\infty}\frac{1}{n^2}\sin\frac{n\pi}{2}\sin n\omega t \\ &= \frac{8V}{\pi^2}\left[\sin\omega t - \frac{1}{9}\sin 3\omega t + \frac{1}{25}\sin 5\omega t - \cdots\right]. \end{aligned} \tag{3.39}$$

Clearly, the Fourier series expansion contains only the odd harmonics, and the accuracy with which the triangular waveform is reproduced increases as the number of terms in Eq. (3.39) is increased.

The amplitude and the phase spectra of the triangular waveform are shown in Figs. 3.6(a) and (b), respectively.

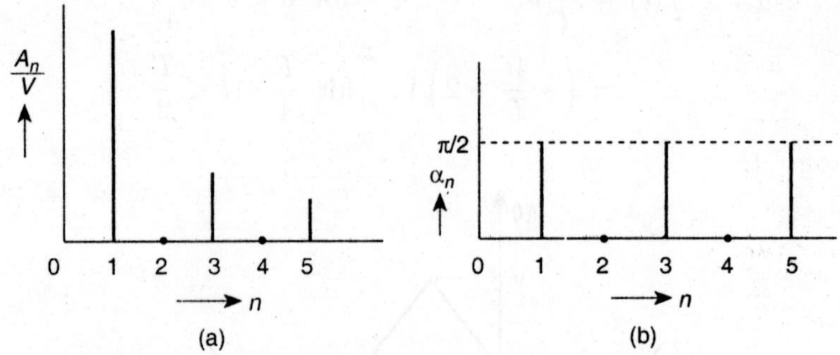

Fig. 3.6 (a) Amplitude and (b) phase spectra for the triangular waveform

(c) Half sinusoidal waveform: Figure 3.7 gives a half sinusoidal waveform of period T and amplitude V. Such a waveform is obtained at the output of a full-wave rectifier driven by a sinusoidal voltage. The function

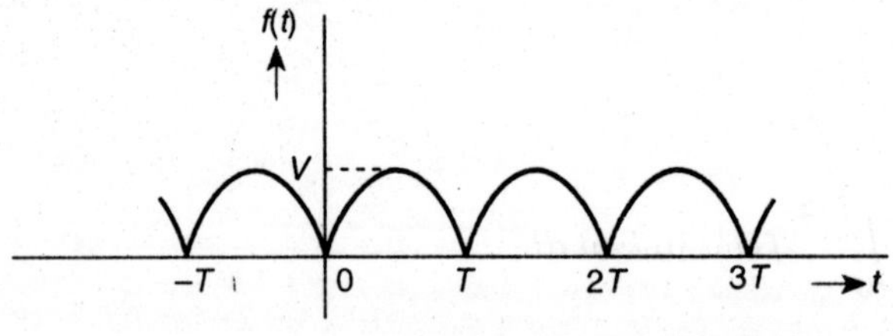

Fig. 3.7 Half sinusoidal wave form

$f(t)$ is an even periodic function here, so that the sine terms disappear in the Fourier series. We have

$$
\begin{aligned}
a_0 &= \frac{2}{T}\int_0^{T/2} f(t)dt = \frac{2}{T}\int_0^{T/2} V\sin\omega t\, dt = \frac{2V}{\pi}.\\
a_k &= \frac{4}{T}\int_0^{T/2} f(t)\cos k\omega t\, dt = \frac{4V}{T}\int_0^{T/2} \sin\omega t\cos k\omega t\, dt\\
&= \frac{2V}{T}\int_0^{T/2} [\sin(k+1)\omega t - \sin(k-1)\omega t]\, dt\\
&= \frac{V}{\pi}\left[\frac{1}{k+1} - \frac{\cos(k+1)\pi}{k+1} + \frac{\cos(k-1)\pi}{k-1} - \frac{1}{k-1}\right]\\
&= \frac{V}{\pi(k^2-1)}[-2-(k-1)\cos(k+1)\pi + (k+1)\cos(k-1)\pi].
\end{aligned}
$$

If k is odd, $a_k = 0$.

If k is even, $a_k = \frac{-4V}{\pi(k^2-1)}$.

So, the Fourier series is

$$f(t) = \frac{2V}{\pi} - \frac{4V}{\pi} \sum_{n=2,4,6,\cdots}^{\infty} \frac{\cos n\omega t}{n^2 - 1} \tag{3.40}$$

$$= \frac{2V}{\pi} - \frac{4V}{\pi}\left(\frac{1}{3}\cos 2\omega t + \frac{1}{15}\cos 4\omega t + \frac{1}{35}\cos 6\omega t + \cdots\right) \tag{3.41}$$

(d) Sawtooth waveform: A sawtooth waveform of time period T and peak value A is depicted in Fig. 3.8. Such a voltage waveform is applied to the time base of a cathode-ray oscilloscope (CRO). Here

$$f(t) = \frac{A}{T}t, \quad \text{for } 0 < t < T.$$

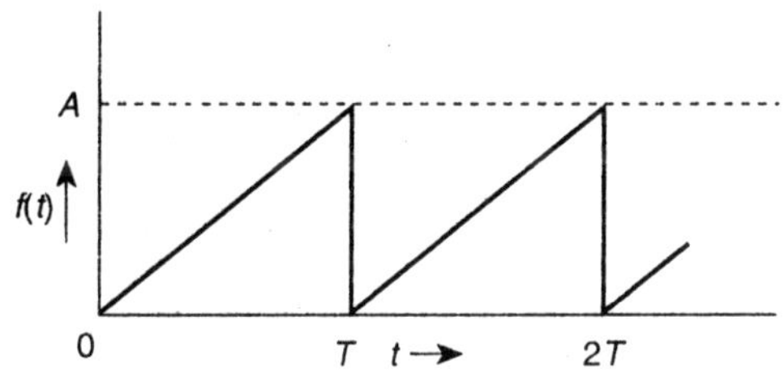

Fig. 3.8 Sawtooth waveform

So,

$$a_0 = \frac{1}{T}\int_0^T f(t)dt = \frac{A}{T^2}\int_0^T t\,dt = \frac{A}{2}.$$

$$a_k = \frac{2}{T}\int_0^T f(t)\cos k\omega t\,dt = \frac{2A}{T^2}\int_0^T t\,\cos k\frac{2\pi}{T}t\,dt$$

$$= \frac{2A}{T^2}\left[t\frac{T}{2k\pi}\sin k\frac{2\pi}{T}t\Big|_0^T - \int_0^T \frac{T}{2k\pi}\sin k\frac{2\pi}{T}t\,dt\right]$$

(Integrating by parts)

$$= \frac{2A}{T^2}\left(\frac{T}{2k\pi}\right)^2 \cos k\frac{2\pi}{T}t\Big|_0^T = 0,$$

whence the cosine terms disappear.

$$b_k = \frac{2}{T}\int_0^T f(t)\sin k\omega t\,dt = \frac{2A}{T^2}\int_0^T t\sin k\frac{2\pi}{T}t\,dt$$

$$= \frac{2A}{T^2}\left[-\frac{tT}{2k\pi}\cos k\frac{2\pi}{T}t\Big|_0^T + \int_0^T \frac{T}{2k\pi}\cos k\frac{2\pi}{T}t\,dt\right]$$

$$= \frac{2A}{T^2}\left[-\frac{T^2}{2k\pi} + \left(\frac{T}{2k\pi}\right)^2 \sin k\frac{2\pi}{T}t\Big|_0^T\right]$$

$$= -\frac{A}{k\pi}.$$

Therefore, the Fourier series for the sawtooth waveform is

$$\begin{aligned} f(t) &= \frac{A}{2} - \frac{A}{\pi}\sum_{n=1}^{\infty}\frac{1}{n}\sin n\omega t \\ &= \frac{A}{2} - \frac{A}{\pi}\left(\sin\omega t + \frac{1}{2}\sin 2\omega t + \frac{1}{3}\sin 3\omega t + \cdots\right). \end{aligned} \tag{3.42}$$

The addition of the first few terms of the series [Eq. (3.42)] is shown graphically in Fig. 3.9. The superposition of more and more terms clearly shows the convergence of the series, i.e., the inclusion of more and more terms in the series improves the resemblance between the resultant curve and the curve under test.

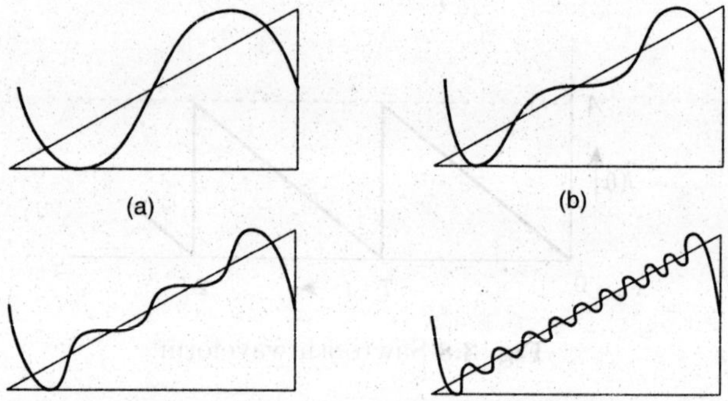

(a) (b) (c) (d)

Fig. 3.9 Superposition of the average and (a) the first harmonic, (b) the first two harmonics, (c) the first three harmonics, and (d) the first ten harmonics in the Fourier series of the sawtooth waveform

Note that at the discontinuity (i.e., at $t = 0$ or $t = T$) the series attains the mean of the values on the two sides of the discontinuity. This value is $A/2$.

3.6 Expansion of the Periodic Function $f(x)$

Here we consider a function $f(x)$ of a real variable x which may not involve time. Suppose that $f(x)$ is periodic of period $(x_2 - x_1)$, and satisfies Dirichlet's conditions; i.e., $f(x)$ is finite and single valued, has a finite number of discontinuities, and a finite number of maxima and minima in the interval $x_1 < x < x_2$. Then a Fourier series can be written for $f(x)$:

$$f(x) = a_0 + \sum_{n=1}^{\infty}(a_n \cos n\omega x + b_n \sin n\omega x), \tag{3.43}$$

where

$$\omega = \frac{2\pi}{x_2 - x_1}. \tag{3.44}$$

If the interval is $(0, 2\pi)$ or $(-\pi, \pi)$, then the period is $x_2 - x_1 = 2\pi$. Hence $\omega = 1$, and Eq. (3.43) becomes

$$f(x) = a_0 + \sum_{n=1}^{\infty} (a_n \cos nx + b_n \sin nx). \tag{3.45}$$

In analogy with Eqs. (3.15), (3.16) and (3.17), the coefficients of Eq. (3.45) can be expressed as

$$a_0 = \frac{1}{2\pi} \int_{-\pi}^{\pi} f(x) dx \tag{3.46}$$

$$a_k = \frac{1}{\pi} \int_{-\pi}^{\pi} f(x) \cos kx \, dx \tag{3.47}$$

and $$b_k = \frac{1}{\pi} \int_{-\pi}^{\pi} f(x) \sin kx \, dx, \tag{3.48}$$

where the interval is $(-\pi, \pi)$. If instead, the interval is $(0, 2\pi)$, the limits of integration in Eqs. (3.46) to (3.48) are taken to be 0 and 2π in place of $-\pi$ and π.

If the function $f(x)$ is given from $x = -l$ to $x = l$, the period is $x_2 - x_1 = 2l$, so that Eq. (3.44) gives $\omega = \frac{\pi}{l}$. Hence Eq. (3.43) reduces to

$$f(x) = a_0 + \sum_{n=1}^{\infty} \left(a_n \cos \frac{n\pi}{l} x + b_n \sin \frac{n\pi}{l} x \right). \tag{3.49}$$

Here the coefficients are given by

$$a_0 = \frac{1}{2l} \int_{-l}^{l} f(x) dx \tag{3.50}$$

$$a_k = \frac{1}{l} \int_{-l}^{l} f(x) \cos \frac{k\pi}{l} x \, dx \tag{3.51}$$

and $$b_k = \frac{1}{l} \int_{-l}^{l} f(x) \sin \frac{k\pi}{l} x \, dx. \tag{3.52}$$

Observation

If $f(x)$ is *not* periodic, the Fourier expansion holds only within the given interval ($-l$ to $+l$, say). Outside this interval, the Fourier series is periodic having the period $2l$, but it does not agree with $f(x)$.

3.7 Half-Range Series

Sometimes a function $f(x)$ in the range $(0, l)$ which is half the period of $2l$, needs to be expanded in a Fourier series. Since the nature of the function outside the range $0 < x < l$ is immaterial, we extend the function in the range $-l < x < 0$ so that the function is odd or even. The Fourier series of $f(x)$ thus contains either the sine series or the cosine series.

If $f(x)$ is to be expanded in a *half-range sine series* in $0 < x < l$, we take $f(x) = -f(x)$ in the range $-l < x < 0$. The function is then *odd* in the entire period $(-l, l)$. The Fourier expansion of $f(x)$ is

$$f(x) = \sum_{n=1}^{\infty} b_n \sin \frac{n\pi x}{l} \tag{3.53}$$

where

$$b_n = \frac{2}{l} \int_0^l f(x) \sin \frac{n\pi x}{l} dx. \tag{3.54}$$

On the other hand, if $f(x)$ has to be expressed as a *half-range cosine series* in $0 < x < l$, we choose $f(x) = f(-x)$ in the range $-l < x < 0$, making it *even* over the period of $2l$. Then,

$$f(x) = a_0 + \sum_{n=1}^{\infty} a_n \cos \frac{n\pi x}{l}, \tag{3.55}$$

where

$$a_0 = \frac{1}{l} \int_0^l f(x) dx \tag{3.56}$$

and

$$a_n = \frac{2}{l} \int_0^l f(x) \cos \frac{n\pi x}{l} dx. \tag{3.57}$$

3.8 Solved Problems

Problem 1. *The function $f(x) = x^2$ is defined within the interval $-\pi \le x \le \pi$, and outside it is periodic.* (i) *Show whether the function is even or odd within $-\pi \le x \le \pi$.* (ii) *Expand $f(x)$ in a Fourier series to show*

$$f(x) = \frac{\pi^2}{3} + \sum_{n=1}^{\infty} \frac{4}{n^2}(-1)^n \cos nx.$$ **[cf. Burd. U. 2001]**

(iii) *Use the above result to show that*

$$\sum_{n=1}^{\infty} \frac{1}{n^2} = \frac{\pi^2}{6}.$$ **[C.U. 1997]**

Solution. (i) We have $f(-x) = (-x)^2 = x^2 = f(x)$. Hence the function is even.

(ii) The Fourier series for $f(x)$ is given by Eq. (3.45). So,

$$\begin{aligned}
a_0 &= \frac{1}{2\pi}\int_{-\pi}^{\pi} f(x)dx = \frac{1}{2\pi}\int_{-\pi}^{\pi} x^2 dx = \frac{1}{6\pi}(\pi^3+\pi^3) = \frac{\pi^2}{3} \\
a_k &= \frac{1}{\pi}\int_{-\pi}^{\pi} f(x)\cos kx\, dx = \frac{1}{\pi}\int_{-\pi}^{\pi} x^2 \cos kx\, dx \\
&= \frac{1}{\pi}\left[\frac{x^2}{k}\sin kx\Big|_{-\pi}^{\pi} - \int_{-\pi}^{\pi}\frac{2x}{k}\sin kx\, dx\right] \qquad \text{(integrating by parts)} \\
&= -\frac{2}{\pi k}\int_{-\pi}^{\pi} x\sin kx\, dx = -\frac{2}{\pi k}\left[-\frac{x}{k}\cos kx\Big|_{-\pi}^{\pi} + \frac{1}{k}\int_{-\pi}^{\pi}\cos kx\, dx\right] \\
&= \frac{4}{k^2}\cos k\pi = \frac{4}{k^2}(-1)^k.
\end{aligned}$$

Since $f(x)$ is an even function, the sine terms are absent, i.e., $b_k = 0$. Hence

$$f(x) = \frac{\pi^2}{3} + \sum_{n=1}^{\infty}\frac{4}{n^2}(-1)^n\cos nx \quad \text{(proved)}.$$

(iii) Putting $x = \pi$ in the above series for $f(x)$ we have

$$f(\pi) = \pi^2 = \frac{\pi^2}{3} + 4\sum_{n=1}^{\infty}\frac{1}{n^2}$$

$$\text{whence} \quad \sum_{n=1}^{\infty}\frac{1}{n^2} = \frac{1}{4}\cdot\frac{2\pi^2}{3} = \frac{\pi^2}{6} \quad \text{(proved)}.$$

Problem 2. *Expand $f(x) = A$, where A is a constant, in the range $0 < x < \pi$, in a Fourier series. Hence show that*

$$\frac{\pi}{4} = 1 - \frac{1}{3} + \frac{1}{5} - \frac{1}{7} + \cdots.$$

[cf. C.U. 1987, '96]

Solution. We expand $f(x)$ here in a half-range sine series in the range $0 < x < \pi$. The Fourier series is

$$f(x) = \sum_{n=1}^{\infty} b_n \sin nx,$$

$$\begin{aligned}
\text{where} \quad b_n &= \frac{2}{\pi}\int_0^{\pi} f(x)\sin nx\, dx \\
&= \frac{2A}{\pi}\left(\frac{-\cos nx}{n}\right)_0^{\pi} = \frac{2A}{\pi n}(1-\cos n\pi).
\end{aligned}$$

If n is odd, $\cos n\pi = -1$ and $b_n = \frac{4A}{\pi n}$.

If n is even, $\cos n\pi = 1$ and $b_n = 0$.

Hence

$$f(x) = A = \frac{4A}{\pi} \sum_{n=1,3,5,\cdots}^{\infty} \frac{1}{n} \sin nx$$
$$= \frac{4A}{\pi} \left[\sin x + \frac{1}{3} \sin 3x + \frac{1}{5} \sin 5x + \cdots \right]$$

Putting $x = \frac{\pi}{2}$, we get

$$\frac{\pi}{4} = 1 - \frac{1}{3} + \frac{1}{5} - \frac{1}{7} + \cdots \quad \text{(proved)}.$$

Problem 3. (a) *Show that the function*

$$f(x) = \begin{cases} 0 \text{ for } -\pi \le x < 0 \\ x \text{ for } \quad 0 \le x < \pi \end{cases}$$

can be expanded in a Fourier series as

$$f(x) = \frac{\pi}{4} - \frac{2}{\pi} \sum_{n=1}^{\infty} \frac{\cos(2n-1)x}{(2n-1)^2} - \sum_{n=1}^{\infty} (-1)^n \frac{\sin nx}{n}. \qquad \text{[C.U. 1991, 2000, '05]}$$

(b) *From this expansion show that*

$$1 + \frac{1}{3^2} + \frac{1}{5^2} + \cdots = \frac{\pi^2}{8}. \qquad \text{[C.U. 2000, '05]}$$

Solution. (a) The Fourier series for $f(x)$ is

$$f(x) = a_0 + \sum_{n=1}^{\infty} (a_n \cos nx + b_n \sin nx).$$

Here the interval is $(-\pi, \pi)$. Hence

$$a_0 = \frac{1}{2\pi} \int_{-\pi}^{\pi} f(x) dx = \frac{1}{2\pi} \left[\int_{-\pi}^{0} 0 \cdot dx + \int_0^{\pi} x\, dx \right] = \frac{\pi}{4}$$
$$a_k = \frac{1}{\pi} \int_{-\pi}^{\pi} f(x) \cos kx\, dx = \frac{1}{\pi} \int_0^{\pi} x \cos kx\, dx.$$

We use here Kronecker's formula

$$\int_0^a f(x)\phi(x) dx = \left[f(x)\phi_1(x) - f'(x)\phi_2(x) + f''(x)\phi_3(x) - \cdots \right]_0^a,$$

where f', f'' etc. are the first, second etc. derivative of $f(x)$; and ϕ_1, ϕ_2, etc. are the first, second etc. integral of $\phi(x)$. The function $f(x)$ is a polynomial of x and $\phi(x)$ is $\sin kx$ or $\cos kx$. Therefore,

$$a_k = \frac{1}{\pi}\left[x\frac{\sin kx}{k} - 1\left(-\frac{\cos kx}{k^2}\right)\right]_0^{\pi} = \frac{1}{\pi k^2}(\cos k\pi - 1)$$

$$b_k = \frac{1}{\pi}\int_{-\pi}^{\pi} f(x)\sin kx\,dx = \frac{1}{\pi}\int_0^{\pi} x\sin kx\,dx$$

$$= \frac{1}{\pi}\left[x\left(\frac{-\cos kx}{k}\right) - 1\left(\frac{-\sin kx}{k^2}\right)\right]_0^{\pi} = -\frac{\cos k\pi}{k}.$$

If k is odd, $\cos k\pi = -1$ and $b_k = \frac{1}{k}$.
If k is even, $\cos k\pi = 1$ and $b_k = -\frac{1}{k}$.
Hence $b_k = (-1)^k(-1/k)$.
So,

$$f(x) = \frac{\pi}{4} + \frac{1}{\pi}\sum_{n=1}^{\infty}\frac{\cos n\pi - 1}{n^2}\cos nx - \sum_{n=1}^{\infty}(-1)^n\frac{\sin nx}{n}.$$

We have

$$\sum_{n=1}^{\infty}(\cos n\pi - 1)\frac{\cos nx}{n^2} = -\frac{2\cos x}{1} - \frac{2\cos 3x}{9} - \frac{2\cos 5x}{25} - \cdots$$

$$= -2\sum_{n=1}^{\infty}\frac{\cos(2n-1)x}{(2n-1)^2}.$$

Thus

$$f(x) = \frac{\pi}{4} - \frac{2}{\pi}\sum_{n=1}^{\infty}\frac{\cos(2n-1)x}{(2n-1)^2} - \sum_{n=1}^{\infty}(-1)^n\frac{\sin nx}{n} \quad \text{(proved).}$$

(b) Putting $x = 0$ we get

$$f(0) = 0 = \frac{\pi}{4} - \frac{2}{\pi}\left(\frac{1}{1^2} + \frac{1}{3^2} + \frac{1}{5^2} + \cdots\right)$$

$$\text{or,} \quad 1 + \frac{1}{3^2} + \frac{1}{5^2} + \cdots = \frac{\pi^2}{8}, \quad \text{(proved).}$$

Problem 4. *Obtain the Fourier series expansion of the following function with period 2π.*

$$f(x) = \begin{cases} x(\pi - x) \text{ for } & 0 \le x \le \pi \\ x(\pi + x) \text{ for } & -\pi \le x \le 0. \end{cases}$$

Hence prove that

$$1 - \frac{1}{3^3} + \frac{1}{5^3} - \frac{1}{7^3} + \cdots = \frac{\pi^3}{32}.$$

[C.U. 1992]

Solution. The Fourier series is

$$f(x) = a_0 + \sum_{n=1}^{\infty}(a_n \cos nx + b_n \sin nx).$$

We have

$$a_0 = \frac{1}{2\pi}\int_{-\pi}^{\pi} f(x)\,dx = \frac{1}{2\pi}\left[\int_{-\pi}^{0} x(\pi + x)dx + \int_0^{\pi} x(\pi - x)dx\right]$$

$$= \frac{1}{2\pi}\left[\left(\pi\frac{x^2}{2} + \frac{x^3}{3}\right)_{-\pi}^{0} + \left(\pi\frac{x^2}{2} - \frac{x^3}{3}\right)_{0}^{\pi}\right] = 0$$

$$a_k = \frac{1}{\pi}\int_{-\pi}^{\pi} f(x)\cos kx\,dx$$

$$= \frac{1}{\pi}\left[\int_{-\pi}^{0} x(\pi + x)\cos kx\,dx + \int_0^{\pi} x(\pi - x)\cos kx\,dx\right]$$

$$= \int_{-\pi}^{0} x\cos kx\,dx + \frac{1}{\pi}\int_{-\pi}^{0} x^2\cos kx\,dx$$

$$+ \int_0^{\pi} x\cos kx\,dx - \frac{1}{\pi}\int_0^{\pi} x^2\cos kx\,dx$$

$$= \int_{\pi}^{0}(-z)\cos k(-z)(-dz) + \frac{1}{\pi}\int_{\pi}^{0}(-z)^2\cos k(-z)(-dz)$$

$$+ \int_0^{\pi} x\cos kx\,dx - \frac{1}{\pi}\int_0^{\pi} x^2\cos kx\,dx,$$

where we have put $x = -z$ in the first two integrals with consequent changes in the limits. Putting now $z = x$, we get

$$a_k = -\int_0^{\pi} x\cos kx\,dx + \frac{1}{\pi}\int_0^{\pi} x^2\cos kx\,dx + \int_0^{\pi} x\cos kx\,dx$$

$$-\frac{1}{\pi}\int_0^{\pi} x^2\cos kx\,dx = 0$$

$$b_k = \frac{1}{\pi}\int_{-\pi}^{\pi} f(x)\sin kx\,dx$$

$$= \frac{1}{\pi}\left[\int_{-\pi}^{0} x(\pi + x)\sin kx\,dx + \int_0^{\pi} x(\pi - x)\sin kx\,dx\right]$$

$$= \int_{-\pi}^{0} x\sin kx\,dx + \frac{1}{\pi}\int_{-\pi}^{0} x^2\sin kx\,dx$$

$$+ \int_0^{\pi} x\sin kx\,dx - \frac{1}{\pi}\int_0^{\pi} x^2\sin kx\,dx$$

$$= 2\int_0^{\pi} x\sin kx\,dx - \frac{2}{\pi}\int_0^{\pi} x^2\sin kx\,dx,$$

since

$$\int_{-\pi}^{0} x \sin kx \, dx = \int_{0}^{\pi} x \sin kx \, dx$$

$$\text{and} \quad \int_{-\pi}^{0} x^2 \sin kx \, dx = -\int_{0}^{\pi} x^2 \sin kx \, dx.$$

Using Kronecker's formula, we obtain

$$\begin{aligned} b_k &= 2\Bigg[x\left(\frac{-\cos kx}{k}\right) - 1\left(\frac{-\sin kx}{k^2}\right) - \frac{1}{\pi}\Bigg\{x^2\left(\frac{-\cos kx}{k}\right) \\ &\qquad -2x\left(\frac{-\sin kx}{k^2}\right) + 2\left(\frac{\cos kx}{k^3}\right)\Bigg\}\Bigg]_0^{\pi} \\ &= \frac{4}{\pi k^3}(1 - \cos k\pi). \end{aligned}$$

If k is odd, $\cos k\pi = -1$ and $b_k = \frac{8}{\pi k^3}$.
If k is even, $\cos k\pi = 1$ and $b_k = 0$.
Hence

$$f(x) = \frac{8}{\pi} \sum_{n=1,3,5,\text{etc.}}^{\infty} \frac{1}{n^3} \sin nx = \frac{8}{\pi}\left[\sin x + \frac{1}{3^3}\sin 3x + \frac{1}{5^3}\sin 5x + \cdots\right].$$

Putting $x = \frac{\pi}{2}$, we have

$$f\left(\frac{\pi}{2}\right) \doteq \frac{\pi}{2}\left(\pi - \frac{\pi}{2}\right) = \frac{8}{\pi}\left[\sin\frac{\pi}{2} + \frac{1}{3^3}\sin\frac{3\pi}{2} + \frac{1}{5^3}\sin\frac{5\pi}{2} + \cdots\right]$$

$$\text{or,} \quad 1 - \frac{1}{3^3} + \frac{1}{5^3} - \frac{1}{7^3} + \cdots = \frac{\pi^3}{32} \text{ proved.}$$

Problem 5. *Expand in a Fourier series the periodic function $f(x)$ with period $2l$ where*

$$f(x) = |x| \text{ for } -l < x < l.$$

Hence show that

$$\frac{1}{1^2} + \frac{1}{3^2} + \frac{1}{5^2} + \cdots = \frac{\pi^3}{8}.$$ **[cf. Burd. U. 1995, C.U. 1999]**

Solution. The Fourier series is

$$f(x) = a_0 + \sum_{n=1}^{\infty}\left(a_n \cos\frac{n\pi x}{l} + b_n \sin\frac{n\pi x}{l}\right).$$

Since $f(x)=|x|$ is an even function, all the sine terms disappear, i.e., $b_n=0$.

$$a_0 = \frac{1}{2l}\int_{-l}^{l} f(x)dx = \frac{2}{2l}\int_0^l x\,dx = \frac{1}{l}\cdot\frac{l^2}{2} = \frac{l}{2}$$

$$a_k = \frac{1}{l}\int_{-l}^{l} f(x)\cos\frac{k\pi}{l}x\,dx = \frac{2}{l}\int_0^l x\cos\frac{k\pi x}{l}dx$$

$$= \frac{2}{l}\left[x\frac{\sin\frac{k\pi x}{l}}{\frac{k\pi}{l}} - 1\frac{\left(-\cos\frac{k\pi x}{l}\right)}{\left(\frac{k\pi}{l}\right)^2}\right]_0^l = \frac{2l}{k^2\pi^2}(\cos k\pi - 1).$$

If k is even, $\cos k\pi = 1$, and $a_k = 0$.
If k is odd, $\cos k\pi = -1$, and $a_k = -\frac{4l}{k^2\pi^2}$.
So,

$$f(x) = |x| = \frac{l}{2} - \frac{4l}{\pi^2}\sum_{n=1,3,5,\cdots}^{\infty}\frac{1}{n^2}\cos\frac{n\pi x}{l}$$

$$= \frac{l}{2} - \frac{4l}{\pi^2}\left[\cos\frac{\pi x}{l} + \frac{1}{3^2}\cos\frac{3\pi x}{l} + \frac{1}{5^2}\cos\frac{5\pi x}{l} + \cdots\right].$$

Putting $x = 0$, it follows that

$$\frac{1}{1^2} + \frac{1}{3^2} + \frac{1}{5^2} + \cdots = \frac{\pi^2}{8}. \quad \text{(Proved)}$$

Problem 6. *Express $f(x) = x$ as a half-range sine series in $0 < x < 2$.*

[Burd. U. 1996]

Solution. Here $f(x) = x = \sum_{n=1}^{\infty} b_n \sin\frac{n\pi x}{2}$

where

$$b_n = \frac{2}{2}\int_0^2 f(x)\sin\frac{n\pi x}{2}dx = \int_0^2 x\sin\frac{n\pi x}{2}dx$$

$$= \left[x\frac{\left(-\cos\frac{n\pi x}{2}\right)}{\frac{n\pi}{2}} - 1\frac{\left(-\sin\frac{n\pi x}{2}\right)}{\left(\frac{n\pi}{2}\right)^2}\right]_0^2 = -\frac{4}{n\pi}\cos n\pi.$$

When n is odd, $\cos n\pi = -1$ and $b_n = \frac{4}{n\pi}$.
When n is even, $\cos n\pi = 1$ and $b_n = -\frac{4}{n\pi}$.
Hence

$$f(x) = x = -\frac{4}{\pi}\sum_{n=1}^{\infty}\frac{(-1)^n}{n}\sin\frac{n\pi x}{2}$$

$$= \frac{4}{\pi}\left[\sin\frac{\pi x}{2} - \frac{1}{2}\sin\pi x + \frac{1}{3}\sin\frac{3}{2}\pi x - \cdots\right].$$

REVIEW QUESTIONS

1. (a) State Fourier's theorem. **[C.U. 1996, 2000, '04; N. Beng. U. 2001]**
   (b) What are the limitations under which a Fourier series expansion is possible? **[cf. Burd. U. 1996]**
   (c) What are Dirichlet's conditions for a function to be piece-wise regular in a given interval? **[C.U. 1987, 2004]**
   (d) Is Fourier's theorem applicable at a point of discontinuity of the function? **[C.U. 2000]**

2. (a) Write the Fourier series for a periodic function $f(t)$ in a general form.
   (b) How are the Fourier coefficients determined? **[cf. N.B.U. 2001]**

3. Can you expand the following functions in a Fourier series? Explain, (i) $\tan x$. **[C.U. 1997, 2004]**, (ii) $\operatorname{cosec} x$, (iii) $\sin(1/x)$.
   [**Hints.** Check Dirichlet's conditions.]

4. What do you mean by the amplitude spectrum and phase spectrum of a periodic function $f(t)$?

5. What simplifications are obtained in the Fourier series if the function is (i) even **[cf. C.U. 2004]** and (ii) odd?

6. Write the exponential form of Fourier series.

7. What do you mean by half-range sine series and half-range cosine series?

PROBLEMS

1. Express $f(x) = x$ as a Fourier series in the interval $-\pi < x < \pi$.
   [**Ans.** $f(x) = 2(\sin x - \frac{1}{2}\sin 2x + \frac{1}{3}\sin 3x - \frac{1}{4}\sin 4x + \cdots)$] **[C.U. 1995]**

2. Expand $f(x) = x^2$ in the range $0 < x < 2\pi$ in a Fourier series.
   [**Ans.** $f(x) = \frac{4\pi^2}{3} + \sum_{n=1}^{\infty}\left(\frac{4}{n^2}\cos nx - \frac{4\pi}{n}\sin nx\right)$] **[C.U. 1986, '90]**

3. Analyse, with the help of Fourier's theorem, a square periodic wave given by

$$y = A \text{ (constant)} \qquad \text{for } 0 \le t \le \frac{T}{2}$$

$$= 0 \qquad \text{for } \frac{T}{2} \le t \le T.$$

What harmonics are present in the expansion? **[C.U. 1988]**

[**Ans.** $y = \frac{A}{2} + \frac{2A}{\pi}\left(\sin\omega t + \frac{1}{3}\sin 3\omega t + \frac{1}{5}\sin 5\omega t + \cdots\right)$]

4. If $f(x) = x - x^2$ in the range $-\pi < x < \pi$, show that

$$f(x) = -\frac{\pi^2}{3} + 4\left[\frac{\cos x}{1^2} - \frac{\cos 2x}{2^2} + \frac{\cos 3x}{3^2} - \cdots\right]$$
$$+2\left[\frac{\sin x}{1} - \frac{\sin 2x}{2} + \frac{\sin 3x}{3} - \cdots\right].$$

Hence show that

$$\frac{1}{1^2} - \frac{1}{2^2} + \frac{1}{3^2} - \frac{1}{4^2} + \cdots = \frac{\pi^2}{12}.$$

[**Hints.** Put $x = 0$ in the Fourier series]

5. If $f(x) = e^{-x}$ for $0 < x < 2\pi$, expand $f(x)$ in a Fourier series.

[**Ans.** $e^{-x} = \frac{1 - e^{-2\pi}}{\pi}\left[\frac{1}{2} + \left(\frac{1}{2}\cos x + \frac{1}{5}\cos 2x + \frac{1}{10}\cos 3x + \cdots\right)\right.$
$\left. + \left(\frac{1}{2}\sin x + \frac{2}{5}\sin 2x + \frac{3}{10}\sin 3x + \cdots\right)\right]$]

6. A periodic function $f(t)$ is defined as follows:

$$f(t) = 0 \quad \text{for } -\frac{\pi}{2} \le \omega t \le 0$$
$$= A\sin\omega t \quad \text{for } 0 \le \omega t \le \frac{\pi}{2},$$

where $\omega T = 2\pi$, T being the time period. Show that the Fourier series for $f(t)$ is

$$f(t) = A\left(\frac{1}{\pi} + \frac{1}{2}\sin\omega t - \frac{2}{1.3\pi}\cos 2\omega t - \frac{2}{3.5\pi}\cos 4\omega t - \cdots\right).$$

Hence show that

$$\frac{1}{2} = \frac{1}{1.3} + \frac{1}{3.5} + \frac{1}{5.7} + \cdots.$$

7. Express $f(x) = x$ as a half-range cosine series in $0 < x < 2$. Hence show that

$$\frac{1}{1^2} + \frac{1}{3^2} + \frac{1}{5^2} + \cdots = \frac{\pi^2}{8}.$$

[**Ans.** $f(x) = 1 - \frac{8}{\pi^2}\left(\frac{\cos\frac{\pi x}{2}}{1^2} + \frac{\cos\frac{3\pi x}{2}}{3^2} + \frac{\cos\frac{5\pi x}{2}}{5^2} + \cdots\right)$]

Chapter 4

Anharmonic Vibrations

4.1 Large-Amplitude Oscillations

For small amplitudes of vibration of a simple harmonic oscillator, the restoring force is proportional to the displacement x from the mean position. For large amplitudes of vibration, the restoring force often contains additional terms involving higher powers of x. Thus, in general, the restoring force is expressed as

$$F(x) = (1 + a_1 x + a_2 x^2 + \cdots)sx, \tag{4.1}$$

where s and a_1, a_2, etc. are constants.

The oscillator then becomes *nonlinear* and the vibrations contain a fundamental frequency and higher harmonics. The oscillations of a nonlinear oscillator are referred to as *anharmonic oscillations.* Assuming that the nonlinearity is small we shall analyse below the vibrations of an anharmonic oscillator.

The differential equation [Eq. (1.2)] for a linear harmonic oscillator is a linear homogeneous differential equation: linear because the equation contains terms depending only on the first power of x and $\frac{d^2x}{dt^2}$, and homogeneous because the equation does not contain any term independent of x.

Linear homogeneous differential equations allow superposition of solutions, i.e., the sum of any two solutions is also a solution. This is known as the *superposition principle.* When the restoring force is given by Eq. (4.1), the differential equation of motion is nonlinear. The superposition principle does not hold for such nonlinear equations.

4.2 Anharmonic Oscillator with a Symmetric Restoring Force

The restoring force $F(x)$ is said to be *symmetric* if its magnitude remains the same but the direction is reversed when the displacement changes from $+x$ to $-x$. Thus $F(x)$ will contain only *odd* powers of x, and for small nonlinearity we approximate Eq. (4.1) as

$$F(x) = sx + bx^3 . \tag{4.2}$$

If m is the mass of the oscillating partic e, the equation of motion is

$$m\frac{d^2x}{dt^2} + F(x) = 0$$

$$\text{or,} \quad \frac{d^2x}{dt^2} + \omega^2 x + \alpha x^3 = 0, \tag{4.3}$$

where

$$\omega = \sqrt{\frac{s}{m}} \tag{4.4}$$

$$\text{and} \quad \alpha = \frac{b}{m}. \tag{4.5}$$

The quantity α is the *nonlinearity parameter* and can be both positive and negative. If α is positive, $F(x)$ increases over the linear value sx for large displacements, so that the spring is *hard.* When α is negative, $F(x)$ decreases below the linear value sx and the spring is *soft.*

We take the boundary conditions to be the following:
At $t = 0$,

$$x = a \quad \text{and} \quad \dot{x} = 0. \tag{4.6}$$

The condition of small nonlinearity is satisfied when the amplitude of oscillation a is such that $|\alpha a^2| \ll \omega^2$.
Let

$$x = x_0 + \alpha x_1 \tag{4.7}$$

$$\text{and} \quad \omega^2 = \omega_0^2 + c_1\alpha. \tag{4.8}$$

Substituting Eqs. (4.7) and (4.8) into Eq. (4.3), and retaining terms up to the linear in α, we get

$$\left(\frac{d^2x_0}{dt^2} + \omega_0^2 x_0\right) + \alpha\left(\frac{d^2x_1}{dt^2} + c_1x_0 + \omega_0^2x_1 + x_0^3\right) = 0. \tag{4.9}$$

Equation (4.9) holds for any arbitrary value of α. So the coefficients of the different powers of α must vanish separately, giving

$$\frac{d^2x_0}{dt^2} + \omega_0^2 x_0 = 0 \tag{4.10}$$

and

$$\frac{d^2x_1}{dt^2} + c_1 x_0 + \omega_0^2 x_1 + x_0^3 = 0. \tag{4.11}$$

The solution of Eq. (4.10) consistent with the boundary condition (4.6) is

$$x_0 = a\cos\omega_0 t. \tag{4.12}$$

Substituting for x_0 from Eq. (4.12) into Eq. (4.11) gives

$$\frac{d^2x_1}{dt^2} + \omega_0^2 x_1 = -\left(c_1 + \frac{3}{4}a^2\right) a\cos\omega_0 t - \frac{a^3}{4}\cos 3\omega_0 t, \tag{4.13}$$

where we have used the trigonometric identity

$$\cos 3\omega_0 t = 4\cos^3\omega_0 t - 3\cos\omega_0 t. \tag{4.14}$$

The first term on the right-hand side of Eq. (4.13) leads to resonance giving an indefinitely large amplitude for x. To avoid this situation, we put the coefficient of $\cos\omega_0 t$ on the right-hand side of Eq. (4.13) to zero. Thus,

$$c_1 = -\frac{3}{4}a^2, \tag{4.15}$$

whence Eq. (4.13) reduces to

$$\frac{d^2x_1}{dt^2} + \omega_0^2 x_1 = -\frac{a^3}{4}\cos 3\omega_0 t. \tag{4.16}$$

The solution of Eq. (4.16) contains the complementary function $x_1 = A\cos\omega_0 t$ and the particular integral $x_1 = B\cos 3\omega_0 t$. Substitution of the particular integral in Eq. (4.16) gives $B = a^3/(32\omega_0^2)$. The complete solution of Eq. (4.16) is then

$$x_1 = A\cos\omega_0 t + \frac{a^3}{32\omega_0^2}\cos 3\omega_0 t. \tag{4.17}$$

The boundary condition shows that at $t = 0$, $x_1 = 0$. So, $A = -a^3/(32\omega_0^2)$. Hence Eq. (4.17) becomes

$$x_1 = \frac{a^3}{32\omega_0^2}(\cos 3\omega_0 t - \cos\omega_0 t). \tag{4.18}$$

The displacement x is obtained from Eq. (4.7):

$$x = a\cos\omega_0 t + \frac{\alpha a^3}{32\omega_0^2}(\cos 3\omega_0 t - \cos\omega_0 t). \qquad (4.19)$$

Thus the oscillations contain a fundamental frequency ω_0 and a third harmonic frequency $3\omega_0$ (Fig. 4.1).

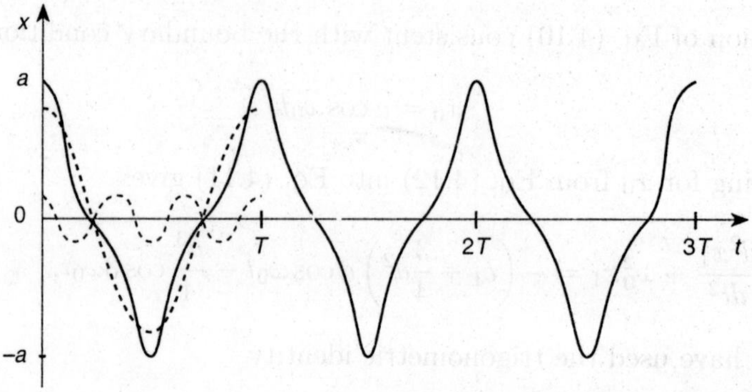

Fig. 4.1 Plot of x versus t as a superposition of fundamental and third harmonic components. The individual components are plotted in the first period : $0<t<T$

The fundamental frequency is obtained from Eq. (4.8):

$$\omega_0^2 = \omega^2 - c_1\alpha = \frac{s}{m} + \frac{3}{4}a^2\alpha,$$

$$\text{or,} \quad \omega_0 = \sqrt{\frac{s}{m} + \frac{3}{4}a^2\alpha}. \qquad (4.20)$$

Equation (4.20) gives the relationship between the frequency ω_0 and the amplitude of oscillation a. If α is positive, the fundamental frequency is enhanced over the linear value $\sqrt{s/m}$; otherwise it decreases below the linear value.

Example. For a simple pendulum of length l, the equation of motion is

$$\frac{d^2\theta}{dt^2} + \frac{g}{l}\sin\theta = 0, \qquad (4.21)$$

where θ is the angular displacement at time t, and g is the acceleration due to gravity. Since $\sin\theta \simeq \theta - \frac{\theta^3}{6}$, Eq. (4.21) reduces to

$$\frac{d^2\theta}{dt^2} + \frac{g}{l}\theta - \frac{g}{6l}\theta^3 = 0. \qquad (4.22)$$

On comparing with Eq. (4.3), we obtain $\omega^2 = g/l$ and $\alpha = -g/(6l)$. As α is negative, a simple pendulum is a soft system. On replacing a by the angular amplitude θ_0, Eq. (4.20) gives

$$\omega_0 = \sqrt{\frac{g}{l}\left(1 - \frac{\theta_0^2}{8}\right)}. \tag{4.23}$$

The time period is

$$\begin{aligned} T &= \frac{2\pi}{\omega_0} = 2\pi\sqrt{\frac{l}{g}}\left(1 - \frac{\theta_0^2}{8}\right)^{-1/2} \\ &\simeq 2\pi\sqrt{\frac{l}{g}}\left(1 + \frac{\theta_0^2}{16}\right). \end{aligned} \tag{4.24}$$

Clearly, the natural frequency decreases and the time period increases with increasing amplitude. When θ_0 is negligibly small, Eq. (4.24) goes over to the familiar form $T = 2\pi\sqrt{l/g}$.

4.3 Anharmonic Oscillator with an Asymmetric Restoring Force

For small nonlinearity, an *asymmetric* restoring force is of the quadratic form:

$$F(x) = sx + cx^2. \tag{4.25}$$

If m is the mass of the oscillating particle, the equation of motion is

$$\begin{aligned} m\frac{d^2x}{dt^2} + sx + cx^2 &= 0 \\ \text{or,} \quad \frac{d^2x}{dt^2} + \omega^2 x + \beta x^2 &= 0, \end{aligned} \tag{4.26}$$

where

$$\omega = \sqrt{\frac{s}{m}} \tag{4.27}$$

$$\text{and} \quad \beta = \frac{c}{m}. \tag{4.28}$$

If the *nonlinearity parameter* β is positive, the spring is stiffer for positive values of the displacement x but it is less stiff for negative values of x. If β is negative, the reverse is true.

The boundary condition is assumed to be as follows:

At $t = 0$,

$$x = a \quad \text{and} \quad \dot{x} = 0. \tag{4.29}$$

The condition of small nonlinearity holds when the amplitude of oscillation is such that $|\beta a| \ll \omega^2$.

Let

$$x = x_0 + \beta x_1 \tag{4.30}$$

$$\text{and} \quad \omega^2 = \omega_0^2 + d_1 \beta. \tag{4.31}$$

Substituting Eqs. (4.30) and (4.31) into Eq. (4.26) and retaining terms up to the linear in β, we obtain

$$\left(\frac{d^2 x_0}{dt^2} + \omega_0^2 x_0\right) + \beta\left(\frac{d^2 x_1}{dt^2} + \omega_0^2 x_1 + d_1 x_0 + x_0^2\right) = 0. \tag{4.32}$$

Since Eq. (4.32) holds for any arbitrary value of β, the coefficients of the different powers of β can be separately equated to zero.

Hence

$$\frac{d^2 x_0}{dt^2} + \omega_0^2 x_0 = 0 \tag{4.33}$$

$$\text{and} \quad \frac{d^2 x_1}{dt^2} + \omega_0^2 x_1 + d_1 x_0 + x_0^2 = 0. \tag{4.34}$$

Consistent with the boundary conditions, the solution of Eq. (4.33) is

$$x_0 = a \cos \omega_0 t. \tag{4.35}$$

Substituting for x_0 from Eq. (4.35) into Eq. (4.34) yields

$$\frac{d^2 x_1}{dt^2} + \omega_0^2 x_1 = -d_1 a \cos \omega_0 t - \frac{a^2}{2}(1 + \cos 2\omega_0 t), \tag{4.36}$$

where we have put $\cos^2 \omega_0 t = \frac{1}{2}(1 + \cos 2\omega_0 t)$. Putting

$$z_1 = x_1 + \frac{a^2}{2\omega_0^2}, \tag{4.37}$$

Eq. (4.36) becomes

$$\frac{d^2 z_1}{dt^2} + \omega_0^2 z_1 = -d_1 a \cos \omega_0 t - \frac{a^2}{2} \cos 2\omega_0 t. \tag{4.38}$$

To avoid resonance, we put

$$d_1 = 0. \tag{4.39}$$

Hence Eq. (4.38) simplifies to

$$\frac{d^2 z_1}{dt^2} + \omega_0^2 z_1 = -\frac{a^2}{2} \cos 2\omega_0 t. \tag{4.40}$$

The solution of Eq. (4.40) has the complementary function $z_1 = A\cos\omega_0 t$ and the particular integral $z_1 = B\cos 2\omega_0 t$. Substituting the particular integral in Eq. (4.40) yields $B = a^2/(6\omega_0^2)$. The complete solution for x_1 is then

$$x_1 = -\frac{a^2}{2\omega_0^2} + A\cos\omega_0 t + \frac{a^2}{6\omega_0^2}\cos 2\omega_0 t. \tag{4.41}$$

The boundary condition shows that at $t = 0$, $x_1 = 0$. Hence $A = a^2/(3\omega_0^2)$. So, Eq. (4.41) becomes

$$x_1 = -\frac{a^2}{2\omega_0^2} + \frac{a^2}{3\omega_0^2}\left(\cos\omega_0 t + \frac{1}{2}\cos 2\omega_0 t\right). \tag{4.42}$$

From Eq. (4.30) we have for the displacement

$$x = a\cos\omega_0 t - \frac{\beta a^2}{2\omega_0^2} + \frac{\beta a^2}{3\omega_0^2}\left(\cos\omega_0 t + \frac{1}{2}\cos 2\omega_0 t\right). \tag{4.43}$$

Therefore, the oscillations contain a fundamental frequency ω_0 and a second harmonic frequency $2\omega_0$. The fundamental frequency of oscillation is obtained from Eq. (4.31). Since $d_1 = 0$ we have

$$\omega_0 = \omega = \sqrt{\frac{s}{m}}. \tag{4.44}$$

Thus there is *no frequency shift* in contrast to the case of the symmetric restoring force.

There is another difference from the symmetric case. The average position of the vibrating mass in the asymmetric case is

$$\bar{x} = -\frac{\beta a^2}{2\omega_0^2}, \tag{4.45}$$

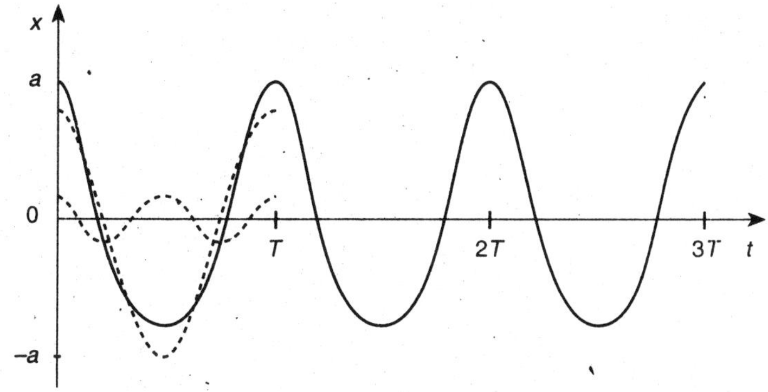

Fig. 4.2 Plot of x versus t as a superposition of fundamental and second harmonic components. The individual components are plotted in the first period : $0 < t < T$

since the average of each cosine term in Eq. (4.43) is zero. Equation (4.45) shows that $\bar{x}$ has the opposite sign to β. Note that for the symmetric restoring force, the mean position of the mass is the same as its equilibrium position.

The plot of the displacement x versus time t for the asymmetric case is displayed in Fig. 4.2.

The thermal expansion of a crystalline solid can be understood from the nonlinear force acting between the atoms. The restoring force between a pair of atoms is asymmetric, the value of β being negative. Hence the average distance between two atoms is

$$\bar{r} = R + \bar{x} = R - \frac{\beta a^2}{2\omega_0^2}, \tag{4.46}$$

where R is the equilibrium separation between the atoms. As β is negative, $\bar{r}$ increases with the amplitude of vibration a, i.e., with a rise of temperature.

4.4 Combination Tones

For any slightly nonlinear spring, the displacement x can be written as a power series in the restoring force F. To visualise this, we consider Eq. (4.25) and rewrite it as

$$cx^2 + sx - F = 0. \tag{4.47}$$

The solution of this quadratic equation is

$$x = \frac{-s \pm \sqrt{s^2 + 4Fc}}{2c}$$

$$= \frac{-s \pm s\left(1 + \frac{4Fc}{s^2}\right)^{1/2}}{2c}. \tag{4.48}$$

Since the nonlinearity term c is small, one can binomially expand the radical and retain terms up to the quadratic in c. Thus,

$$x = \frac{1}{2c}\left[-s \pm s\left(1 + \frac{2Fc}{s^2} - \frac{2F^2c^2}{s^4}\right)\right]. \tag{4.49}$$

For $c = 0$, we have $x = F/s$. In order to agree with this value, the negative sign before the first bracket in Eq. (4.49) must be rejected. So,

$$x = \frac{F}{s}\left(1 - \frac{Fc}{s^2}\right). \tag{4.50}$$

Extending Eq. (4.50), we can generally write

$$x = AF + BF^2 + CF^3 + \cdots \tag{4.51}$$

where A, B and C are constants. For stiffness controlled motion, the driving force is nearly equal in magnitude to the restoring force. So, F can be taken to be the driving force in Eq. (4.51).

Consider two simple harmonic forces with different angular frequencies ω_1 and ω_2 acting simultaneously. Then

$$F = F_1 \cos \omega_1 t + F_2 \cos \omega_2 t. \tag{4.52}$$

For simplicity, we assume that only the first two terms in Eq. (4.51) are significant. Substituting Eq. (4.52) then yields

$$x = A(F_1 \cos \omega_1 t + F_2 \cos \omega_2 t) + B(F_1 \cos \omega_1 t + F_2 \cos \omega_2 t)^2. \tag{4.53}$$

The coefficient of A gives the linear response, showing the presence of the exciting frequencies ω_1 and ω_2; it will produce beats if $\omega_1 \simeq \omega_2$. The coefficient of B can be expanded as follows:

$$\begin{aligned}(F_1 \cos \omega_1 t + F_2 \cos \omega_2 t)^2 &= F_1^2 \cos^2 \omega_1 t + F_2^2 \cos^2 \omega_2 t \\ &\quad + 2F_1 F_2 \cos \omega_1 t \cos \omega_2 t \\ &= \frac{1}{2} F_1^2 (\cos 2\omega_1 t + 1) + \frac{1}{2} F_2^2 (\cos 2\omega_2 t + 1) \\ &\quad + F_1 F_2 \left[\cos(\omega_1 + \omega_2)t + \cos(\omega_1 - \omega_2)t\right].\end{aligned} \tag{4.54}$$

Thus we have the second harmonics of both the exciting frequencies apart from a constant term, and the sum frequency $(\omega_1 + \omega_2)/(2\pi)$ and the difference frequency $(\omega_1 - \omega_2)/(2\pi)$, assuming $\omega_1 > \omega_2$. If the F^3 term is retained in Eq. (4.51), we have the additional frequencies $(2\omega_1 \pm \omega_2)/(2\pi)$ and $(2\omega_2 \pm \omega_1)/(2\pi)$.

Thus, when two pure, strong tones of frequencies f_1 and f_2 are simultaneously impressed on a nonlinear system, a series of new tones of frequencies $a_1 f_1 \pm a_2 f_2$, where a_1 and a_2 are small integers, are produced. These frequencies are termed the *combination frequencies*, and the corresponding tones are known as *combination tones*. The tones of frequencies $f_1 \pm f_2$ are the *first-order combination tones*; $f_1 - f_2$ is the *difference tone* and $f_1 + f_2$ is the *summation tone*. The remaining combination tones of frequencies $2f_1 \pm f_2$, $2f_2 \pm f_1$ etc. are the *higher order tones*. Usually, the difference tone is most intense and readily perceived. The combination frequencies are increasingly important as the driving force amplitudes F_1 and F_2 are increased. A *subsidiary resonance* is expected if a combination frequency equals the natural frequency of the system.

The combination tones that are produced outside the ear are known as the *objective combination tones*. The objective nature of the combination tone can be demonstrated if two intense notes of frequencies f_1 and f_2 superimpose on a resonator tuned to the difference frequency $f_1 - f_2$. The nonlinearity of the resonator produces a difference frequency and the resonator responds vigorously to this frequency.

The combination tones produced only in the ear are referred to as the *subjective combination tones.* The ear itself has an intrinsic nonlinearity. The eardrum may respond linearly to harmonic pressure variations caused by the sound reaching it. These vibrations are transmitted to the sensor organ (the basilar membrane) through a fluid in a coiled tube called the cochlea. This part of the system is nonlinear and gives rise to the difference frequency. If the sound reaching the eardrum contains the frequencies f and $\frac{3}{2}f$, the brain receives sound sensation for notes of frequencies f, $\frac{3}{2}f$ and to the difference frequency $\frac{1}{2}f$. The sum frequency $\left(f + \frac{3}{2}f\right)$ is not heard for psychophysical reasons not yet clearly understood. The musical interval for any two frequencies in the ratio 3 : 2 is a *fifth,* and a note having the frequency $f/2$ is an *octave* below one of the frequencies, f. A violin player, while tuning two strings to an exact fifth, listens for this octave.

The difference tone is clearly heard when a police whistle or a referee's whistle (as used in a soccer match) is blown. These whistles have two slightly unequal short-barrels, each generating a high frequency note. These notes can be heard if one of the barrels is closed by a finger and the whistle is blown. However, when the whistle is blown with both the barrels open, a strong low-pitched note corresponding to the difference of the frequencies of the barrels, is heard.

The presence of the combination tone is *undesirable* in a sound reproduction system, e.g. a loudspeaker. It is difficult to construct a loudspeaker in which the moving part whose forced vibration emits the sound, gives a strictly linear restoring force. Also, the electromagnetic driving force may itself be anharmonic, either because it is not exactly proportional to the signal current or because the current is not a true replica of the source signal. Since the nonlinear effects are important at large amplitudes, the obvious solution to get rid of the combination frequencies is to operate the device for displacements which are small compared to the design range.

The theory of the combination tone outlined above is sometimes called *Helmholtz intensity theory.* The theory demands that the primary tones must be very intense to produce combination tones. However, weak primaries are also found to produce combination tones in asymmetric systems. Such a system was realised by Waetzmann who loaded a membrane with a central weight on one side of the membrane only. If the membrane is subjected to two harmonic tones, the vibration of the membrane shows the presence of the combination tones. The theory developed by Waetzmann in this connection is known as *Waetzmann's general asymmetry theory.* He argued that the eardrum with its load ossicles on one side would behave as the loaded membrane discussed above, and so produce the subjective combination tone.

4.5 Aural Harmonics

For a nonlinear system, the displacement can be expressed as a power series of the force [vide Eq. (4.51)]. Putting $F = F_0 \cos \omega t$, we obtain from Eq. (4.51)

$$x = AF_0 \cos \omega t + BF_0^2 \cos^2 \omega t, \tag{4.55}$$

neglecting the higher order terms. Since $\cos^2 \omega t = \frac{1}{2}(1 + \cos 2\omega t)$, Eq. (4.55) gives

$$x = \frac{1}{2}BF_0^2 + AF_0 \cos \omega t + \frac{1}{2}BF_0^2 \cos 2\omega t \tag{4.56}$$

Thus the inclusion of the F^2 term gives a second harmonic of frequency 2ω apart from a constant term in x. Consideration of the F^3 term also in x would give a additional frequency 3ω.

In the ear and a majority of the acoustic devices, $B \ll A$, $C \ll B$, etc. So the effects of the higher order terms are unimportant except when the exciting intensity is high. For the ear, this exciting intensity level is about 50 dB above threshold for a 1.2 kHz tone.

When a pure tone of high intensity is received by the ear, higher harmonics are also heard in addition to the fundamental. These harmonics are called *aural harmonics*. Their origin lies in the nonlinearity of the response of the ear. For instance, for an intense 200 Hz pure tone, a careful listener will hear in addition to 200 Hz, weaker tones of frequencies 400 and 600 Hz. An aural harmonic can be recognized by superposing in the ear another tone which will generate about 3 or 4 beats with the suspected aural harmonic. The intensity of the superimposed tone is adjusted till the beats are clearly perceived.

4.6 Solved Problems

Problem 1. *An asymmetric nonlinear vibrator is subjected simultaneously to two harmonic tones of frequencies* 50 Hz *and* 60 Hz, *respectively. Assuming that the resonant frequency is very high, give all the frequencies present in the forced vibration. Ignore the third and the higher powers of the force in the series expansion of the displacement.*

Solution. If f_1 and f_2 are the two exciting frequencies, the forced vibration of the system consists of (i) a steady component of the displacement, i.e., a frequency zero; (ii) the frequencies f_1 and f_2, i.e., 50 Hz and 60 Hz; (iii) the frequencies $2f_1$ and $2f_2$, i.e., 100 Hz and 120 Hz; and (iv) the frequencies $(f_1 + f_2)$ and $(f_1 - f_2)$, i.e., 110 Hz and 10 Hz.

Problem 2. *A tuning fork of frequency 64 Hz is acted on by two harmonic tones, one of which has the frequency 200 Hz. If the tuning fork responds to the difference frequency, find the frequency of the other tone.*

Solution. If the desired frequency is f_1, we have $f_1 = 200 \pm 64$, i.e., $f_1 = 264$ Hz or $f_1 = 136$ Hz.

Problem 3. *Determine the first-order percentage correction in the time period of a simple pendulum when the angular amplitude of oscillation is* (i) $4°$ *and* (ii) $50°$.

Solution. The first-order corrected formula for the time period of a pendulum is $T = 2\pi\sqrt{l/g}\left(1 + \frac{\theta_0^2}{16}\right)$.

(i) If $\theta_0 = 4° = \frac{\pi}{45}$ rad, we have $\frac{\theta_0^2}{16} = 3.05 \times 10^{-4}$. So, the percentage correction in T is 0.03%. (ii) When $\theta_0 = 50° = \frac{\pi \times 50}{180}$ rad, we have $\frac{\theta_0^2}{16} = 4.76 \times 10^{-2}$. Hence the desired percentage correction is 4.76%.

REVIEW QUESTIONS

1. (a) What are anharmonic oscillations?
   (b) Distinguish between a linear and a nonlinear oscillator.
   (c) State the characteristics of an anharmonic oscillator.

2. An anharmonic oscillator is characterized by a restoring force $F = sx + bx^3$. Show that for small nonlinearity the oscillations contain a fundamental frequency and a third harmonic frequency. Also, find the fundamental frequency as a function of the amplitude of oscillation.

3. The force of restitution of an oscillator is $F = sx + cx^2$. Assuming small nonlinearity show that the average position of a vibrating mass is shifted from its equilibrium position and that the oscillations contain a fundamental frequency and a second harmonic. Does the fundamental frequency depend on the amplitude of oscillation?

4. How do you account for the thermal expansion of a crystalline solid assuming that the restoring force between the vibrating atoms is nonlinear?

5. What are combination tones? How are they produced? **[C.U. 2005]**

6. Give a theory to show that combination tones are produced as a result of the double forcing of a nonlinear system. **[cf. Burd.U. 1996]**

7. Discuss the conditions under which combination tones are produced. "Combination tones are both subjective and objective." — Explain.

8. Give examples where the combination tone is useful and where it is undesirable.

9. What are aural harmonics? How do they originate?

PROBLEMS

1. The frequency of a tuning fork is 400 Hz. It is acted on simultaneously by two harmonic tones one of which has the frequency 50 Hz and the other has the frequency in the range 360 to 500 Hz. If the tuning fork responds to the difference tone, calculate the exact frequency of the other tone.
[**Ans.** 450 Hz]

2. An asymmetric nonlinear oscillator is driven by two harmonic tones of frequencies 90 Hz and 100 Hz, respectively. The resonant frequency of the oscillator is very high. Considering only the terms up to the quadratic of the force in the series expansion of the displacement, find the frequencies present in the forced vibration.
[**Ans.** 0, 10, 90, 100, 180, 190, 200 Hz]

Chapter 5

Wave Motion

5.1 Wave Motion in an Elastic Medium

If a particle of a material medium is slightly displaced from its equilibrium position and then released, it starts to oscillate about the equilibrium position by virtue of inertial and elastic properties. The vibrating particle being bound to the neighbouring particles of the medium by cohesive forces, the vibratory motion is imparted to the neighbouring particles. Such a process of transference of the vibration from particle to particle, is known as a *wave motion*. Through a wave motion, the disturbance and the energy are carried from one point to another of a material medium without any physical transport of the medium. In a wave motion, the oscillations of the particles are similar, but a phase difference exists between the oscillations of the different particles. The wave motion is generated by the tensile or the elastic forces between the particles of the medium. Such waves propagating from one point to another in a material medium due to its elasticity, are called *elastic waves*. Sound waves are elastic waves.

5.2 Types of Waves

In general, there are two types of wave motion: *transverse* and *longitudinal*.

When the particles of the medium vibrate at right angles to the direction of propagation of the wave, the wave is said to be a *transverse wave*. Waves along a stretched string are transverse waves.

When the particles of the medium vibrate parallel to the direction of propagation of the wave, the wave is termed a *longitudinal wave*. Sound waves in solids, liquids and gases, compressional waves along a string, etc. are examples of longitudinal waves.

In a transverse wave, the layers of the medium vibrate tangentially with respect to each other, and so one layer exerts a *shearing force* to its adjacent layer. Such forces exist in solid and to some extent in liquids, but not in gases. So, transverse waves cannot be produced in a gas. However, all material media are more or less compressible. Therefore, longitudinal waves can propagate through solids, liquids and gases.

In wave motion, the disturbance is a function of both position and time. If the disturbance at a point is a periodic function of time, the wave is said to have time periodicity. If at a given instant of time, the disturbance has the same value at equidistant points on the path of the wave, the wave has a space periodicity as well. Waves having both time and space periodicity are called *periodic waves.* In actual practice, the magnitude of the disturbance decays gradually resulting in a change of the shape of the disturbance. To start with, one considers the waves which do not change shape and have a constant magnitude, i.e., the waves which have *no attenuation* and are of a *constant type.* Thus *periodic waves of constant type* form the basis of the study of wave motion.

5.3 Terms Related to Periodic Waves of Constant Type

The curve giving the variation of the displacement of an oscillating particle in the wave as a function of time, is called the *time-displacement curve.* In the simplest case, the time-displacement curve is a sine curve (termed a *sinusoid*). For a wave of constant type, all the particles on the path of propagation will have identical time-displacement curves. But a particle farther from the source will attain at a later time the displacement of a particle nearer the source. This *phase lag* in a homogeneous medium is proportional to the distance between the particles.

The curve showing, at a particular instant of time, the displacements of the particles on the line of propagation as a function of their distance from a suitably chosen point on the line of propagation, is referred to as the *space-displacement curve.* In the simplest case, such a curve is a sine curve. Clearly, the space-displacement curve gives the shape or the form of the wave, and is therefore also known as the *waveform* or the *wave profile.* For a wave of constant type, the waveform does not change.

The maximum displacement of a particle on the path of the wave is called the *amplitude* of the wave. The space-displacement curve of Fig. 5.1 shows that the amplitude of the wave is A.

The *period* of the wave, denoted by T is the time needed for the generation of a complete wave. Since a complete wave is produced by a full cycle of oscillation of a particle on the line of propagation, the period of the wave can be alternatively defined as the time needed for a vibrating particle on the line of propagation to undergo a complete cycle of oscillation.

The *frequency*, denoted by n, of a wave is defined as the number of complete waves produced in the medium per second. Alternatively, the frequency of a wave is the number of complete oscillations per second of a vibrating particle on the line of propagation. Since the time for the generation of a complete wave is T, the number of complete waves generated per second is $1/T$. Thus the frequency is given by $n = 1/T$. The SI unit of frequency is hertz (Hz).

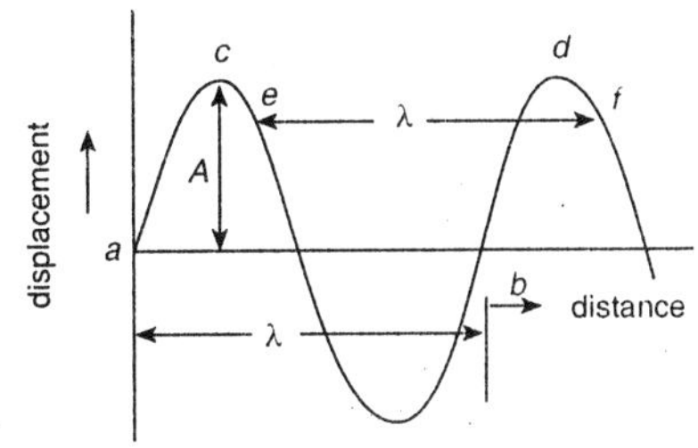

Fig. 5.1 Space-displacement curve

The shortest distance between any two particles on the path of the wave having the same phase of vibration at any instant of time, is termed the *wavelength* (λ). The distance λ is shown in Fig. 5.1. Note that the period gives the time periodicity and the wavelength gives the space periodicity of the wave.

The velocity of propagation of an unchanging waveform is referred to as the *wave velocity* or *phase velocity*. A particle on the line of propagation at x_2 has at a later instant of time t_2 the same phase of vibration as another particle at x_1 nearer to the source at an earlier instant of time t_1. Hence, a given phase has propagated from x_1 to x_2 in time $(t_2 - t_1)$, so that the phase velocity is

$$c = \frac{x_2 - x_1}{t_2 - t_1}.$$

When a wave is produced in a medium, the plane passing through the particles of the medium in the same phase of vibration is referred to as the *wavefront*. The propagation of the wave through the medium can be looked upon as the movement of the wavefront with a velocity equal to the wave velocity. In a homogeneous isotropic medium, the wave velocity is the same in all directions. So, for waves generated from a point source in the medium, the wavefronts are concentric spherical surfaces with the source at the centre. Such waves are known as *spherical waves*. At a long distance from the source, a small portion of a spherical wavefront can be considered to be a plane. Therefore, that small portion of the wavefront can be termed a *plane wavefront*. A wave with plane wavefronts is referred to as a *plane wave*. A plane wave moves in a fixed direction without spreading laterally.

The direction of propagation of a wave is called a *ray*. Rays are normal to the wavefronts. They are shown by straight lines with arrowheads pointing to the direction of the propagation of the waves.

Observations

(i) Let T, n, λ and c be the period, frequency, wavelength and the wave velocity, respectively. By definition the distance traversed by the wave in time T is λ, so that the distance traversed in unit time is λ/T. But this distance is the wave velocity c. Hence $c = \lambda/T = n\lambda$, because $nT = 1$.

(ii) The phase difference between two oscillating particles on the line of propagation separated by a distance of one wavelength (λ) is 2π radian or 360°. Consequently, the phase difference between two oscillating particles a unit distance apart is $2\pi/\lambda$ radian or $(360/\lambda)°$. So, the phase difference between two vibrating particles separated by a distance x is $2\pi x/\lambda$ radian or $(360x/\lambda)°$.

5.4 Characteristics of Progressive Waves

A wave propagating from one point to another in a medium without being subjected to any boundary condition, is termed a *progressive wave*. A progressive wave has the following characteristics :

(i) The waves are produced by the periodic vibrations of the particles in the medium.

(ii) Each particle of the medium oscillates about its mean position with the same amplitude and period. In the simplest case, the vibrations are simple harmonic.

(iii) The phase difference between two vibrating particles on the line of propagation is proportional to the distance between the particles.

(iv) The medium as a whole does not move along with the wave. Only the energy is carried by the advancement of the waveform in the direction of propagation of the wave.

(v) The wave velocity in a given medium is a constant determined by the density and the elasticity of the medium.

5.5 Mathematical Representation of a Plane Progressive Wave

A plane progressive wave moves in a fixed direction (say, the positive x-direction) without spreading laterally and without being hampered by any boundary condition.

Let the disturbance associated with the wave be denoted by ψ; it may be the particle displacement. ψ is termed the *wave field parameter*. Its value changes at a particular point along the direction of propagation with time t; at a given instant of time, ψ changes from point to point along the line of propagation.

Let at time $t = 0$, the values of ψ at points along the line of propagation be given by $\psi(X, 0) = f(X)$, where X is the position coordinate of a point measured from an arbitrary origin O (Fig. 5.2). If the waveform remains constant as the wave moves, the same function gives the values of ψ when the origin moves with the wave. The distance travelled by the origin in time t is ct, where c is the wave velocity. The value $\psi(X, 0)$ which the wave field parameter had at $t = 0$ at a point distant X from the origin O now appears at a distance $x = ct + X$ from the origin O (Fig. 5.2). So,

$$\psi(X, 0) = f(X) = f(x - ct) = \psi(x, t). \tag{5.1}$$

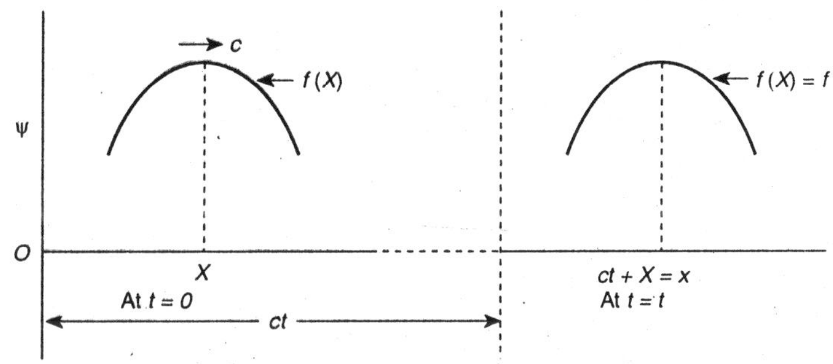

Fig. 5.2 Propagation of the disturbance ψ

The function $f(x - ct)$ represents a plane progressive wave propagating with the velocity c in the positive x-direction. We could interchange x and ct, so that the function $f(ct - x)$ also represents the same plane progressive wave. To visualize this, we take X on the other side of the origin in Fig. 5.2 and hence get $ct - X = x$, giving $X = ct - x$. In the subsequent discussion, we shall retain the form $f(ct - x)$ in keeping with the common practice.

The function $f(ct - x)$ has the following *characteristic features* :

(i) At a particular instant of time, the quantity $ct - x$ and hence $f(ct - x)$ remains the same at all points on a plane perpendicular to the x-axis. Thus the wave fronts are plane, so that $f(ct - x)$ *represents a plane wave propagating in the positive x-direction.* The quantity $(ct - x)$ is the *phase*.

(ii) Let $\psi(x, t)$ denote the value of the field parameter at x at time t. As the wave moves in the positive x-direction, the same value occurs at

$x + dx$ at time $t + dt$. Therefore,

$$\psi(x,t) = \psi(x + dx, t + dt)$$

$$\text{or,} \quad f(ct - x) = f[c(t + dt) - (x + dx)].$$

This shows that $cdt - dx = 0$,

$$\text{or,} \quad c = \frac{dx}{dt}, \tag{5.2}$$

i.e., c is the *phase velocity* in the positive x-direction.

Observations

(i) With increasing time, the phase $(ct - x)$ can remain constant if x increases. Thus $f(ct - x)$ must propagate in the positive x-direction. For a plane wave propagating in the negative x-direction, the phase must remain constant with increasing t if x decreases. Thus the phase must be $(ct + x)$, and $f(ct + x)$ *represents a plane progressive wave propagating in the negative x-direction with the velocity c.*

(ii) If the direction of propagation of a plane progressive wave is not along a co-ordinate axis, its equation can be written as $\psi = f(ct - \vec{r} \cdot \hat{a})$, where $\vec{r} = \hat{\imath}x + \hat{\jmath}y + \hat{k}z$. Here, $\hat{\imath}, \hat{\jmath}$ and $\hat{k}$ are unit vectors along the X, Y, Z axes, respectively, and $\hat{a}$ is a unit vector in the direction of wave propagation. Since $\hat{a} = \hat{\imath}l + \hat{\jmath}m + \hat{k}n$ where l, m, n are the direction cosines of the propagation line with respect to the co-ordinate axes, we have

$$\psi(x, y, z, t) = f(ct - lx - my - nz). \tag{5.3}$$

For a given value of t, ψ is the same for all (x, y, z) points if $lx + my + nz$ = constant. The equation $lx + my + nz$ = constant, represents a plane, the normal to which has the direction cosines l, m and n with respect to the coordinate axes. Here the wavefronts are planes of which the normal possesses the direction cosines l, m, n.

5.6 Differential Wave Equation in One Dimension

A plane progressive wave propagating in the positive x-direction with unaltered waveform is represented by $\psi = f(ct - x)$. Here $f(ct - x)$ is the mathematical expression for the wave field parameter ψ. Putting $z = ct - x$, we have $\frac{\partial z}{\partial x} = -1$ and $\frac{\partial z}{\partial t} = c$. Hence

$$\frac{\partial \psi}{\partial x} = \frac{d\psi}{dz}\frac{\partial z}{\partial x} = -\frac{d\psi}{dz}, \tag{5.4}$$

$$\text{and} \quad \frac{\partial^2 \psi}{\partial x^2} = -\frac{\partial}{\partial x}\left(\frac{d\psi}{dz}\right) = -\frac{d}{dz}\left(\frac{d\psi}{dz}\right)\frac{\partial z}{\partial x} = \frac{d^2\psi}{dz^2}. \tag{5.5}$$

Again

$$\frac{\partial \psi}{\partial t} = \frac{d\psi}{dz}\frac{\partial z}{\partial t} = c\frac{d\psi}{dz} \tag{5.6}$$

$$\text{and} \quad \frac{\partial^2 \psi}{\partial t^2} = c\frac{\partial}{\partial t}\left(\frac{d\psi}{dz}\right) = c\frac{d}{dz}\left(\frac{d\psi}{dz}\right)\frac{\partial z}{\partial t} = c^2\frac{d^2\psi}{dz^2}. \tag{5.7}$$

Equations (5.5) and (5.7) give

$$\frac{\partial^2 \psi}{\partial t^2} = c^2\frac{\partial^2 \psi}{\partial x^2}, \tag{5.8}$$

where c is the phase velocity of the wave.

Starting with $\psi = f(ct + x)$ we also arrive at the differential Eq. (5.8). Equation (5.8) is known as *the differential equation in one dimension for plane waves of constant type.* The general solution of Eq. (5.8) is

$$\psi = f_1(ct - x) + f_2(ct + x), \tag{5.9}$$

where f_1 and f_2 are any two functions. f_1 represents a plane wave moving in the positive x-direction with the velocity c, and f_2 represents a plane wave moving in the negative x-direction with the same velocity. As f_1 and f_2 are, in general, different functions, the waveforms of the two waves need not be the same.

For small displacements of the particles on the line of propagation of the wave, the restoring force is proportional to the displacement. Hence for small amplitudes, the oscillations of the particle are simple harmonic. So, the displacement of a particle at the position x at time t can be written as

$$\xi = A\cos k(ct - x) = \mathrm{Re}\left[Ae^{jk(ct-x)}\right] \tag{5.10}$$

$$\text{or,} \quad \xi = A\sin k(ct - x) = \mathrm{Im}\left[Ae^{jk(ct-x)}\right], \tag{5.11}$$

if the wave propagates in the positive x-direction. 'Re' and 'Im' denote respectively the real and the imaginary part of the complex quantity. Note that the physical displacement is either a sine or a cosine function, and *not* a complex quantity. For a wave moving in the negative x-direction, we write

$$\xi = A\cos k(ct + x) = \mathrm{Re}\left[Ae^{jk(ct+x)}\right] \tag{5.12}$$

$$\text{or,} \quad \xi = A\sin k(ct + x) = \mathrm{Im}\left[Ae^{jk(ct+x)}\right]. \tag{5.13}$$

Note that here the wave field parameter is the particle displacement ξ.

In Eqs. (5.10) through (5.13), A is the amplitude of the wave. If ω is the angular frequency of the simple harmonic motion we have $kc = \omega$. If n is the frequency and T is the period, then $\omega = \frac{2\pi}{T} = 2\pi n$. The distance covered by the wave in time T is the wavelength λ. Hence $\lambda = cT = \frac{\omega T}{k} = \frac{2\pi}{k}$, or $k = \frac{2\pi}{\lambda}$. Thus we get the following two important relationships :

$$c = \frac{\omega}{k} \tag{5.14}$$

$$\text{and} \quad k = \frac{2\pi}{\lambda}. \tag{5.15}$$

The quantity k represents the number of waves in a distance of 2π, and is called the *propagation constant.* Therefore, a plane harmonic wave propagating in the positive x-direction can be expressed in any of the following forms :

$$\xi = A\cos k(ct - x) \tag{5.16a}$$

$$\xi = A\cos(\omega t - kx) \tag{5.16b}$$

$$\xi = A\cos\frac{2\pi}{\lambda}(ct - x) \tag{5.16c}$$

$$\xi = \text{Re}\left[Ae^{j(\omega t - kx)}\right]. \tag{5.16d}$$

Instead of cosine functions, sine functions could also be used in these equations.

Observation

Equation (5.8) holds only for plane waves of constant type. It is not applicable to waves of large amplitudes and attenuated waves.

5.7 Velocity of Propagation of Plane Longitudinal Waves in an Elastic Fluid

We shall derive here an expression for the velocity of plane longitudinal waves propagating in an elastic fluid medium on the basis of the following assumptions :

(i) The medium is homogeneous and isotropic.

(ii) Dissipative forces originating from viscosity and thermal conduction are absent.

(iii) The effect of gravity is negligible, so that, in equilibrium, the pressure and the density are the same everywhere in the medium.

(iv) The strain produced by the wave in the medium is so small that Hooke's law holds.

We consider a cylinder of the fluid of cross sectional area α, the axis of the cylinder coinciding with the direction of propagation of the wave. Let A_1 and B_1 be two closely spaced transverse plane sections of the cylinder (Fig. 5.3). Suppose that x and $x + \delta x$ be the equilibrium positions of the planes A_1 and B_1 with respect to an arbitrarily chosen origin, δx being much smaller than the wavelength λ of the propagating wave. The particles on the planes A_1 and B_1 are displaced due to the excess pressure produced by the progressive longitudinal wave. On the elapse of a short time interval δt (which is much smaller than the period T of the wave), let the particles on the plane A_1 be displaced parallel to the cylinder axis by ξ to A_2, where $\xi \ll \delta x$. The corresponding displacement of the particles on B_1 in time δt is B_1B_2, where $B_1B_2 = \xi + \delta\xi = \xi + \frac{\partial \xi}{\partial x}\delta x$.

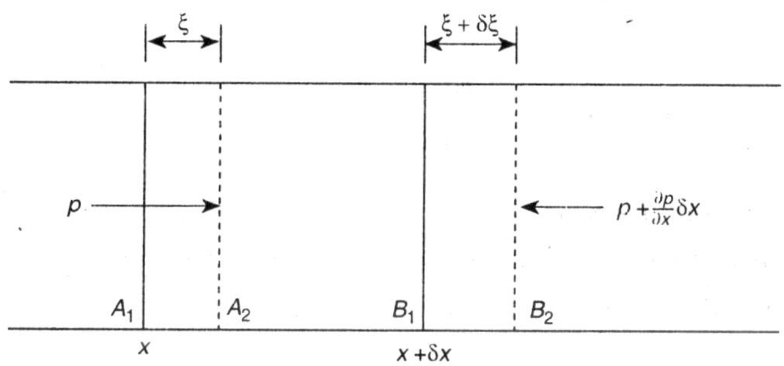

Fig. 5.3 A fluid cylinder with transverse plane sections

The displacements of the particles change the volume of the fluid between the planes. The initial volume V_0 of the fluid between A_1 and B_1 is

$$V_0 = \alpha\delta x. \tag{5.17}$$

The final volume of the fluid between A_2 and B_2 is

$$V_f = \alpha(\delta x + \delta\xi) = \alpha\left(\delta x + \frac{\partial \xi}{\partial x}\delta x\right). \tag{5.18}$$

So, the increase in volume is $\delta V = V_f - V_0 = \alpha\frac{\partial \xi}{\partial x}\delta x$. The ratio of the increase in volume δV to the initial volume V_0 is defined to be the *dilatation* or the volume strain Δ. Thus

$$\Delta = \frac{\delta V}{V_0} = \frac{\partial \xi}{\partial x}. \tag{5.19}$$

When a layer of a fluid is compressed by a disturbance, the pressure in the layer increases from the equilibrium value P_0 by an amount p, referred to as the *excess pressure*. When the layer suffers a rarefaction, p is negative. For sound waves, p is known as the *sound pressure* or the *acoustic pressure*.

The mass of the fluid between the planes A_1 and B_1 is the same as that between the planes A_2 and B_2. This mass is $\rho_0 \alpha \delta x$, where ρ_0 is the equilibrium fluid density. The acoustic pressure difference between A_1 and A_2 is negligible since $\xi \ll \delta x$. Therefore, we can take the excess pressure at either A_1 or A_2 to be p, and that at either B_1 or B_2 as $p + \frac{\partial p}{\partial x}\delta x$. These two pressures act on the fluid slab A_2B_2 in opposite directions (Fig. 5.3), and produce two effects :

(a) Equal and opposite excess pressures p develop a stress on the fluid between A_2 and B_2. The stress p, if compressional, produces the volume strain $-\frac{\delta V}{V_0}$. By Hooke's law the *bulk modulus* K is defined to be the ratio between the volume stress to the volume strain :

$$K = -\frac{p}{\delta V/V_0} = -\frac{p}{\Delta} = -\frac{p}{\left(\frac{\partial \xi}{\partial x}\right)}$$

$$\text{or,} \quad p = -K\frac{\partial \xi}{\partial x}. \tag{5.20}$$

(b) The resultant force $\alpha\frac{\partial p}{\partial x}\delta x$ exerted on the fluid between B_2 and A_2 in the direction from B_2 to A_2, produces by Newton's second law of motion, an acceleration $\frac{\partial^2 \xi}{\partial t^2}$ of the fluid. Since force = mass × acceleration, we have

$$-\alpha\frac{\partial p}{\partial x}\delta x = (\rho_0 \alpha \delta x)\frac{\partial^2 \xi}{\partial t^2}, \tag{5.21}$$

the negative sign accounting for the fact that the unbalanced force is in the negative x-direction. Equation (5.21) simplifies to

$$-\frac{\partial p}{\partial x} = \rho_0\frac{\partial^2 \xi}{\partial t^2}. \tag{5.22}$$

Substituting for p from Eq. (5.20) into Eq. (5.22), we obtain

$$K\frac{\partial^2 \xi}{\partial x^2} = \rho_0\frac{\partial^2 \xi}{\partial t^2}$$

$$\text{or,} \quad \frac{\partial^2 \xi}{\partial t^2} = \frac{K}{\rho_0}\frac{\partial^2 \xi}{\partial x^2}. \tag{5.23}$$

Comparing Eq. (5.23) with Eq. (5.8), we find that ξ satisfies the differential wave equation for plane waves, the wave velocity for ξ being

$$c = \sqrt{\frac{K}{\rho_0}}. \tag{5.24}$$

Observation

The ratio of the increase in density $\delta\rho$ of a layer to the initial density ρ_0 is defined to be the *condensation*, s. Thus $s = \delta\rho/\rho_0$, or $\delta\rho = \rho_0 s$. The final density is $\rho = \rho_0 + \delta\rho = \rho_0(1+s)$. If V_0 and V be the initial and the final volumes of the fluid slab, respectively, we have $\rho V = \rho_0 V_0$, since the mass of the slab is a constant. As $V = V_0(1+\Delta)$, we obtain $(1+s)(1+\Delta) = 1$. The quantities s and Δ are small fractions so that the product $s\Delta$ is negligible. Therefore, $1 + s + \Delta = 1$, or

$$s = -\Delta. \tag{5.25}$$

Using Eq. (5.25) in the expression $K = -p/\Delta$, we get $K = p/s$, or

$$p = Ks \tag{5.26}$$

$$= c^2\rho_0 s, \tag{5.27}$$

where Eq. (5.24) has been used.

5.8 Velocity of Sound in Gases

When a sound wave propagates in a gas, the pressure changes, i.e., the compressions and rarefactions occur so rapidly that there is no exchange of heat between the layers and the surroundings. The process is thus *adiabatic* obeying the equation PV^γ = constant, where γ is the ratio of the specific heat of the gas at constant pressure to that at constant volume. The differentiation of the above equation gives

$$\gamma PV^{\gamma-1}dV + V^\gamma dP = 0$$

$$\text{or,} \quad \gamma P dV = -V dP$$

$$\text{or,} \quad \frac{dP}{dV} = -\frac{\gamma P}{V}. \tag{5.28}$$

The bulk modulus is

$$K = -\frac{p}{\frac{\delta V}{V_0}} = -V\frac{dP}{dV} \tag{5.29}$$

since $p = dP$, $\delta V = dV$ and $V_0 = V$. Substituting for $\frac{dP}{dV}$ from Eq. (5.28) into Eq. (5.29) gives $K = \gamma P$, so that the velocity of sound in the gas is

$$c = \sqrt{\frac{K}{\rho_0}} = \sqrt{\frac{\gamma P}{\rho}}. \tag{5.30}$$

For air at NTP, $\rho = 1.293$ kg/m^3, $P = 1.013 \times 10^5$ Pa and $\gamma = 1.41$. Substituting these values in Eq. (5.30), we have $c = 332.3$ m/s, in good agreement with the experimental value. The velocity of sound in a gas depends on the density, temperature, and humidity. It travels faster in a lighter gas. The velocity increases with a increase of temperature and humidity of the gas. Furthermore, since γ is greater for a monatomic gas, the velocity of sound in a monatomic gas is greater than that in a diatomic gas. For a fuller discussion on the effects of these various factors on the velocity of sound in a gaseous medium, the reader is referred to elsewhere[1].

A relationship can be found between the velocity of sound and the root-mean square (rms) speed of the gas molecules. The kinetic theory of gases predicts that the gas pressure is $P = \frac{1}{3}\rho v^2$, where ρ is the density of the gas and v is the rms speed of the gas molecules. Thus $v = \sqrt{3P/\rho}$, and the speed of sound is

$$c = \sqrt{\frac{\gamma P}{\rho}} = \left(\frac{\gamma}{3}\right)^{\frac{1}{2}} v. \tag{5.31}$$

For air or a diatomic gas, $\gamma = 1.41$, so that $c = 0.69v$. Thus *c and v are of the same order of magnitude.* This can be understood physically as follows : The compression and rarefaction are handed over from one layer of the gas to an adjacent layer through the collisions between the gas molecules moving with their thermal speeds. The speed of propagation of compression and rarefaction, i.e., the speed of sound is thus expected to be of the same order as the rms speed v of the gas molecules.

We shall now find the velocity of sound in a mixture of several gases. Let P_1, P_2, P_3, etc. be the partial pressures of the gases in the mixture giving the total pressure P at a temperature T. Suppose that the densities of the constituent gases under these conditions be ρ_1, ρ_2, ρ_3, etc. Then the density of the mixture is

$$\rho = \frac{P_1\rho_1 + P_2\rho_2 + \cdots}{P} = \frac{1}{P}\sum_i P_i\rho_i. \tag{5.32}$$

If γ is the effective ratio of the specific heats, we have

$$\frac{P}{\gamma - 1} = \frac{P_1}{\gamma_1 - 1} + \frac{P_2}{\gamma_2 - 1} + \cdots = \sum_i \frac{P_i}{\gamma_i - 1}. \tag{5.33}$$

With ρ and γ given by Eqs. (5.32) and (5.33), respectively, the speed of sound in the gas mixture is obtained from the expression $c = \sqrt{\gamma P/\rho}$.

5.9 Velocity of Sound in Liquids

The speed of sound in liquids is given by Eq. (5.24). Since the bulk modulus K is larger for liquids than in gases, the speed of sound in liquids in generally

[1] 'Elements of Physics' (for Higher secondary students) Vol. 1, by D. Chattopadhyay and P.C. Rakshit (New Age International, New Delhi).

greater than that in gases. For example, the sound speed in water at 30°C is 1510 m/s. There is some controversy as to whether the bulk modulus for liquids is to be taken under adiabatic or isothermal conditions. In many cases, the difference between the two values is not significant. A measurement of c yields a value for K, referred to as the *acoustic bulk modulus* of the liquid.

The speed of sound in liquids depends on temperature and to some extent on pressure, since both K and ρ_0 are affected by these factors. The sound speed in sea water increases with the salinity and with the depth below the surface.

5.10 Velocity of Longitudinal Waves in a Solid Bar

We consider a solid bar of cross sectional area α, the length of the bar being comparable to the wavelength of the longitudinal wave propagating along the bar. The bar is assumed to be thin enough so that a longitudinal stress produces the same displacement of the molecules over a given cross section. Let the density and Young's modulus of the material of the bar be ρ and Y, respectively. We choose two planes A_1 and B_1 inside the bar perpendicular to its axis, the initial distances of A_1 and B_1 from an arbitrary origin being x and $x + \delta x$, respectively (Fig. 5.4). The passage of the longitudinal wave causes the molecules in the rod to be displaced, thus shifting the planes A_1 and B_1. Let at any instant of time, the plane A_1 be displaced by ξ to the position A_2. Correspondingly the plane B_1 is displaced by $\xi + \delta\xi$ to B_2. The distance A_2B_2 is $\delta x + \delta\xi = \delta x + \frac{\partial \xi}{\partial x}\delta x$. Thus the distance between the planes increases by $\delta\xi = \frac{\partial \xi}{\partial x}\delta x$. Therefore, the elongation of the bar per unit length at this point is $\frac{\partial \xi}{\partial x}$. Consequently, the compressive strain is $-\frac{\partial \xi}{\partial x}$.

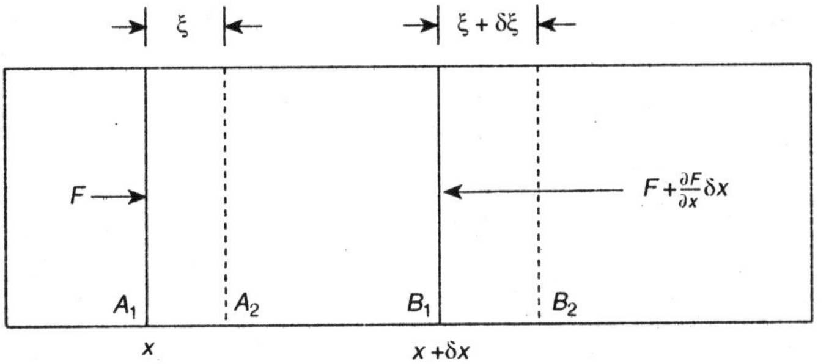

Fig. 5.4 Propagation of longitudinal waves in a thin solid rod

Since $\xi \ll \delta x$, the forces on A_1 and A_2 and those on B_1 and B_2 can be taken to be equal. If F is the force on A_1 (or A_2) acting in the positive x-direction, the force on B_1 (or B_2) acting in the opposite direction is $F + \frac{\partial F}{\partial x}\delta x$. These two forces produce two effects:

(i) A compressive stress F/α on the slab A_1B_1 develops the compressive strain $-\frac{\partial \xi}{\partial x}$. From Hooke's law, we have

$$Y = \frac{\text{stress}}{\text{strain}} = \frac{F/\alpha}{\left(-\frac{\partial \xi}{\partial x}\right)}$$

$$\text{or,} \quad F = -Y\alpha\frac{\partial \xi}{\partial x}. \tag{5.34}$$

(ii) An unbalanced resultant force $\frac{\partial F}{\partial x}\delta x$ causes, by Newton's second law of motion, an acceleration $\frac{\partial^2 \xi}{\partial t^2}$ of the slab of mass $\rho\alpha\delta x$. Since Force = mass × acceleration, we have

$$-\frac{\partial F}{\partial x}\delta x = \rho\alpha\delta x\frac{\partial^2 \xi}{\partial t^2}, \tag{5.35}$$

the negative sign implying that the force is in the negative x-direction. Equation (5.35) simplifies to

$$-\frac{\partial F}{\partial x} = \rho\alpha\frac{\partial^2 \xi}{\partial t^2}. \tag{5.36}$$

Differentiating Eq. (5.34) with respect to x and substituting in Eq. (5.36), we obtain

$$Y\frac{\partial^2 \xi}{\partial x^2} = \rho\frac{\partial^2 \xi}{\partial t^2}$$

$$\text{or,} \quad \frac{\partial^2 \xi}{\partial t^2} = \frac{Y}{\rho}\frac{\partial^2 \xi}{\partial x^2}. \tag{5.37}$$

Comparison of Eq. (5.37) with Eq. (5.8) shows that ξ satisfies the differential wave equation for plane waves. The wave velocity for ξ is therefore given by

$$c = \sqrt{\frac{Y}{\rho}}. \tag{5.38}$$

Note that the analysis given above is parallel to the one given in Sec. 5.7. In fact, Eq. (5.38) is obtained from Eq. (5.24) by replacing bulk modulus K with Young's modulus Y.

Since, for metals, Y is of the order of 10^{11}Pa and ρ is about 10^4kg/m^3, the velocity of sound in thin metal rods is a few times 10^3 m/s. This velocity is greater than that in liquids or gases.

Observations

If, instead of in a long thin bar, longitudinal waves propagate in an extended solid medium, there would be no lateral expansion or contraction associated with the longitudinal contraction or expansion during the passage of the wave, as we had for a long thin bar. Therefore, the axial modulus X should replace Young's modulus Y in Eqs. (5.37) and (5.38), so that in an extended solid the velocity of sound is given by

$$c = \sqrt{\frac{X}{\rho}}. \tag{5.39}$$

The axial modulus X can be written as

$$X = \frac{3K + 4n}{3} = \frac{Y(1-\sigma)}{(1+\sigma)(1-2\sigma)} \tag{5.40}$$

where K is the bulk modulus, n is the rigidity modulus, and σ is Poisson's ratio. So, Eq. (5.39) gives

$$c = \sqrt{\frac{3K + 4n}{3\rho}} = \sqrt{\frac{Y(1-\sigma)}{\rho(1+\sigma)(1-2\sigma)}}. \tag{5.41}$$

Waves travelling through the earth's crust during an earthquake are approximate examples of waves travelling in an extended solid.

5.11 Properties of the Differential Wave Equation

We have seen that the particle displacement ξ obeys the differential equation for plane waves, i.e.,

$$\frac{\partial^2 \xi}{\partial t^2} = c^2 \frac{\partial^2 \xi}{\partial x^2}. \tag{5.42}$$

ξ is a function of the independent variables x and t. We shall show here that the particle velocity, dilatation, condensation, acoustic pressure etc. satisfy the same differential wave equation as ξ and propagate with the same phase velocity c.

(i) The particle velocity u is given by

$$u = \frac{\partial \xi}{\partial t}. \tag{5.43}$$

Differentiating both sides of Eq. (5.42) with respect to t, we obtain

$$\frac{\partial}{\partial t}\left(\frac{\partial^2 \xi}{\partial t^2}\right) = c^2 \frac{\partial}{\partial t}\left(\frac{\partial^2 \xi}{\partial x^2}\right). \tag{5.44}$$

Since x and t are independent variables, the order of differentiation with respect to x and t can be interchanged in Eq. (5.44) to yield

$$\frac{\partial^2}{\partial t^2}\left(\frac{\partial \xi}{\partial t}\right) = c^2 \frac{\partial^2}{\partial x^2}\left(\frac{\partial \xi}{\partial t}\right)$$

or,

$$\frac{\partial^2 u}{\partial t^2} = c^2 \frac{\partial^2 u}{\partial x^2},$$

which is the same as Eq. (5.42). Thus the particle velocity u obeys the wave equation. In a similar way, it is easily shown that the particle acceleration $\frac{\partial u}{\partial t} = \frac{\partial^2 \xi}{\partial t^2}$ satisfies the wave equation.

(ii) The dilatation in plane waves has been already shown to be given by

$$\Delta = \frac{\partial \xi}{\partial x} = -s \tag{5.45}$$

where s is the condensation. Differentiating Eq. (5.42) with respect to x gives

$$\frac{\partial}{\partial x}\left(\frac{\partial^2 \xi}{\partial t^2}\right) = c^2 \frac{\partial}{\partial x}\left(\frac{\partial^2 \xi}{\partial x^2}\right)$$

$$\text{or,}\quad \frac{\partial^2}{\partial t^2}\left(\frac{\partial \xi}{\partial x}\right) = c^2 \frac{\partial^2}{\partial x^2}\left(\frac{\partial \xi}{\partial x}\right)$$

$$\text{or,}\quad \frac{\partial^2 \Delta}{\partial t^2} = c^2 \frac{\partial^2 \Delta}{\partial x^2}. \tag{5.46}$$

So, Δ (and hence s) obeys the wave equation.

(iii) The acoustic pressure p is related to the condensation s by $p = Ks$, where K is the bulk modulus. Since s satisfies the wave equation, it follows that p also satisfies the same, and so does the pressure gradient $\partial p/\partial x$. Furthermore, since the excess density is $\delta\rho = \rho_0 s$, obviously the excess density obeys the wave equation.

5.12 Simple Harmonic Solution of the Wave Equation

We have seen that the differential equation for the wave field parameter ψ associated with a plane wave propagating with the phase velocity c, is

$$\frac{\partial^2 \psi}{\partial t^2} = c^2 \frac{\partial^2 \psi}{\partial x^2}. \tag{5.47}$$

The results of Sec. 5.11 show that apart from the particle displacement ξ, the particle velocity, dilatation, acoustic pressure etc. can serve as the wave field parameter. For small displacements when Hooke's law is valid, i.e., the restoring force is proportional to the displacement, the particle displacement ξ varies simple harmonically. Any of the forms given by Eqs. (5.16a) through (5.16d) for ξ is acceptable. By direct substitution, it is readily verified that these forms satisfy the differential wave equation.

Taking

$$\xi = A\cos(\omega t - kx), \tag{5.48}$$

similar harmonic solutions can be written for any of the other wave field parameters. For example, we write for the particle velocity

$$u = \frac{\partial \xi}{\partial t} = -\omega A \sin(\omega t - kx)$$

$$= \omega A \cos\left(\omega t - kx + \frac{\pi}{2}\right). \tag{5.49}$$

The dilatation is

$$\Delta = \frac{\partial \xi}{\partial x} = kA \sin(\omega t - kx) = kA \cos\left(\omega t - kx - \frac{\pi}{2}\right). \tag{5.50}$$

The condensation is

$$s = -\Delta = -kA \sin(\omega t - kx) = kA \cos\left(\omega t - kx + \frac{\pi}{2}\right). \tag{5.51}$$

The acoustic pressure is

$$p = Ks = -KkA \sin(\omega t - kx) = -c^2 \rho_0 kA \sin(\omega t - kx)$$

$$= c^2 \rho_0 kA \cos\left(\omega t - kx + \frac{\pi}{2}\right). \tag{5.52}$$

The excess density is

$$d = \rho_0 s = -\rho_0 kA \sin(\omega t - kx) = \rho_0 kA \cos\left(\omega t - kx + \frac{\pi}{2}\right). \tag{5.53}$$

The above relationships show that u, s, p and d are in phase and all of them lead the particle displacement ξ by $\pi/2$ in time. Only the dilatation Δ lags ξ by $\pi/2$ in time.

Observation

The maximum acoustic pressure is $c^2 \rho_0 kA$, where A is the maximum particle displacement. The plots of ξ and p against t for a particular value of x are shown in Fig. 5.5. The curves show the harmonic variations given by Eqs. (5.48) and (5.52). Because of the phase difference of $\pi/2$ between ξ and p, the excess pressure is zero at maximum or minimum displacement; the excess pressure is a maximum or a minimum when the displacement is zero.

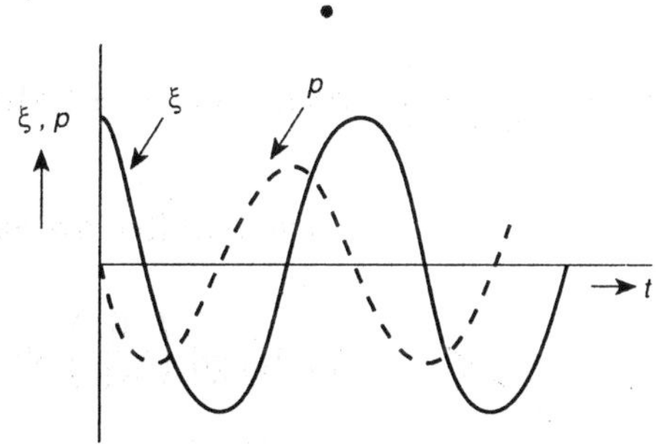

Fig. 5.5 Plots of ξ (solid line) and p (broken lines) against t for a monochromatic sound wave

The root-mean square sound pressure $p_{\rm rms}$ is given by

$$p_{\rm rms}^2 = \frac{1}{T}\int_0^T p^2 dt = \frac{(c^2\rho_0 kA)^2}{T}\int_0^T \sin^2(\omega t - kx)dt$$

$$= \frac{(c^2\rho_0 kA)^2}{2},$$

so that
$$p_{\rm rms} = \frac{c^2\rho_0 kA}{\sqrt{2}} = \frac{\sqrt{2}\pi c^2\rho_0 A}{\lambda}. \tag{5.54}$$

5.13 Energy Density of Plane Progressive Wave

When sound waves propagate in a medium, the particles in the medium start vibrating and so the medium acquires an energy which is partly kinetic and partly potential. The average energy contained in a unit volume of the medium is termed the *energy density.*

Suppose that a plane, harmonic, longitudinal wave is travelling in the positive x-direction. The kinetic energy of a layer of the medium having the thickness δx and a unit cross sectional area perpendicular to the direction of propagation of the wave, is

$$dE_K = \frac{1}{2}\rho_0\delta x\left(\frac{\partial\xi}{\partial t}\right)^2. \tag{5.55}$$

The thickness δx is so small that all the particles in the layer have the same velocity $(\partial\xi/\partial t)$. For a harmonic wave,

$$\xi = A\cos(\omega t - kx),$$

so that
$$dE_K = \frac{1}{2}\rho_0\delta x A^2\omega^2\sin^2(\omega t - kx).$$

The time average of the $\sin^2$ term over a period is 1/2, so that the average kinetic energy of the layer is

$$\overline{dE}_K = \frac{1}{4}\rho_0\delta x A^2\omega^2. \tag{5.56}$$

The time average of the kinetic energy density is

$$\bar{E}_K = \frac{1}{4}\rho_0 A^2\omega^2. \tag{5.57}$$

The same expression for $\bar{E}_K$ is obtained by taking the space average.

To calculate the potential energy, we note that the increase in the potential energy dE_p is the work done in compressing the layer by an increase of the pressure from P_0 to P. Thus

$$dE_P = -\int_{P_0}^{P} P \cdot dV \qquad (5.58)$$

where dV is the change in volume. The negative sign in Eq. (5.58) appears since an increase of V is associated with a decrease in P. If K is the bulk modulus, we have $K = -V_0\frac{dP}{dV}$, or $dV = -\frac{V_0}{K}dP$. Hence Eq. (5.58) gives

$$\begin{aligned} dE_P &= \frac{V_0}{K}\int_{P_0}^{P} PdP = \frac{V_0}{2K}(P^2 - P_0^2) \\ &= \frac{V_0}{2K}(P+P_0)(P-P_0) = \frac{V_0}{2K}(2P_0+p)p \\ &= \frac{V_0}{2K}(2P_0p + p^2), \end{aligned} \qquad (5.59)$$

since $P = P_0 + p$, where p is the excess pressure. As p varies sinusoidally following Eq. (5.52), its average value over a full cycle is zero, whereas the average value of p^2 is $\frac{1}{2}(c^2\rho_0 kA)^2$. The space averages of p and p^2 are the same as their time averages. So, the average potential energy of the layer is

$$\overline{dE_P} = \frac{V_0}{4K}(c^2\rho_0 kA)^2 = \frac{V_0}{4}c^2\rho_0 k^2 A^2 = \frac{V_0}{4}\rho_0\omega^2 A^2, \qquad (5.60)$$

since $K = \rho_0 c^2$ and $c^2k^2 = \omega^2$. As V_0, the volume of the layer, is δx, we obtain $\overline{dE_P} = \overline{dE_K}$. The average potential energy density is

$$\bar{E}_P = \frac{1}{4}\rho_0 A^2\omega^2. \qquad (5.61)$$

The total average energy density is

$$\bar{E} = \bar{E}_K + \bar{E}_P = \frac{1}{2}\rho_0 A^2\omega^2. \qquad (5.62)$$

Equations (5.62) shows that the energy density is proportional to the density of the medium ρ_0, the square of the amplitude A, and the square of the angular frequency ω of the wave.

5.14 Intensity

The intensity of sound or the *acoustic intensity* is the time rate of flow of average energy per unit area of the wavefront in a direction normal to the wavefront. Thus, if dW is the average energy crossing an area dS on the wavefront normally per unit time, the intensity is

$$I = \frac{dW}{dS}. \qquad (5.63)$$

The acoustic intensity is clearly the average acoustic power flowing normally through a unit area of the wavefront.

For plane progressive waves, the energy crossing a unit area of the wavefront in a unit time along the line of propagation is the energy contained in a cylinder of unit cross sectional area and length c, where c is the phase velocity. Thus the intensity is the product of the average energy density and the phase velocity. In other words,

$$I = \bar{E}c = \frac{1}{2}\rho_0 A^2\omega^2 c. \tag{5.64}$$

From Eq. (5.52), we note that the maximum acoustic pressure is

$$p_{\max} = c^2\rho_0 kA = c\rho_0\omega A, \tag{5.65}$$

since $c = \omega/k$. Using Eq. (5.65) we rewrite Eq. (5.64) as

$$I = \frac{p_{\max}^2}{2\rho_0 c} = \frac{p_{\rm rms}^2}{\rho_0 c}. \tag{5.66}$$

Equation (5.66) expresses the intensity of a plane wave in terms of the acoustic pressure (maximum or rms).
The root mean square particle velocity $u_{\rm rms}$ is obtained from Eq. (5.49). We have

$$\begin{aligned} u_{\rm rms}^2 &= \frac{1}{T}\int_0^T u^2 dt = \frac{\omega^2 A^2}{T}\int_0^T \sin^2(\omega t - kx)dt \\ &= \frac{\omega^2 A^2}{2}, \end{aligned}$$

whence

$$u_{\rm rms} = \frac{\omega A}{\sqrt{2}}. \tag{5.67}$$

Equations (5.54) and (5.67) give

$$p_{\rm rms}u_{\rm rms} = \frac{c\rho_0\omega A}{\sqrt{2}}\frac{\omega A}{\sqrt{2}} = \frac{1}{2}\omega^2 A^2 c\rho_0 = I. \tag{5.68}$$

Thus *the acoustic intensity for a plane wave is the product of the rms sound pressure and the rms particle velocity.*

The SI unit of intensity is W/m^2. It is assumed here that the wave is not attenuated as it travels, i.e., there is no absorption and dissipation of energy in the medium. The energy entering a section of the medium is the same as the energy leaving the section in the same time interval. That is why the intensity given by Eq. (5.64) is independent of x, the position coordinate along the direction of propagation. In the presence of attenuation, the intensity would decrease with increasing x for a forward wave.

Observations

(i) Let the acoustic pressure over a unit area of the wavefront in the direction of propagation be p, and the corresponding particle velocity be u. Then the instantaneous acoustic power is pu. From Eqs. (5.49) and (5.52) we write $pu = c^2\rho_0 k\omega A^2 \sin^2(\omega t - kx)$. The average value of $\sin^2(\omega t - kx)$ over a full cycle being $1/2$, the *average acoustic power* per unit area or the *intensity* is

$$I = \overline{pu} = \frac{1}{2}c^2\rho_0 k\omega A^2 = \frac{1}{2}\rho_0 c\omega^2 A^2,$$

which agrees with Eq. (5.64).

As $\sin^2(\omega t - kx) = \frac{1}{2}[1 - \cos 2(\omega t - kx)]$, we have for the instantaneous acoustic power

$$pu = \tfrac{1}{2}\rho_0 c\omega^2 A^2 [1 - \cos 2(\omega t - kx)]$$
$$= I[1 - \cos 2(\omega t - kx)].$$

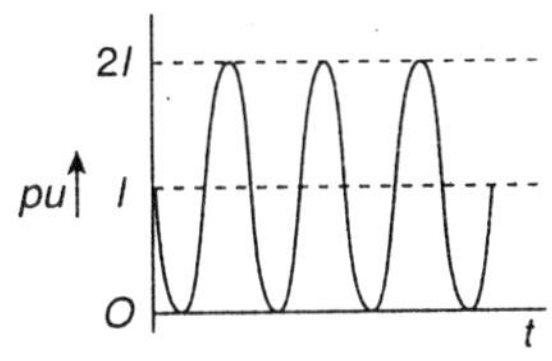

Fig. 5.5A Variation of the instantaneous acoustic power with time for $x = 0$

This shows that for a given value of x, pu oscillates with time with an angular frequency 2ω about the intensity I with an amplitude I (Fig. 5.5A).

(ii) The instantaneous power in an electric circuit is vi where v and i are the instantaneous voltage and current, respectively. This expression is analogous to the expression pu for the instantaneous acoustic power. Thus the acoustic pressure p is analogous to the voltage v, and the particle velocity u to the current i. Just as the ratio v/i is termed the impedance, the analogous ratio p/u is called the *specific acoustic impedance*, Z_s :

$$Z_s = \frac{p}{u}.$$

Using Eqs. (5.49), (5.52) and (5.30) we have

$$Z_s = \frac{c^2\rho_0 k}{\omega} = \rho_0 c = \sqrt{K\rho_0}.$$

Note that Z_s depends only on the parameters K and ρ_0 of the medium, i.e., it is a characteristic of the medium. Hence Z_s is often referred to as the *characteristic acoustic impedance* of the medium.

From Eq. (5.66) we can express the intensity in terms of the acoustic impedance :

$$I = \frac{p_{\text{rms}}^2}{Z_s}. \tag{5.68a}$$

The CGS unit of $\rho_0 c$ is g.cm^{-2}.s^{-1}, called *rayls*, after Lord Rayleigh. For air at 20°C and standard atmospheric pressure, $\rho_0 = 1.21 \times 10^{-3}$ g/cm^3 and $c = 3.43 \times 10^4$ cm/s, so that $\rho_0 c = 41.5$ rayls. In MKS units, $\rho_0 c$ is expressed in mks rayls ($\equiv$ kg.m^{-2}.s^{-1}). Thus for air at 20°C and standard atmospheric pressure, $\rho_0 c = 415$ mks rayls.

(iii) The sound waves incident on a surface exert on it a pressure, known as the *acoustic radiation pressure.* This pressure equals the average rate of change of momentum of the particles incident on a unit area of the surface. If u is the particle velocity and m is the mass per unit area, then the rate of change of momentum in question is $\frac{d(mu)}{dt} = m\frac{du}{dt} + u\frac{dm}{dt}$. Since u is a sinusoid, the time average of $\frac{du}{dt}$ is zero. Also, $\frac{dm}{dt} = \rho_0 u$. So, the time average of $u\frac{dm}{dt}$ is $\frac{1}{2}\rho_0\omega^2 A^2 = \bar{E} = I/c$. Thus the radiation pressure equals the average energy density of the wave, which is intensity divided by the phase velocity.

5.15 Bel and Phon

In practice, intensities of sound waves are expressed in relative values in terms of some standard intensity. The auditory sensation of loudness increases or decreases as the intensity increases or decreases. The response of the human ear is not proportional to the intensity; rather, the sensation is governed by the logarithm of the intensity. The intensity is a physical quantity and has an objective existence. But loudness is a sensation dictated by the human ear and judged by the brain, and is therefore partly subjective.

A relationship between the intensity and the loudness is given by the *Weber-Fechner law,* which states that the increase in loudness dL for an average listener is proportional to dI/I, where dI is a small increase of the intensity I. Thus,

$$dL = C_0 \frac{dI}{I},$$

where C_0 is a constant. Integrating, we obtain

$$L = C_0 \ln I + C_1 = C \log_{10} I + C_1, \tag{5.69}$$

where C and C_1 are constants. So, if L_1 and L_2 refer to the loudness for the intensities I_1 and I_2, respectively, the loudness change corresponding to the intensity change from I_1 to I_2 is

$$L_2 - L_1 = C \log_{10} \frac{I_2}{I_1}. \tag{5.70}$$

Equation (5.70) shows that a doubling of the intensity does not give a doubling of the loudness.

The quantity $\log_{10}(I_2/I_1)$ gives the intensity level in bel (B) of I_2 relative to the standard intensity I_1. The term 'bel' is coined from Graham Bell, the inventor of telephone. Since 'bel' is a large unit, the intensity level is often expressed in decibel (dB). Since $1\,\text{B} = 10\,\text{dB}$, the intensity level in dB is $10\log_{10}(I_2/I_1)$.

The threshold audibility, i.e., the lowest audibility for a note of frequency 1 kHz corresponds to an intensity of 10^{-12}W/m^2. This is taken as the standard intensity. A sound having an intensity 10 times the threshold value has an intensity level of 1 B or 10 dB.

The loudness level of a sound is measured with respect to that of the standard reference tone of frequency 1 kHz and intensity 10^{-12}W/m^2. The sound whose loudness level is to be determined, and the standard reference tone are alternately heard by an average listener with both ears from the same distance. The intensity of the reference tone is slowly increased till the two sounds appear to be equally loud. If, under this condition, the intensity level of the reference tone is n dB above threshold, the sound under test is said to have an equivalent loudness level of n phon.

Observations

(1) We have noted that the intensity level (abbreviated IL) of a sound of intensity I is expressed in dB by

$$\text{IL} = 10\log_{10}\left(\frac{I}{I_0}\right),$$

where I_0 is the reference intensity corresponding to the threshold of audibility, i.e., $I_0 = 10^{-12}\ \text{W/m}^2$. From Eq. (5.66), we find that the intensity I is proportional to p^2_{rms}. So,

$$\text{IL} = 10\log_{10}\left(\frac{p_{\text{rms}}}{p_0}\right)^2 = 20\log_{10}\frac{p_{\text{rms}}}{p_0},$$

where p_0 is the rms value of the reference pressure for the threshold of audibility in air. Note that for air, $\rho_0 = 1.293\ \text{kg/m}^3$ and $c = 332.3\ \text{m/s}$. Substituting these values in Eq. (5.66) together with the reference intensity $10^{-12}\ \text{W/m}^2$, we obtain for the corresponding reference rms acoustic pressure $p_0 = 2\times10^{-5}\ \text{N/m}^2 = 2\times10^{-5}\,\text{Pa}$.

Many detectors respond to the changes in the pressure of the acoustic wave rather than to the intensity. Hence often the acoustic pressure is used instead of intensity in sound measurements. Consequently, the quantity $20\log_{10}(p_{\text{rms}}/p_0)$ is referred to as the *sound pressure level* (abbreviated SPL) in dB for the effective acoustic pressure p_{rms}. Thus

$$\text{SPL} = 20\log_{10}\left(\frac{p_{\text{rms}}}{p_0}\right).$$

For plane progressive waves in air, the above equations for IL and SPL, give the same numerical results. But in more complex sound fields, viz., standing wave patterns, intensity and pressure are not simply related by Eq. (5.66), and so the above expressions for IL and SPL will not give identical values.

(2) The intensity level (in dB) of a sound of intensity I_2 with respect to another sound of intensity I_1 is

$$
\begin{aligned}
N &= 10\log_{10}\frac{I_2}{I_1} = 10\,(\log_{10} I_2 - \log_{10} I_1) \\
&= 10\,[(\log_{10} I_2 - \log_{10} I_0) - (\log_{10} I_1 - \log_{10} I_0)] \\
&= 10\log_{10}\frac{I_2}{I_0} - 10\log_{10}\frac{I_1}{I_0},
\end{aligned}
$$

where I_0 is the standard threshold intensity. Hence the intensity level of I_2 relative to I_1 is the *difference* of the intensity levels of I_2 and I_1 with respect to the standard threshold.

If $I_2 > I_1$, then N is positive. N is negative when $I_2 < I_1$. If $N = 1$, $I_2 : I_1 = 10^{0.1} : 1 = 1.259 : 1$. Thus 1 dB corresponds to an arithmetical intensity ratio of 1.259 : 1. This is nearly the *minimum* intensity change that a normal ear can perceive.

(3) From the detectable feeblest sound to the loudest sound irritating to the ear, the intensity changes by about 10^{14} or 140 dB. Thus the use of decibels makes it possible to express large intensity ratios by relatively small numbers.

5.16 Phase Velocity and Group Velocity

A plane progressive harmonic wave propagating in the positive x-direction can be expressed as

$$\xi = A\sin(\omega t - kx), \tag{5.71}$$

where ξ is the particle displacement at the position coordinate x at time t; A is the amplitude, ω is the angular frequency, and k is the propagation constant of the wave. If, at time $t+dt$, the constant phase $(\omega t - kx)$ advances to the point $x + dx$, we have

$$\omega t - kx = \omega(t + dt) - k(x + dx)$$

whence the velocity c with which the phase moves is given by

$$c = \frac{dx}{dt} = \frac{\omega}{k}, \tag{5.72}$$

c is referred to as the *phase velocity* or the *wave velocity*.

If two or more plane simple harmonic waves of the same amplitude but different frequencies superimpose, a *group* of waves is formed. The amplitude of the group changes with distance, and the velocity with which the maximum of the wave group travels is referred to as the *group velocity*. The energy is transmitted with the group velocity.

Interestingly, a monochromatic wave train of finite length can be treated as a wave group, since it is formed by the superposition of a theoretically infinite number of plane harmonic waves of continuously differing frequencies. In practice, the frequencies are within a band whose width is determined by the length of the wave train. The shorter the length, the wider the frequency band.

For small amplitude acoustic waves moving in an isotropic homogeneous elastic medium, the wave velocity does not depend on the frequency. The velocities of the component waves being equal, the wave group does not change its form as it advances; here the group velocity is the same as the phase velocity. If the medium is dispersive, i.e., if the phase velocity depends on the frequency, then the group velocity will be different from the phase velocity.

To find an expression for the group velocity, we consider that the group is formed by the combination of two waves of equal amplitude A and slightly different angular frequencies ω and $\omega + d\omega$, travelling with the propagation constants k and $k + dk$. The resultant displacement of a particle at x at time t under the simultaneous action of the two waves is

$$\xi = A\sin(\omega t - kx) + A\sin\left[(\omega + d\omega)t - (k + dk)x\right]. \tag{5.73}$$

Resorting to the trigonometric identity $\sin A + \sin B = 2\sin\frac{A+B}{2}\cos\frac{B-A}{2}$, we can write Eq. (5.73) as

$$\begin{aligned}\xi &= 2A\sin\tfrac{1}{2}[(\omega t - kx) + (\omega + d\omega)t - (k + dk)x] \times \\ &\qquad \cos\tfrac{1}{2}[(\omega + d\omega)t - (k + dk)x - (\omega t - kx)] \\ &= 2A\cos\left(\frac{td\omega - xdk}{2}\right)\sin\left[\left(\omega + \frac{d\omega}{2}\right)t - \left(k + \frac{dk}{2}\right)x\right]. \end{aligned} \tag{5.74}$$

In Eq. (5.74), the slowly varying cosine term can be taken as the amplitude of the resultant wave represented by the sine term. Looking at the sine term we find that the angular frequency and the propagation constant of the resultant wave are $\omega + d\omega/2$ and $k + dk/2$, respectively. Note that these are the mean values for the component waves. The phase velocity of the composite wave is

$$c = \frac{\omega + (d\omega/2)}{k + (dk/2)} \simeq \frac{\omega}{k}, \tag{5.75}$$

i.e., nearly the phase velocity of the component waves.

The amplitude represented by the cosine term advances with the group velocity v. If the phase $(td\omega - xdk)/2$ associated with it progresses to $(x + dx)$ at time $(t + dt)$, we have

$$td\omega - xdk = (t + dt)d\omega - (x + dx)dk,$$

so that the group velocity is

$$v = \frac{dx}{dt} = \frac{d\omega}{dk}. \tag{5.76}$$

Equation (5.72) gives $\omega = ck$, or $d\omega = cdk + kdc$

$$\text{or,} \quad v = \frac{d\omega}{dk} = c + k\frac{dc}{dk}. \tag{5.77}$$

Again, $k = 2\pi/\lambda$, or, $\lambda = 2\pi/k$, or, $\frac{d\lambda}{dk} = -\frac{2\pi}{k^2}$. Hence Eq. (5.77) gives

$$v = c + k\frac{dc}{d\lambda} \cdot \frac{d\lambda}{dk} = c - \frac{2\pi}{k}\frac{dc}{d\lambda} = c - \lambda\frac{dc}{d\lambda}. \tag{5.78}$$

Equations (5.77) and (5.78) relate the group velocity and the phase velocity.

In a nondispersive medium, the phase velocity c is independent of the frequency ω or the wavelength λ, i.e., $\frac{dc}{d\lambda} = 0$. Hence $v = c$, i.e., the group velocity and the phase velocity are equal. In a dispersive medium where c increases with increasing λ, $\frac{dc}{d\lambda}$ is positive, so that $v < c$, i.e., the group velocity with which the energy propagates, is *less than* the phase velocity.

The two component waves of slightly different wavelengths and their resultant wave, as given by Eq. (5.74), are depicted in Fig. 5.6. The solid curve in Fig. 5.6(c) represents the sine term. The envelope shown by the dashed lines represents the cosine term. The figure clearly shows the amplitude variation of the resultant wave given by the sine term. The amplitude varies simple harmonically with x and t, and the waves are split into groups. The velocity of the wave group or the envelope is v.

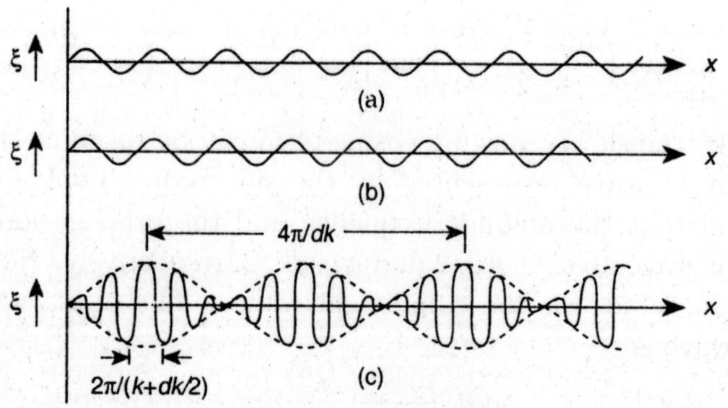

Fig. 5.6 The component waves are given in (a) and (b); the resultant wave in (c)

5.17 Stationary Waves

When two or more waves propagating independently of one another in a medium at the same time superimpose, the resultant displacement, velocity, and acceleration of any particle of the medium in the region of overlap is the vector sum of the displacements, velocities, and accelerations of the particle caused by the individual waves. This is known as the *principle of superposition.* For waves propagating in a material medium, the principle holds for small amplitudes only, because only then the particle displacement is proportional to the restoring force.

When two identical progressive acoustic waves moving in a medium along the same straight line with the same velocity in opposite directions superimpose, the resulting waves alternately wax and wane but do not propagate in either direction. Such waves are known as *stationary waves* or *standing waves.*

The standing waves are confined to the region of the medium where the progressive waves overlap. They do not advance through the medium, but alternately expand and shrink. In the simplest case, the particles in a stationary wave oscillate simple harmonically about their mean positions with different amplitudes. At some equidistant points the amplitude is zero, i.e., at these positions the particles of the medium are permanently at rest. These points are called the *nodes.* The amplitude of vibration of the successive particles from one node to the next at first increases, attains a maximum midway between the two nodes, and then decreases to zero at the next node. The points where the particles oscillate with the maximum amplitude are termed the *antinodes.* All the particles between any two successive nodes oscillate in the same phase. But the particles between a pair of consecutive nodes differ in phase by 180° with the particles between the adjacent pair of nodes. The distance between two successive nodes or two successive antinodes is equal to half the wavelength of the overlapping waves.

To derive the equation of a stationary wave we consider two identical harmonic progressive waves moving in a medium along the x-axis in opposite directions. The particle displacement for the wave propagating in the positive x-direction is

$$\xi_1 = a \sin \frac{2\pi}{\lambda}(ct - x). \tag{5.79}$$

The same for the wave travelling in the negative x-direction is

$$\xi_2 = a \sin \frac{2\pi}{\lambda}(ct + x). \tag{5.80}$$

Here a, λ and c are respectively the amplitude, the wavelength, and the phase velocity of waves. When the waves superpose, the resultant particle displacement is

$$\xi = \xi_1 + \xi_2 = a\left[\sin\frac{2\pi}{\lambda}(ct - x) + \sin\frac{2\pi}{\lambda}(ct + x)\right]$$

$$= 2a\sin\frac{2\pi ct}{\lambda}\cos\frac{2\pi x}{\lambda} = A\sin\frac{2\pi ct}{\lambda}, \tag{5.81}$$

where

$$A = 2a\cos\frac{2\pi x}{\lambda}. \tag{5.82}$$

Equation (5.81) is known as the *equation of the stationary wave*. It shows that the wave is not progressive because the argument of the sine function contains ct and not $(ct-x)$. A particle at a distance x from the chosen origin oscillates simple harmonically with an amplitude A given by Eq. (5.82). This equation reveals that the amplitude A of the stationary wave is not a constant; it is a periodic function of x.

At a node, $A = 0$, i.e., $\cos\frac{2\pi x}{\lambda} = 0$, i.e., $\frac{2\pi x}{\lambda} = (2n+1)\frac{\pi}{2}$, i.e., $x = (2n+1)\frac{\lambda}{4}$, where $n = 0, \pm1, \pm2$ etc. Obviously, the distance between two consecutive nodes is $\lambda/2$.

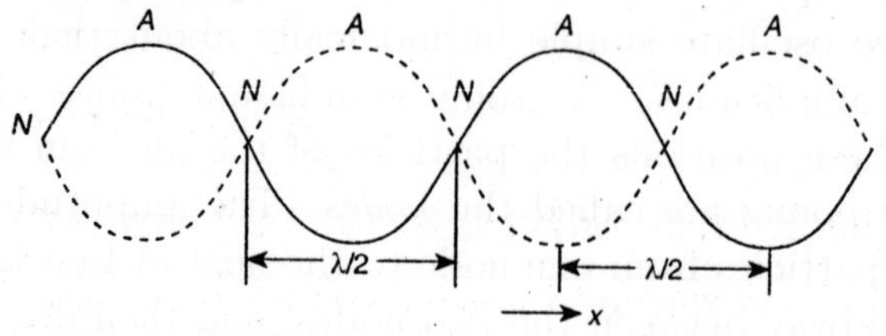

Fig. 5.7 Stationary wave pattern with nodes (N) and antinodes (A)

At an antinode, the amplitude A is a maximum, i.e., $\cos\frac{2\pi x}{\lambda} = \pm1$, i.e., $\frac{2\pi x}{\lambda} = n\pi$ or, $x = \frac{n\lambda}{2}$, where $n = 0, \pm1, \pm2$ etc. The maximum amplitude at an antinode is $\pm 2a$, i.e., numerically two times the amplitude of a component progressive wave. Obviously, the antinodes are midway between the nodes, and the distance between two successive antinodes is $\lambda/2$ (Fig. 5.7).

A. Pressure variation along a stationary wave

The excess pressure in a longitudinal wave is $p = -K\frac{\partial\xi}{\partial x}$, where K is the bulk modulus of the medium and ξ is the particle displacement. For the two waves given by Eqs. (5.79) and (5.80), the acoustic pressures are respectively $p_1 = -K\frac{\partial\xi_1}{\partial x}$ and $p_2 = -K\frac{\partial\xi_2}{\partial x}$. The net excess pressure is $p = p_1 + p_2 = -K\frac{\partial}{\partial x}(\xi_1 + \xi_2) = -K\frac{\partial\xi}{\partial x}$, where the resultant particle displacement ξ is given by Eq. (5.81). Hence

$$p = K\frac{4\pi a}{\lambda}\sin\frac{2\pi x}{\lambda}\sin\frac{2\pi ct}{\lambda} = A_1\sin\frac{2\pi ct}{\lambda}, \tag{5.83}$$

where

$$A_1 = \frac{4\pi Ka}{\lambda} \sin \frac{2\pi x}{\lambda}. \tag{5.84}$$

The quantity A_1 gives the pressure amplitude varying harmonically with x. The pressure amplitude is zero when $\sin \frac{2\pi x}{\lambda} = 0$, i.e. $\frac{2\pi x}{\lambda} = n\pi$, i.e., $x = \frac{n\lambda}{2}$, where $n = 0, \pm 1, \pm 2$ etc. These points are known as *pressure nodes.* Note that *the pressure nodes coincide with the displacement antinodes.*

The pressure amplitude is a maximum when $\sin \frac{2\pi x}{\lambda} = \pm 1$, i.e., $\frac{2\pi x}{\lambda} = (2n+1)\frac{\pi}{2}$, i.e., $x = (2n+1)\frac{\lambda}{4}$, where $n = 0, \pm 1, \pm 2$ etc. These points are called the *pressure antinodes*, and *they coincide with the displacement nodes.* Thus at points where the displacement is a maximum the excess pressure is zero, and vice versa.

B. Stationary waves produced by reflection

If a boundary surface is placed in the path of a progressive wave, the wave is reflected from the surface. The reflected wave travels backwards and superimposes on the forward wave, thus producing stationary waves. Let the particle displacement for the incident wave moving in the positive x-direction be $\xi_1 = a \sin \frac{2\pi}{\lambda}(ct-x)$. The reflected wave moves in the negative x-direction, and so the corresponding particle displacement is $\xi_2 = ra \sin \frac{2\pi}{\lambda}(ct+x)$, where $r(\leq 1)$ is the ratio of the reflected amplitude to the incident amplitude. The quantity r is known as the *displacement reflection coefficient.* Suppose that the reflector is positioned at $x = 0$. At this point, the resultant particle displacement is

$$\xi = \xi_1 + \xi_2 = (1+r)a \sin \frac{2\pi}{\lambda} ct. \tag{5.85}$$

From the formula for the excess pressure $p = -K\frac{\partial \xi}{\partial x}$, the excess pressure for the incident and the reflected waves are $p_1 = \frac{2\pi Ka}{\lambda} \cos \frac{2\pi}{\lambda}(ct - x)$ and $p_2 = -\frac{2\pi Kar}{\lambda} \cos \frac{2\pi}{\lambda}(ct + x)$. The resultant excess pressure at $x = 0$ is

$$p = p_1 + p_2 = \frac{2\pi Ka}{\lambda}(1-r) \cos \frac{2\pi ct}{\lambda}. \tag{5.86}$$

Case 1. If the reflector is perfectly rigid, the particle displacement ξ given by Eq. (5.85) is zero at $x = 0$ for all values of t. Hence $r = -1$. This shows that at the point of reflection, there is a phase shift of π between the particle displacement of the incident wave and that of the reflected wave. The *boundary is a displacement node and consequently a pressure antinode.*

Case 2. If the reflector can move freely, the resultant excess pressure p given by Eq. (5.86) vanishes at $x = 0$ for all values of t. Hence $r = 1$, so that the particle displacement suffers no phase change upon reflection. Here *the boundary is a pressure node and a displacement antinode.*

The above two cases represent extreme situations. If the reflecting wall is not perfectly rigid, r lies between 0 and -1. Then the reflected wave has a smaller amplitude and carries less energy than the incident wave. The resulting stationary wave will contain imperfect nodes. The amplitude at the nodes will be $(1+r)a$ and that at the antinodes will be $(1-r)a$, r being negative. The intensity being proportional to the square of the amplitude, the intensity of the reflected wave is r^2 times the intensity of the incident wave. So, the *absorption coefficient* of the reflector is $(1-r^2)$. This quantity is considerably important in the acoustics of auditoria.

The ratio of the maximum to the minimum particle displacement in a stationary wave is termed the *standing wave ratio* (abbreviated SWR). The ratio r^2 between the intensity of the reflected wave to that of the incident wave is known as the *power reflection coefficient.* This coefficient is clearly related to the SWR.

C. Energy of stationary waves

The excess pressure p over a unit area in a plane perpendicular to the direction of propagation of the forward component wave is given by Eq. (5.83). The corresponding particle displacement is expressed by Eq. (5.81). The particle velocity is

$$u = \frac{\partial \xi}{\partial t} = \frac{4\pi ac}{\lambda} \cos\frac{2\pi ct}{\lambda} \cos\frac{2\pi x}{\lambda}. \tag{5.87}$$

So, the instantaneous acoustic power flow is

$$pu = \left(\frac{4\pi a}{\lambda}\right)^2 Kc \sin\frac{2\pi ct}{\lambda} \cos\frac{2\pi ct}{\lambda} \sin\frac{2\pi x}{\lambda} \cos\frac{2\pi x}{\lambda}.$$

Since the average value of $\sin\frac{2\pi ct}{\lambda}\cos\frac{2\pi ct}{\lambda}$ over a complete period (0 to T) is zero, it follows that the average transmission of power through any area for a stationary wave is zero.

To calculate the energy density in a stationary wave, we consider a layer of the medium of thickness δx and unit cross sectional area perpendicular to the x-direction. The kinetic energy of the layer is

$$dE_K = \frac{1}{2}\rho_0 u^2 \delta x.$$

Using Eq. (5.87), we obtain

$$dE_K = \frac{1}{2}\rho_0 \delta x \left(\frac{4\pi ac}{\lambda}\right)^2 \cos^2\frac{2\pi ct}{\lambda} \cos^2\frac{2\pi x}{\lambda}. \tag{5.88}$$

The time average of $\cos^2 \frac{2\pi ct}{\lambda}$ over a period is 1/2. Hence the average kinetic energy of the layer is

$$\overline{dE_K} = \frac{1}{4}\rho_0 \delta x \left(\frac{4\pi ac}{\lambda}\right)^2 \cos^2 \frac{2\pi x}{\lambda}. \tag{5.89}$$

The increase in the potential energy dE_p of the layer is given by Eq. (5.59). For the stationary wave, the excess pressure p is obtained from Eq. (5.83). The time average of p over a complete cycle is zero, and that of p^2 is $\frac{1}{2}(4\pi Ka/\lambda)^2 \sin^2(2\pi x/\lambda)$. Hence the average potential energy of the layer is

$$\overline{dE_p} = \frac{V_0}{4K}\left(\frac{4\pi Ka}{\lambda}\right)^2 \sin^2\frac{2\pi x}{\lambda} = \frac{1}{4}\rho_0 \delta x \left(\frac{4\pi ac}{\lambda}\right)^2 \sin^2 \frac{2\pi x}{\lambda}, \tag{5.90}$$

since $V_0 = \delta x$ and $K = \rho_0 c^2$. The total average energy of the layer is

$$\overline{dE} = \overline{dE_K} + \overline{dE_p} = \frac{1}{4}\rho_0 \delta x \left(\frac{4\pi ac}{\lambda}\right)^2. \tag{5.91}$$

The average energy density, i.e., the average energy per unit volume is

$$\bar{E} = \frac{4\pi^2 a^2 c^2}{\lambda^2}\rho_0 = \rho_0 \omega^2 a^2. \tag{5.92}$$

Note that the average energy density contained in a stationary wave is two times that for a progressive wave [see Eq. (5.62)].

Observation

Let two plane progressive harmonic waves propagating along the same line in opposite directions overlap. The waves have unequal amplitudes and are represented by $y_1 = a_1 \sin(\omega t - kx)$ and $y_2 = a_2 \sin(\omega t + kx)$; $a_1 > a_2$. Let $a_1 = a_2 + a$. Then $y = y_1 + y_2 = 2a_2 \sin \omega t \cos kx + a \sin(\omega t - kx)$. Thus, the resultant wave consists of a stationary wave (represented by the first term on the right-hand side) and a progressive wave travelling along the same line (represented by the second term).

Progressive and stationary waves : a comparison

Progressive waves	Stationary waves
1. Progressive waves are generated by the periodic vibrations (which are simple harmonic in the simplest case) of the particles of the medium.	1. Stationary waves are formed by the superposition of two identical progressive waves propagating along the same line in opposite directions.

(contd. on next page)

(contd. from previous page)

Progressive waves	Stationary waves
2. Each particle of the medium on the path of the wave oscillates about its mean position wih the same amplitude and period.	2. The particles oscillate with the same period but different amplitudes. The amplitude is a maximum at the antinodes, and is always zero at the nodes.
3. The wave moves due to the elasticity of the medium.	3. Stationary waves do not move through the medium, but they alternately expand and shrink in segments in the region of the medium where the progressive waves overlap.
4. The instantaneous phase difference between two vibrating particles on the path of the wave is proportional to the equilibrium distance between the particles.	4. All the vibrating particles between two successive nodes are in the same phase. But the particles between a pair of consecutive nodes differ in phase by 180° with the particles between the adjacent pair of nodes.
5. All the particles do not cross their equilibrium positions simultaneously.	5. All the particles cross their equilibrium positions simultaneously two times in a period.
6. The pressure and density at each point periodically suffer similar changes.	6. For longitudinal waves, pressure and density undergo maximum changes at the nodes. There is no change at the antinodes.
7. The average energy density is distributed equally between the kinetic and potential forms.	7. The average energy density is the sum of the average energy densities of the component progressive waves.
8. The equation of a progressive harmonic wave travelling in the positive x direction is $\xi = a \sin \frac{2\pi}{\lambda}(ct - x)$. The equation shows that with increasing time t, the distance x must increase to keep $(ct-x)$ constant. Thus, planes of constant phase move in the positive x direction, showing that the wave is progressive.	8. The equation of a stationary wave is $\xi = 2a \sin \frac{2\pi ct}{\lambda} \cos \frac{2\pi x}{\lambda}$. The equation does not contain a periodic function of $(ct \mp x)$ which characterizes a progressive wave.

5.18 Doppler Effect

The pitch of a note, i.e., its frequency as perceived by a listener, appears to change when there is a relative motion between the source of sound and the listener. The apparent inequality between the emitted and the perceived frequencies is referred to as the *Doppler effect*. The pitch or the frequency of the whistle of a railway engine or the sound of an aeroplane appears to increase as the source approaches a listener. The pitch or the frequency suddenly drops as the source passes the listener and recedes from him. The closure the listener stands to the line of motion of the source and the higher the speed, the more sudden the change. For an aeroplane flying overhead. the change in pitch occurs rather slowly when the plane is far above the ground, but the change is rapid when it is at a low height.

Doppler effect is common for all wave motions including light. However, for a detectable change, the relative velocity between the source and the observer must be a sizable fraction of the velocity of the wave. Doppler effect is readily detected for sound waves since the velocity of sound is not so high. The effect is more difficult to observe for light waves whose velocity is very high. The slight changes in the frequency of light coming from stars moving towards or away from the earth have been recorded by spectographs and they give a wealth of information about the motion of the stars.

For a moving source, the change in frequency occurs because, although the source emits waves at a constant rate, the waves emitted in a specified time occupy a shorter or a longer length than when the source is stationary. Consequently, the apparent wavelength and hence the frequency changes. If the observer moves, he receives a larger or a smaller number of waves in a specified time than when he was stationary relative to the source. The motion of the medium changes the speed of sound, resulting in a change of apparent frequency. A quantitative calculation of the change in frequency in different cases of interest is given below.

(i) Source moving, observer at rest

Let O be the observer at rest and S be the source moving with a velocity v_s along SO while continuously emitting sound waves of frequency n and

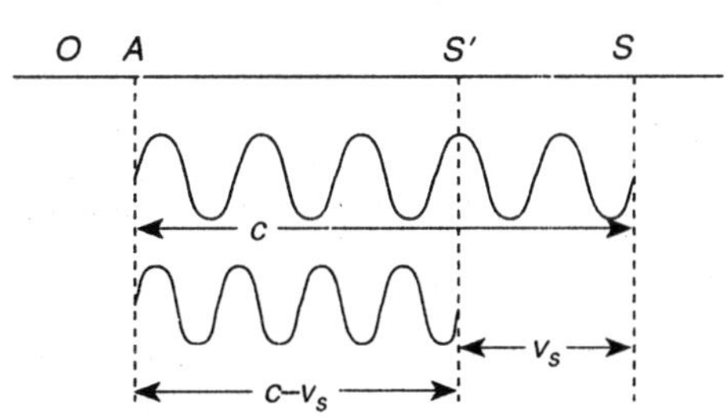

Fig. 5.8 Doppler effect for a moving source and a stationary observer

wavelength λ. The waves propagate towards the observer with a velocity c. All the n waves emitted by the source in 1s will occupy the distance $SA = c$, if the source is stationary (Fig. 5.8). However, if in this period, the source moves to S', where $SS' = v_s$, these n waves will occupy the distance $S'A = c - v_s$. The original wavelength is $\lambda = SA/n = c/n$, whereas the changed wavelength is $\lambda' = S'A/n = \frac{c-v_s}{n}$. Thus, when the source moves towards the observer, the wavelength is shortened. The velocity of sound remaining unchanged, the observer interprets the waves of wavelength λ' as the waves of frequency

$$n' = \frac{c}{\lambda'} = \frac{c}{(c - v_s)/n} = \frac{cn}{c - v_s}. \tag{5.93}$$

The apparent change in frequency of the note is

$$n' - n = n\left(\frac{c}{c - v_s} - 1\right) = \frac{nv_s}{c - v_s}. \tag{5.94}$$

The fractional change in frequency is

$$\frac{n' - n}{n} = \frac{v_s}{c - v_s}. \tag{5.95}$$

If the source moves away from the observer, v_s is negative, and so the apparent frequency of the note is

$$n' = \frac{cn}{c + v_s}. \tag{5.96}$$

When the wind blows with a velocity v_m in the direction of sound, the sound velocity is effectively enhanced to $c + v_m$. In this case, Eq. (5.93) modifies to

$$n' = \frac{(c + v_m)n}{c + v_m - v_s}. \tag{5.97}$$

If the wind blows against the sound, the effective sound velocity is $c - v_m$, so that Eq. (5.97) becomes

$$n' = \frac{(c - v_m)n}{c - v_m - v_s}. \tag{5.98}$$

(ii) Observer moving, source at rest

Suppose that a stationary source S emits continuously sound waves of frequency n and wavelength λ, the speed of sound being c. The observer O moves with a velocity v_O away from the source (Fig. 5.9).

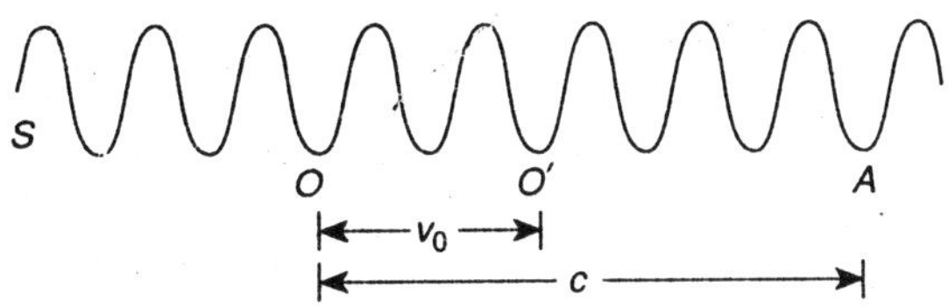

Fig. 5.9 Doppler effect for a moving observer and fixed source

If the observer were at rest, the n waves crossing O in 1s would occupy a length $OA = c$. But during this period the observer moves to O' where $OO' = v_0$. Thus, in 1s he receives the waves occupying the length $O'A = c - v_0$. This number is

$$n' = \frac{c - v_0}{\lambda} = \frac{c - v_0}{c/n},$$

since n waves are contained in the distance c. So, the apparent frequency is

$$n' = \frac{(c - v_0)n}{c}. \tag{5.99}$$

If the observer approaches the source, v_0 is negative so that the apparent frequency becomes

$$n' = \frac{(c + v_0)n}{c}. \tag{5.100}$$

If the wind blows in the direction of sound with the speed v_m, the effective velocity of sound is raised to $c + v_m$. Thus Eq. (5.99) becomes

$$n' = \frac{(c + v_m - v_0)n}{c + v_m}. \tag{5.101}$$

When the wind blows against the sound, Eq. (5.101) is replaced by

$$n' = \frac{(c - v_m - v_0)n}{c - v_m}. \tag{5.102}$$

(iii) Both source and observer moving

Let the observer move with a velocity v_0 towards the source which, in turn, approaches the observer with a speed v_s. When the observer is at rest and the source moves, the apparent frequency of the note is [see Eq. (5.93)]

$$n_1 = \frac{cn}{c - v_s}.$$

The observer is now assumed to approach the source which remains stationary and emits sounds of frequency n_1. So, the apparent frequency of the note is [see Eq. (5.100)]

$$n' = \frac{(c + v_0)n_1}{c} = \frac{(c + v_0)n}{c - v_s}. \tag{5.103}$$

If either the source or the observer recedes from the other, the sign of v_s or v_0 must be changed accordingly.

When the wind blows in the direction of sound, c is to be replaced by $c + v_m$, where v_m is the velocity of the medium. Then Eq. (5.103) modifies to

$$n' = \frac{(c + v_m + v_0)n}{c + v_m - v_s}. \tag{5.104}$$

For the wind blowing against the sound, the sign of v_m is to be reversed.

(iv) Source moving not in line with stationary observer

In most cases, the stationary observer O stands some distance away from the line of motion AB of the source (Fig. 5.10). Suppose that the source travels along AB with a velocity v_s. At the position S_1 of the source, the component of the source velocity along S_1O is $v_s \cos\theta$, and is towards the observer. So, the apparent frequency of the note received by the observer, obtained from Eq. (5.93), is

$$n' = \frac{cn}{c - v_s \cos\theta}. \tag{5.105}$$

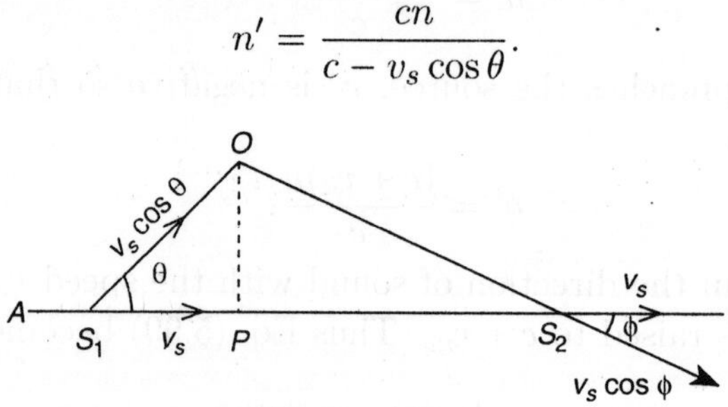

Fig. 5.10 Doppler effect when the stationary observer is not in line with the moving source

As the source approaches P, the foot of the perpendicular from O on AB, θ increases continuously so that $v \cos\theta$ decreases and n' increases. When the source is at P, $\theta = \pi/2$ and $n' = n$. As the source moves past P and comes to S_2 on the other side of P, the velocity component $v_s \cos\phi$ acts along OS_2 away from the observer. Now the apparent frequency of the note, obtained from Eq. (5.96), is

$$n' = \frac{cn}{c + v_s \cos\phi}. \tag{5.106}$$

The apparent frequency decreases as the source moves away. The sudden drop of the frequency or the pitch as the source crosses P is more pronounced, the higher the speed v_s.

(v) Stationary source and observer moving not in line with the source

Suppose that the source S is at rest and the observer moves with a velocity v_0 along AB (Fig. 5.11). Let the observer be located at O_1 at some instant of time and at O_2 after a time interval δt. Hence $O_1O_2 = v_0\delta t$. In this interval δt, the wavefront O_1C travels to DE where $O_1D = CE = c\delta t$, where c is the speed of sound.

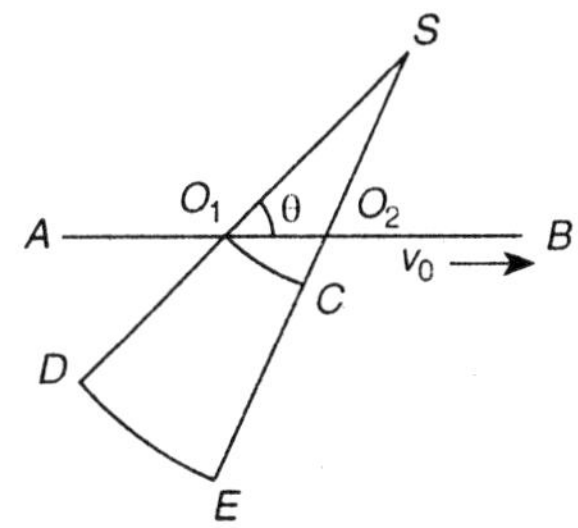

Fig. 5.11 Doppler effect for an observer moving not in line with a fixed source

If the observer were at rest, he would receive in time δt the waves confined within O_1D. But as he travels to O_2, he also receives the waves contained in O_2E. So, if n and n' are the exact and the apparent frequencies of the note, we have $n = O_1D/\lambda = c\delta t/\lambda$ and $n' = O_2E/\lambda = (O_2C + CE)/\lambda = (O_2C + c\delta t)/\lambda$. If $\theta = \angle SO_1B, O_2C = O_1O_2 \cos\theta = v_0 \cos\theta\delta t$. Therefore,

$$n' = (v_0 \cos\theta + c)\delta t/\lambda,$$

and

$$\frac{n'}{n} = \frac{v_0 \cos\theta + c}{c}.$$

The apparent frequency is thus given by

$$n' = \frac{(c + v_0 \cos\theta)n}{c}. \tag{5.107}$$

If $\theta = 0$, Eq. (5.107) gives $n' = \frac{(c+v_0)n}{c}$, agreeing with Eq. (5.100).

If $\theta = \pi$, the same equation yields $n' = \frac{(c-v_0)n}{c}$, which agrees with Eq. (5.99).

If $\theta = \pi/2$, we have $n' = n$.

Observations

(i) If plane sound waves are reflected from a *moving mirror*, the wavelength of the reflected wave will be different from that of the incident wave emitted from a stationary source. Thus the frequency of the reflected wave will differ from that of the incident wave. This phenomenon is occasionally known as *Doppler's principle*.

Let the incidence be normal and the mirror move towards the source with a speed v. The length of the column intercepted by the mirror in $1s$ is $c + v$, where c is the speed of sound. The number of waves in the column is $(c + v)/\lambda = n(c + v)/c$, where n and c are respectively the frequency and the wavelength of the note emitted by the source. The reflected column occupies a length $c - v$ and contains $n(c + v)/c$ waves in it. Therefore, the wavelength of the reflected wave is

$$\lambda' = \frac{c - v}{n(c + v)/c} = \frac{c}{n}\frac{c - v}{c + v} = \lambda\frac{c - v}{c + v}. \tag{5.108}$$

If a listener who is at rest relative to the source, receives the reflected waves, he will sense a frequency

$$n' = \frac{c}{\lambda'} = n\frac{c+v}{c-v}. \tag{5.109}$$

The listener will also receive the direct waves of frequency n. The reflected and the direct wave will produce in his ear *beats* of frequency

$$n' - n = n\left(\frac{c+v}{c-v} - 1\right) = \frac{2vn}{c-v}. \tag{5.110}$$

(ii) Suppose that an observer travels along a straight line connecting two sources S_1 and S_2 which emit notes of the same frequency n (Fig. 5.12). If the observer O moves towards S_2 with a speed v_0, then the apparent frequency of the note received from S_1 is [see Eq. (5.99)]

$$n_1 = \frac{(c - v_0)n}{c}.$$

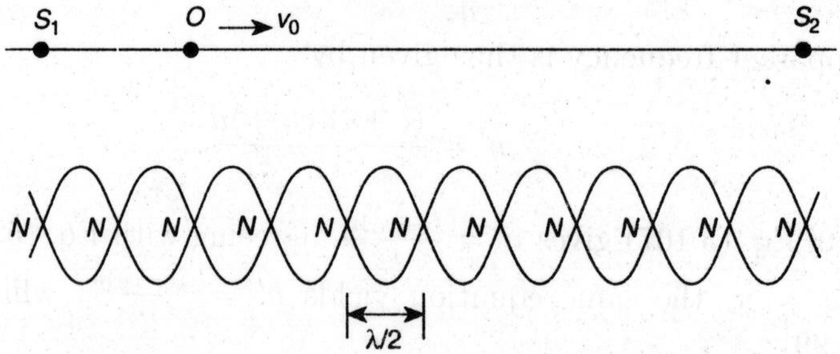

Fig. 5.12 Connection between Doppler effect, stationary waves and beats

The apparent frequency of the note received from S_2 is

$$n_2 = \frac{(c + v_0)n}{c}.$$

The two frequencies n_1 and n_2 will produce beats of frequency

$$n_2 - n_1 = \frac{2v_0 n}{c} = \frac{2v_0}{\lambda}, \tag{5.111}$$

where λ is the wavelength of the emitted note.

Here the observer can be considered to move through the stationary waves formed by the superposition of the waves emitted by S_1 and S_2 and progressing in opposite directions. The distance between two consecutive nodes is $\lambda/2$, so that the number of minima of sound encountered by the observer in 1s is $v_0/(\lambda/2) = 2v_0/\lambda$, agreeing with Eq. (5.111). Here we find an interesting combination of Doppler effect, stationary waves and beats.

(iii) If a sound source travels through a medium with a speed exceeding the phase velocity of waves in the medium, we have $v_s > c$ in the Doppler formulae, whence the formulae lose their significance. In this case, the wavefront assumes the shape of a cone with the moving source at its apex. The bow waves of a speed boat converter, and the shock waves from an aeroplane speeding through the air with a velocity exceeding the sound speed in air, i.e., with a supersonic velocity, furnish examples of such conical wavefronts.

When a bullet moving with a supersonic speed passes an observer three distinct sounds are heard. First, a click is heard due to shock waves as the bullet flies past the observer. Then a breezing sound caused by the eddies due to the passage of the bullet through the air, is heard. The sound reporting the explosion of the gun powder is heard last. The shock wave due to the high-speed bullet or aeroplane is transmitted laterally and not in the forward direction. So the sound in this case is heard only when the projectile crosses the observer.

(iv) A simple *demonstration* of Doppler effect in the laboratory consists in mounting a reed at the end of a rod about 0.6 m long. The other end of the rod is attached to a turntable which rotates at a high speed. The frequency of the reed whistling through the air increases as the reed approaches the observer and decreases as the reed moves away.

Interestingly, the human ear is most sensitive to frequencies of about 3 kHz, where the ear can detect a frequency change of about 3 Hz. Thus, if an observer approaches or recedes from an audio oscillator of frequency 3 kHz with a speed of about 0.4 m/s, he can detect the Doppler shift in frequency.

5.19 Large-Amplitude Waves and Shock Waves

Thus far we discussed small-amplitude waves for which the bulk modulus K could be taken to be a constant, independent of density or pressure. When the condensation s is large, the variation of pressure and density is also large. In such cases where the wave-amplitude is large, K cannot be treated as a constant.

Equation (5.22) shows that $\frac{\partial^2\xi}{\partial t^2} = -\frac{1}{\rho_0}\frac{\partial p}{\partial x}$. Assuming that the change is adiabatic, we have $\frac{P}{P_0} = \left(\frac{V_0}{V}\right)^\gamma$, or $P = P_0\left[\frac{V_0}{V_0(1+\Delta)}\right]^\gamma$, where Δ is the dilatation. In general, we have shown in Sec. 5.7 that $(1+s)(1+\Delta) = 1$, or, $1+\Delta = 1/(1+s)$. Furthermore, Eq. (5.19) gives $\Delta = \partial\xi/\partial x$. Therefore, $\partial\Delta/\partial x = \partial^2\xi/\partial x^2$.

We have $P = P_0(1+\Delta)^{-\gamma}$, so that

$$\frac{\partial p}{\partial x} = \frac{\partial P}{\partial x} = -\gamma P_0(1+\Delta)^{-(\gamma+1)}\frac{\partial \Delta}{\partial x}$$
$$= -\gamma P_0(1+\Delta)^{-(\gamma+1)}\frac{\partial^2 \xi}{\partial x^2}.$$

So,

$$\frac{\partial^2 \xi}{\partial t^2} = \frac{\gamma P_0}{\rho_0}(1+\Delta)^{-(\gamma+1)}\frac{\partial^2 \xi}{\partial x^2} = c^2(1+s)^{\gamma+1}\frac{\partial^2 \xi}{\partial x^2} = c_e^2\frac{\partial^2 \xi}{\partial x^2}. \quad (5.112)$$

Here $c\left(=\sqrt{\frac{\gamma P_0}{\rho_0}}\right)$ is the phase velocity for small-amplitude waves and

$$c_e = c(1+s)^{\frac{\gamma+1}{2}}. \quad (5.113)$$

Equation (5.112) has the same form as the wave equation (5.8) with c_e replacing c. Note that the effective wave velocity c_e depends on s. Thus, each part of the wave disturbance has its own velocity of propagation depending on the local condensation. Condensation is larger at high-pressure regions than at low-pressure regions. Hence the high-pressure regions move faster than the low-pressure regions, so that the wave gets distorted and takes a saw-tooth shape (Fig. 5.13).

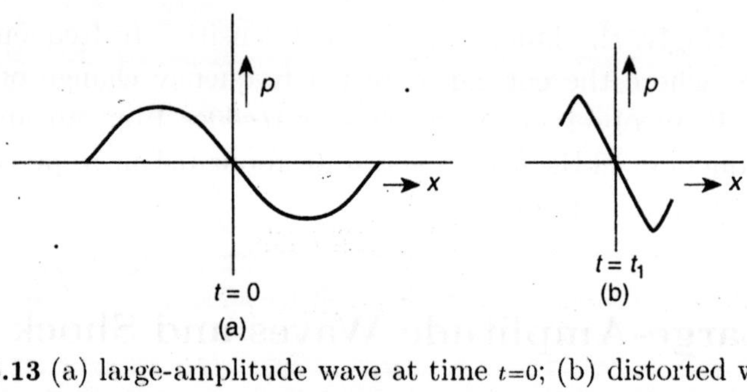

Fig. 5.13 (a) large-amplitude wave at time t=0; (b) distorted wave at a subsequent time t_1 due to the faster moving high-pressure regions

At 1 kHz, the lowest audible intensity in air is 10^{-12}W/m^2. The loudest sound causing a sensation of pain in the ear is about 140 dB above the threshold in intensity. The corresponding rms condensation s_{rms}, calculated from the relationship $I = c^3\rho_0 s_{\text{rms}}^2$, is about 1.4×10^{-3}. Modern techniques can produce compression waves with s comparable to unity in a gas. The study of such compression waves with large s values is the subject matter of *macrosonics*.

For small-amplitude waves, the particle velocity u is much smaller than c, the relationship between the two being $u = cs$. For large-amplitude waves,

u and c are comparable. For shock waves, which we are going to discuss, u exceeds c. A whip makes a cracking sound as its tip moves faster than sound. The study of shock waves is extremely important in relation to ballistics, missiles, explosions, and aerodynamics.

A large-amplitude (i.e., macrosonic) wave taking the saw-tooth shape of Fig. 5.13(b), is a *shock wave.* The velocity of propagation of the shock wave normal to the wavefront is larger than the velocity of the low-amplitude sound wave in the gas. The ratio of the particle velocity to the local sound velocity is called the *Mach number.* The salient features of a shock wave are as follows :

(i) A shock wave has a large change of excess pressure over a small region of space. Such waves originate in rapid compressional disturbances such as in an explosion in a fluid.

(ii) The moving pressure amplitude in a shock wave is much larger than that in an ordinary sound wave.

(iii) The velocity of a shock wave exceeds considerably that of an ordinary sound wave.

Mathematical treatment of shock waves

For an ideal shock wave, one can assume a discontinuity of density, pressure, and particle velocity. Accordingly, we resort to a set of difference equations based on conservation principles, rather than using the differential wave equation for wave propagation. We consider a tube of unit cross sectional area and assume that a shock front moves through it in the positive x-direction (Fig. 5.14). The density, pressure, and the particle velocity behind and in front of the shock front can be taken to be constant. Let these be ρ_1, p_1, u_1, respectively, behind the shock front, and ρ_2, p_2, u_2 in front of the shock front.

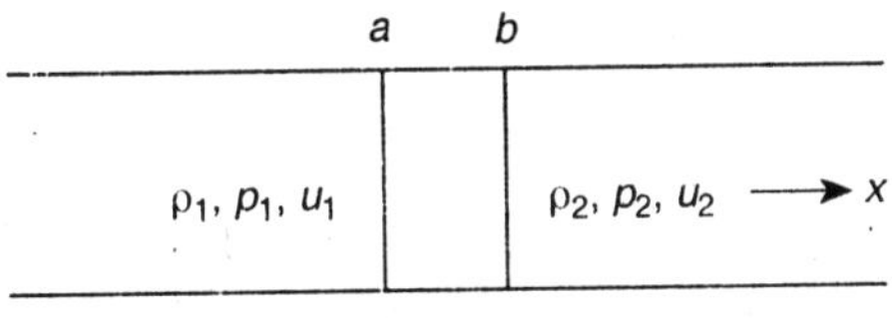

Fig. 5.14 Movement of a shock wave

The equation of continuity or the principle of conservation of mass tells us that the time rate of flow of mass into the front equals that out of it, i.e.,

$$\rho_1 u_1 = \rho_2 u_2. \tag{5.114}$$

The momentum flowing into the front in a unit time is $u_1(\rho_1 u_1) = \rho_1 u_1^2$, and that flowing out of the front in a unit time is $u_2(\rho_2 u_2) = \rho_2 u_2^2$. Therefore, the time rate of increase of momentum of the fluid in the front is $\rho_1 u_1^2 - \rho_2 u_2^2$. This must be equal to the resultant force $p_2 - p_1$ experienced by the front. Thus $\rho_1 u_1^2 - \rho_2 u_2^2 = p_2 - p_1$, or,

$$p_1 + \rho_1 u_1^2 = p_2 + \rho_2 u_2^2. \tag{5.115}$$

Solving Eqs. (5.114) and (5.115) for u_1 and u_2, we obtain

$$u_1 = \frac{1}{\rho_1}\left(\frac{p_1 - p_2}{\frac{1}{\rho_2} - \frac{1}{\rho_1}}\right)^{1/2} \tag{5.116}$$

and

$$u_2 = \frac{1}{\rho_2}\left(\frac{p_1 - p_2}{\frac{1}{\rho_2} - \frac{1}{\rho_1}}\right)^{1/2}. \tag{5.117}$$

So,

$$u_2 - u_1 = \left[(p_1 - p_2)\left(\frac{1}{\rho_2} - \frac{1}{\rho_1}\right)\right]^{1/2}. \tag{5.118}$$

If $p_1 - p_2 \ll p_2$ and $\rho_1 - \rho_2 \ll \rho_2$, Eqs. (5.116) and (5.117) give

$$u_1 = u_2 = \sqrt{\Delta p / \Delta \rho}. \tag{5.119}$$

Actually, u_1 is the velocity of the shock front itself. For small pressure variations, Eq. (5.119) shows that both u_1 and u_2 reduce to the usual sound velocity in the medium.

The time rate of work done (i.e., the power) on the shock-front fluid is $p_2 u_2 - p_1 u_1$. This is consumed in the rate of increase of the kinetic energy $\frac{1}{2}\rho_2 u_2 (u_1^2 - u_2^2)$ and the internal energy $\rho_2 u_2 \Delta U$, where ΔU is the rate of increase of the internal energy per unit mass. That is,

$$p_2 u_2 - p_1 u_1 = \rho_2 u_2 \left[\frac{1}{2}(u_1^2 - u_2^2) + \Delta U\right] \tag{5.120}$$

For a perfect gas, $p/\rho = RT$, where T is the temperature, and $\Delta U = C_v(T_1 - T_2)$. Noting that $p/(R\rho) = T$ and $R = C_v(\gamma - 1)$, we have

$$\Delta U = \frac{1}{\gamma - 1}\left(\frac{p_1}{\rho_1} - \frac{p_2}{\rho_2}\right). \tag{5.121}$$

Substituting for ΔU from Eq. (5.121) and for u_1 and u_2 from Eqs. (5.116) and (5.117) into Eq. (5.120), we obtain

$$\frac{p_1}{p_2} = \frac{\rho_1(\gamma + 1) - \rho_2(\gamma - 1)}{\rho_2(\gamma + 1) - \rho_1(\gamma - 1)} \tag{5.122}$$

and

$$\frac{\rho_1}{\rho_2} = \frac{p_1(\gamma+1)+p_2(\gamma-1)}{p_2(\gamma+1)+p_1(\gamma-1)}. \tag{5.123}$$

The quantities p_1/p_2 and ρ_1/ρ_2 give a feeling of the shock strength. The ratio of the temperatures on the two sides of the shock front is

$$\frac{T_1}{T_2} = \frac{p_1/\rho_1}{p_2/\rho_2} = \frac{c_1^2}{c_2^2}, \tag{5.124}$$

where $c_1(=\sqrt{\gamma p_1 \rho_1})$ and $c_2(=\sqrt{\gamma p_2/\rho_2})$ are the sound speeds.

We introduce here the Mach numbers $M_1 = u_1/c_1$ and $M_2 = u_2/c_2$. Then Eq. (5.114) gives $\rho_1^2 u_1^2 = M_1^2 c_1^2 \rho_1^2 = \gamma M_1^2 p_1 \rho_1 = \gamma M_2^2 p_2 \rho_2$, so that

$$\frac{\rho_1}{\rho_2} = \frac{M_2^2 p_2}{M_1^2 p_1}. \tag{5.125}$$

Again Eq. (5.115) yields

$$p_1 + M_1^2 \gamma p_1 = p_2 + M_2^2 \gamma p_2. \tag{5.126}$$

Equation (5.126) can be used to eliminate M_1 or M_2 in the ratio ρ_1/ρ_2 given by Eq. (5.125). Substituting the result in Eq. (5.122) gives

$$\frac{p_1}{p_2} = \frac{\gamma(2M_2^2-1)+1}{\gamma+1} = \frac{\gamma+1}{\gamma(2M_1^2-1)+1}. \tag{5.127}$$

Using Eq. (5.127) in Eq. (5.123) gives

$$\frac{\rho_1}{\rho_2} = \frac{M_2^2(\gamma+1)}{M_2^2(\gamma-1)+2} = \frac{M_1^2(\gamma-1)+2}{M_1^2(\gamma+1)}. \tag{5.128}$$

These equations immediately yield

$$M_1^2 = \frac{2+(\gamma-1)M_2^2}{2\gamma M_2^2-(\gamma-1)} \tag{5.129}$$

and

$$M_2^2 = \frac{2+(\gamma-1)M_1^2}{2\gamma M_1^2-(\gamma-1)}. \tag{5.130}$$

If $M_1 > 1$, we obtain from Eq. (5.129), $M_2 < 1$. Thus, a supersonic flow velocity on passing through a shock front becomes subsonic.

5.20 Differential Equation for Waves in Three Dimensions

As stated earlier, the acoustic waves generated by a vibrating point source in a homogeneous isotropic medium have wavefronts that are concentric spherical surfaces with the source at the centre. Such waves are referred to as *spherical waves.* At a sufficiently long distance from the source, a small part of the spherical wavefront is approximately a plane, so that this small portion can be treated as a *plane wavefront.* A wave with plane wavefronts is called a *plane wave.* A plane wave propagates in a fixed direction without diverging laterally. So far, we have considered plane waves. We shall now discuss spherical waves, for which a wave equation in three dimensions is required. The beams in spherical waves generated by a point source are divergent. As the energy emitted from the source spreads over a larger spherical surface with increasing distance from the source, the sound pressure and the intensity drop off with increasing distance.

The beams can be made convergent if they are focussed by suitable reflectors to the centre of a sphere.

To derive the wave equation in three dimensions, we consider an arbitrary volume τ in the fluid medium bounded by the surface S (Fig. 5.15). If ρ is the mass density of the fluid and $\vec{V}$ is the particle velocity at an elemental area ds on the surface S, the time rate of out flow (or *efflux*) of the fluid mass through the closed surface S is $\oint_S \rho\vec{V}\cdot\hat{n}\,ds$, where $\hat{n}$ is a unit vector along the outward normal to ds. Since mass is *conserved*, this rate of outflow must be equal to the time rate of decrease of mass in the volume τ, i.e., equal to $-\frac{d}{dt}\int_\tau \rho\,d\tau$. Thus

$$\oint_S \rho\vec{V}\cdot\hat{n}\,ds = -\frac{d}{dt}\int_\tau \rho\,d\tau \tag{5.131}$$

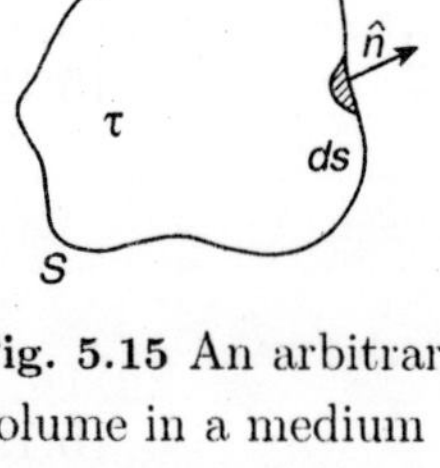

Fig. 5.15 An arbitrary volume in a medium in motion

The time derivative on the right-hand side of Eq. (5.131) can be taken inside the volume integral whence it becomes a partial derivative. Furthermore, the left-hand side of Eq. (5.131) can be converted into a volume integral by applying Gauss's divergence theorem in vector analysis. Thus Eq. (5.131) gives

$$\int_\tau \left[\vec{\nabla}\cdot(\rho\vec{V}) + \frac{\partial\rho}{\partial t}\right] d\tau = 0. \tag{5.132}$$

Since Eq. (5.132) is valid for any arbitrary volume, we must have

$$\vec{\nabla}\cdot\rho\vec{V} + \frac{\partial\rho}{\partial t} = 0. \tag{5.133}$$

Equation (5.133) is referred to as the *continuity equation*. It is a hydrodynamic relationship expressing the conservation of fluid mass.

The quantity $\rho\vec{V}$ can be identified with the *mass current density* $\vec{J}_m$ which gives the rate of flow of mass through a unit area perpendicular to $\vec{V}$. The continuity equation is clearly seen to be analogous to the corresponding equation in electricity with the electric current density $\vec{J}$ replacing $\vec{J}_m$ and the electric charge density replacing the mass density.

If p is the excess pressure, the unbalanced force acting on a volume element $d\tau$ in the fluid is $(\vec{\nabla}p)d\tau$. By Newton's second law of motion, this unbalanced force must be equated to the time rate of change of momentum of the fluid in $d\tau$, i.e., to $-\frac{\partial}{\partial t}(\vec{V}\rho)d\tau$. Hence

$$(\vec{\nabla}p)d\tau = -\frac{\partial}{\partial t}(\vec{V}\rho)d\tau$$

$$\text{or,} \quad \vec{\nabla}p = -\frac{\partial}{\partial t}(\rho\vec{V}). \tag{5.134}$$

Equation (5.134) gives

$$\vec{\nabla}\cdot\vec{\nabla}p = -\frac{\partial}{\partial t}(\vec{\nabla}\cdot\rho\vec{V})$$

$$\text{or,} \quad \nabla^2 p = \frac{\partial^2\rho}{\partial t^2} \tag{5.135}$$

where Eq. (5.133) has been used to substitute for $\vec{\nabla}\cdot\rho\vec{V}$. The operator ∇^2 is known as the *Laplacian operator*.

The *elastic properties* of the medium give for small values of the condensation s, the following relationships for the excess pressure p and the instantaneous density ρ :

$$p = Ks \quad \text{and} \quad \rho = \rho_0(1+s)$$

where K is the bulk modulus and ρ_0 is the equilibrium density. Therefore,

$$\rho = \rho_0 + \rho_0\frac{p}{K} \quad \text{and} \quad \frac{\partial^2\rho}{\partial t^2} = \frac{\rho_0}{K}\frac{\partial^2 p}{\partial t^2} = \frac{1}{c^2}\frac{\partial^2 p}{\partial t^2}.$$

where $c^2 = K/\rho_0$ [see Eq. (5.24)]. Substituting for $\frac{\partial^2\rho}{\partial t^2}$ in Eq. (5.135) gives

$$\frac{\partial^2 p}{\partial t^2} = c^2\nabla^2 p. \tag{5.136}$$

Equation (5.136) is the *differential acoustic wave equation in three dimensions for the acoustic pressure.*

The relationship $p = Ks = K\frac{d\rho}{\rho_0} = c^2 d\rho$ shows that p is proportional to s and $d\rho$. So the wave equation can also be written in terms of the condensation s or the incremental density $d\rho$:

$$\frac{\partial^2 s}{\partial t^2} = c^2 \nabla^2 s \tag{5.137}$$

$$\text{and} \quad \frac{\partial^2 d\rho}{\partial t^2} = c^2 \nabla^2 d\rho. \tag{5.138}$$

5.21 Spherical Wave Equation and its Solution

For *spherical symmetry*,

$$\nabla^2 \equiv \frac{\partial^2}{\partial r^2} + \frac{2}{r}\frac{\partial}{\partial r}$$

where r is the radial distance from the origin. Hence, for spherical waves, the wave equation for the acoustic pressure [Eq. (5.136)] takes the form

$$\frac{\partial^2 p}{\partial t^2} = c^2 \left(\frac{\partial^2 p}{\partial r^2} + \frac{2}{r}\frac{\partial p}{\partial r} \right) = \frac{c^2}{r}\frac{\partial^2}{\partial r^2}(rp) \tag{5.139}$$

As r and t are independent variables, Eq. (5.139) can be rewritten as

$$\frac{\partial^2 (rp)}{\partial t^2} = c^2 \frac{\partial^2 (rp)}{\partial r^2}, \tag{5.140}$$

which is the *spherical wave equation* for the acoustic pressure.

From Eqs. (5.137) and (5.138), the spherical wave equations for s and $d\rho$ can be similarly obtained as

$$\frac{\partial^2 (rs)}{\partial t^2} = c^2 \frac{\partial^2 (rs)}{\partial r^2} \tag{5.141}$$

$$\text{and} \quad \frac{\partial^2 (rd\rho)}{\partial t^2} = c^2 \frac{\partial^2 (rd\rho)}{\partial r^2}. \tag{5.142}$$

Comparing the spherical wave equation [Eq. (5.140)] with the plane wave equation [Eq. (5.8)], we find that they have similar forms with rp replacing ψ in the numerator and r replacing x in the denominator. The solution of Eq. (5.140) can therefore be expressed as

$$rp = f_1(ct - r) + f_2(ct + r)$$

$$\text{or,} \quad p = \frac{1}{r} f_1(ct - r) + \frac{1}{r} f_2(ct + r). \tag{5.143}$$

The first term on the right-hand side of Eq. (5.143) represents a divergent spherical wave propagating radially outward with velocity c. The second term represents a *convergent* spherical wave travelling radially inward with velocity c.

We shall deal with only the divergent wave here. The solution shows that the acoustic pressure p is inversely proportional to the distance r from the origin. The decrease of the acoustic pressure with increasing distance can be visualized as follows. The intensity at a distance r from the origin is $I = W/(4\pi r^2)$, where W is the power emitted by the source. Since W is constant and I varies as p^2, it follows that $p \propto 1/r$.

The solutions for s and $d\rho(= \rho - \rho_0)$ have the same forms as that for p, and so s and $d\rho$ also drop off as $1/r$.

If p varies sinusoidally, we can write

$$p = \frac{A}{r}e^{j(\omega t - Kr)} \tag{5.144}$$

where A is a constant to be determined from some boundary condition. The real or the imaginary part of Eq. (5.144) gives the physical solution. Similar expressions can be written for s and $d\rho$.

The particle velocity for a spherical wave is defined by $V_r = \frac{\partial \xi_r}{\partial t}$, where ξ_r is the particle displacement along r. The particle velocity can be related to the acoustic pressure by noting that for small amplitudes ρ in Eq. (5.134) can be replaced by ρ_0 giving

$$\vec{\nabla} p = -\rho_0 \frac{\partial \vec{V}}{\partial t}. \tag{5.145}$$

For spherical waves, Eq. (5.145) reduces to

$$\frac{\partial p}{\partial r} = -\rho_0 \frac{\partial V_r}{\partial t} \tag{5.146}$$

which is the dynamic equation relating p and V_r. With p given by Eq. (5.144), we get from Eq. (5.146)

$$\frac{\partial V_r}{\partial t} = -\frac{1}{\rho_0}\frac{\partial p}{\partial r} = \frac{A}{\rho_0}\left(\frac{1}{r^2} + \frac{jk}{r}\right)e^{j(\omega t - kr)}$$

So,

$$V_r = \frac{A}{\omega \rho_0}\left(\frac{k}{r} - \frac{j}{r^2}\right)e^{j(\omega t - kr)} \tag{5.147}$$

The *specific acoustic impedance* Z_s at a point is defined as the ratio of the acoustic pressure and the particle velocity at that point. For spherical waves, we obtain

$$Z_s = \frac{p}{V_r} \tag{5.148}$$

Using Eqs. (5.144) and (5.147) we get

$$Z_s = \frac{\omega \rho_0}{k - \frac{j}{r}} = \frac{\rho_0 c}{1 - \frac{j}{kr}}, \tag{5.149}$$

since $c = \omega/k$. Equation (5.149) can be expressed as

$$Z_s = \rho_0 c \frac{1 + \frac{j}{kr}}{1 + \frac{1}{k^2r^2}} = \rho_0 c \left(\frac{k^2r^2}{1 + k^2r^2} + \frac{jkr}{1 + k^2r^2} \right)$$
$$= \rho_0 c \frac{kr}{\sqrt{1 + k^2r^2}} e^{j\theta} \tag{5.150}$$

where $\tan\theta = 1/(kr)$. We have $\cos\theta = kr/\sqrt{1 + k^2r^2}$, so that

$$|Z_s| = \rho_0 c \cos\theta. \tag{5.151}$$

Equation (5.150) reveals that Z_s has a resistive part R_s and a reactive part X_s, i.e.,

$$Z_s = R_s + jX_s \tag{5.152}$$

where

$$R_s = \rho_0 ck^2r^2/(1 + k^2r^2) \quad \text{and} \quad X_s = \rho_0 ckr/(1 + k^2r^2).$$

The complex nature of Z_s reflects that p and V_r are not in phase.

At a sufficiently long distance from the origin, $k^2r^2 \gg 1$. Then $R_s \simeq \rho_0 c$ and $X_s \simeq 0$. Now p and V_r will be in phase. Thus at a long distance from the origin a spherical wave can be treated as a plane wave with the specific acoustic impedance equal to the characteristic impedance $\rho_0 c$ of the medium (see Sec. 5.14).

5.22 Intensity for Spherical Waves

Acoustic intensity is defined as the flow of the average acoustic power normally through a unit area of the wavefront.

If P is the instantaneous pressure and V_r is the instantaneous velocity for a spherical wave, the instantaneous power flow normally through a unit area of the wavefront is PV_r. The average value of the integral of the product PV_r over a full cycle gives the acoustic intensity I. If P_0 is the steady equilibrium pressure, we have $P = P_0 + p$. The integration of P_0V_r over a complete cycle is zero, so that the static pressure contributes nothing to the acoustic intensity. Therefore,

$$I = \frac{1}{T} \int_0^T pV_r dt \tag{5.153}$$

where T is the time period of the pressure variation.

If $p_{\max}$ is the acoustic pressure amplitude and $V_{\max}$ is the particle velocity amplitude, we can write

$$p = p_{\max} \sin\omega t \quad \text{and} \quad V_r = V_{\max} \sin(\omega t - \theta)$$

where $\tan\theta = 1/(kr)$ (see Sec. 5.21). Hence

$$\begin{aligned} I &= \frac{p_{\max}V_{\max}}{T}\int_0^T \sin\omega t \sin(\omega t - \theta)dt \\ &= \frac{p_{\max}V_{\max}}{2T}\int_0^T [\cos\theta - \cos(2\omega t - \theta)]\,dt \\ &= \frac{1}{2}p_{\max}V_{\max}\cos\theta = p_{\rm rms}V_{\rm rms}\cos\theta \end{aligned} \tag{5.154}$$

where $p_{\rm rms}(= p_{\max}/\sqrt{2})$ is the root-mean-square or the effective pressure and $V_{\rm rms}(= V_{\max}/\sqrt{2})$ is the root-mean-square or the effective particle velocity.

Equation (5.151) shows that the magnitude of p/V_r is $\rho_0 c\cos\theta$, i.e., $p_{\rm rms}/V_{\rm rms} = \rho_0 c\cos\theta$, or $\cos\theta = p_{\rm rms}/(\rho_0 c V_{\rm rms})$. Hence Eq. (5.154) yields

$$I = \frac{p_{\rm rms}^2}{\rho_0 c} = \frac{p_{\max}^2}{2\rho_0 c}. \tag{5.155}$$

Equation (5.155) is the same as Eq. (5.66) obtained for plane waves. The intensity I can be expressed in terms of $V_{\rm rms}$ or $V_{\max}$ by noting that $p_{\rm rms} = \rho_0 c V_{\rm rms}\cos\theta$. Equation (5.155) then gives

$$I = \rho_0 c V_{\rm rms}^2\cos^2\theta = \frac{1}{2}\rho_0 c V_{\max}^2\cos^2\theta. \tag{5.156}$$

If W is the total power emitted by the sound source located at the origin and I is the intensity at a distance r from the origin we have

$$W = 4\pi r^2 I = \frac{4\pi r^2 p_{\rm rms}^2}{\rho_0 c} = 2\pi r^2\frac{p_{\max}^2}{\rho_0 c}. \tag{5.157}$$

5.23 Solved Problems

Problem 1. *A police car driving at* 1.5 m/s *sounds a siren emitting a note of frequency* 550 Hz. *The sound is reflected from a stationary object ahead of the car. What will be the number of beats heard per second by an observer in the car? Assume that the velocity of sound is* 330 m/s.

[**C.U. 1996; N. Beng. U. 2001**]

Solution. When the observer approaches a source with a speed v_0, and the source approaches the observer with a speed v_s, the apparent frequency of the tone is

$$n' = n\frac{c + v_0}{c - v_s}.$$

In this problem, the reflected sound can be thought to be emitted from the image of the source formed by the reflector. As the source approaches the reflector with a velocity v_s, the image moves towards the source with the same velocity v_s. So, here $v_0 = v_s$ and the apparent frequency of the reflected sound is

$$n' = n\frac{c+v_s}{c-v_s}.$$

Hence, the number of beats is

$$n' - n = \frac{2v_s n}{c-v_s} = \frac{2\times 1.5\times 550}{330-1.5} = \frac{2\times 1.5\times 550}{328.5} = 5.02\,\text{s}^{-1}.$$

Problem 2. *The intensity level in a conversation is* 70 dB *above the threshold of* $10^{-12}\,\text{W/m}^2$. *Calculate the amplitude of vibration of the air particles in the sound wave. Given, velocity of sound* = 350 m/s, *density of air* = 1.25 g/litre, *mean frequency* = 500 Hz. **[C.U. 1993]**

Solution. If I is the intensity level of the conversation, we have from the formula : Intensity level in dB $= 10\log_{10}(I/I_0)$,

$$70 = 10\log_{10}\left(I/10^{-12}\right)$$

whence

$$I = 10^7 \times 10^{-12} = 10^{-5}\,\text{W/m}^2.$$

Again

$$I = \frac{1}{2}\rho_0 A^2\omega^2 c,$$

where ρ_0 = air density = 1.25 g/litre = 1.25 kg/m^3,

c = velocity of sound = 350 m/s,

ω = angular frequency $= 2\pi f = 2\times\pi\times 500 = 1000\pi$ rad/s.

Substituting the values we have for the amplitude A of vibration of the air particles

$$A = \left(\frac{2\times 10^{-5}}{1.25\times(1000\pi)^2\times 350}\right)^{1/2} = 6.8\times 10^{-8}\,\text{m} = 0.068\,\mu\text{m}.$$

Problem 3. *For a plane harmonic longitudinal wave of frequency* 1000 Hz *in air at STP, the rms excess pressure is* 200 dyne/cm^2, *the phase velocity being* 350 m/s. *Calculate the maximum particle displacement and particle velocity. Given,* $\gamma_{\text{air}} = 1.40$. **[C.U. 1992]**

Solution. The maximum acoustic pressure is

$$p_{\max} = \sqrt{2}p_{\text{rms}} = \sqrt{2}\times 200 = 282.8\,\text{dyne/cm}^2.$$

The wavelength is

$$\lambda = \frac{\text{phase velocity}}{\text{frequency}} = \frac{350}{1000}\,\text{m} = 35\,\text{cm}.$$

The propagation constant is $k = 2\pi/\lambda = 2 \times 3.14/35 = 0.1795\,\text{cm}^{-1}$. The angular frequency is $\omega = 2\pi f = 2000\pi = 6283$ rad/s. Since $p = Ks$, we have $p_{\text{max}} = Ks_{\text{max}}$, where s_{max} is the maximum condensation. For air at STP, $K = \gamma P_0 = 1.40 \times 1.013 \times 10^6\,\text{dyne/cm}^2$. Hence,

$$s_{\text{max}} = \frac{282.3}{1.40 \times 1.013 \times 10^6} = 1.994 \times 10^{-4}.$$

If the maximum particle displacement is A, we have

$$A = \frac{s_{\text{max}}}{k} = \frac{1.994 \times 10^{-4}}{0.1795} = 1.111 \times 10^{-3}\,\text{cm}.$$

If the maximum particle velocity is u_{max},

$$u_{\text{max}} = \omega A = 6283 \times 1.111 \times 10^{-3} = 6.98\,\text{cm/s}.$$

Problem 4. *An increase of pressure of* 100 kPa *causes a certain volume of water to decrease by* 5×10^{-3} *per cent of its original volume. Show that the speed of sound in water is about* 1414 m/s. **[C.U. 1991]**

Solution. The speed of sound in water is

$$c = \sqrt{\frac{K}{\rho_0}}$$

where K = Bulk modulus $= p/\left|\frac{\delta v}{v_0}\right| = 100 \times 10^3/(5 \times 10^{-5}) = 2 \times 10^9$ Pa, and ρ_0 = density $= 1\,\text{g/cm}^3 = 10^3\,\text{kg/m}^3$. Hence

$$\begin{aligned} c &= \sqrt{\frac{2 \times 10^9}{10^3}} \\ &= \sqrt{2} \times 10^3 = 1.414 \times 10^3 \\ &= 1414\,\text{m/s}. \end{aligned}$$

Problem 5. *Calculate the average energy density in a plane progressive harmonic wave in air if the intensity level of sound is* 140 dB *relative to the threshold intensity of* $10^{-16}\,\text{W/cm}^2$. *Given, speed of sound in air* = 340 m/s.

[cf. Burd. U. 1994]

Solution. If I be the intensity of the sound, we have

$$140 = 10 \log_{10}\left(\frac{I}{10^{-16}}\right),$$

since dB $= 10\log_{10}(I/I_0)$. So, $I = 10^{14} \times 10^{-16} = 10^{-2}\,\text{W/cm}^2 = 10^2\,\text{W/m}^2$. If $\bar{E}$ is the average energy density, $\bar{E} = I/c = 10^2/340 = 0.294\,\text{J/m}^3$.

Problem 6. *The phase velocity v of deep water waves of wavelength λ is given by $v^2 = \frac{g\lambda}{2\pi} + \frac{2\pi S}{\rho\lambda}$, where g is the acceleration due to gravity, S is the surface tension and ρ is the density of the liquid. Determine the wavelength λ_0 of the waves which do not disperse in water. Show that for wavelengths $\lambda \ll \lambda_0$, the group velocity v_g is $3v/2$ and for $\lambda \gg \lambda_0$, $v_g \simeq v/2$.*

[Burd. U. 1990]

Solution. For waves that do not disperse, we have $\frac{dv}{d\lambda} = 0$. Differentiating the expression for v^2, we get

$$2v\frac{dv}{d\lambda} = \frac{g}{2\pi} - \frac{2\pi S}{\rho\lambda^2} = 0,$$

$$\text{or} \quad \lambda^2 = \lambda_0^2 = \frac{4\pi^2 S}{\rho g},$$

$$\text{or} \quad \lambda_0 = 2\pi\sqrt{\frac{S}{\rho g}}.$$

If v_g is the group velocity, then $v_g = v - \lambda\frac{dv}{d\lambda}$

$$= v - \lambda\left(\frac{g}{4\pi v} - \frac{\pi S}{\rho v\lambda^2}\right) = v - \lambda\left(\frac{\pi S}{\rho v\lambda_0^2} - \frac{\pi S}{\rho v\lambda^2}\right)$$

$$= v - \frac{\pi S}{\rho v\lambda_0}\left(\frac{\lambda}{\lambda_0} - \frac{\lambda_0}{\lambda}\right).$$

If $\lambda \ll \lambda_0$, we obtain

$$v_g \simeq v + \frac{\pi S}{\rho v\lambda} = v + \frac{v}{2} - \frac{g\lambda}{4\pi v} = \frac{3}{2}v - \frac{\pi S}{\rho v\lambda_0}\left(\frac{\lambda}{\lambda_0}\right) \simeq \frac{3}{2}v.$$

If $\lambda \gg \lambda_0$, we get

$$v_g \simeq v - \frac{\pi S}{\rho v}\frac{\lambda}{\lambda_0^2} = v - \frac{\pi S}{\rho v}\cdot\frac{\lambda\rho g}{4\pi^2 S}$$

$$= v - \frac{\lambda g}{4\pi v} = v - \frac{v}{2} - \frac{\pi S}{\rho\lambda v} = \frac{v}{2} - \frac{\lambda_0 g}{4\pi v}\left(\frac{\lambda_0}{\lambda}\right) \simeq \frac{v}{2}.$$

Problem 7. *If two sound waves, one in air and the other in water, have equal intensities, what is the ratio of the pressure amplitude of the wave in water to that of the wave in air? (Density of air = $0.00129\,\mathrm{g/cm^3}$, velocity of sound in water and air are 1.45×10^5 cm/s and 3.3×10^4 cm/s, respectively.)*

[Burd. U. 1989, cf. C.U. 2000]

Solution. The intensity I is related to the pressure amplitude $p_{\max}$ by the relationship $I = p_{\max}^2/(2\rho_0 c)$ where ρ_0 is the density of the medium and c is the speed of sound. When the intensity in air and that in water are equal,

$(p_{\max}^2)_w/(p_{\max}^2)_a = (\rho_0 c)_w/(\rho_0 c)_a$ where the subscripts 'w' and 'a' stand for water and air, respectively. Substituting the given values, we get

$$\frac{(p_{\max})_w}{(p_{\max})_a} = \left(\frac{1 \times 1.45 \times 10^5}{0.00129 \times 3.3 \times 10^4}\right)^{1/2} = 58.4.$$

Problem 8. *A certain loudspeaker has a circular opening with a diameter of* 15 cm. *Assume that the sound it emits is uniform and outward through the entire opening. If the sound intensity at the opening is* $100\mu\mathrm{W/m^2}$, *show that the power being radiated as sound is about* $1.77\mu\mathrm{W}$.

[**C.U. 1991**]

Solution. The area of the opening is $\alpha = \pi d^2/4 = 3.14 \times 15^2/4\,\mathrm{cm}^2$. The intensity at the opening is $I = 100\mu\mathrm{W/m^2} = 10^{-2}\mu\mathrm{W/cm^2}$. Hence power $= I\alpha = 10^{-2} \times 3.14 \times 225/4 = 1.77\mu\mathrm{W}$.

Problem 9. *A microphone emits a* 1 kHz *pure tone having an intensity level of* 65 dB. *Calculate the actual intensity* (*reference* $= 2 \times 10^{-5}\,\mathrm{N/m^2}$) *and the loudness level of the sound. When another source of the same intensity is switched on, calculate the increase in the intensity level in* dB. [**C.U. 1998**]

Solution. For the reference pressure of $2 \times 10^{-5}\,\mathrm{N/m^2}$ the reference intensity is $10^{-12}\,\mathrm{W/m^2}$. Hence, if I is the actual intensity of the initial source, we have

$$65 = 10\log_{10}\left(\frac{I}{10^{-12}}\right), \quad \text{whence } I = \sqrt{10}\,\mu\,\mathrm{W/m^2} = 3.16\,\mu\,\mathrm{W/m^2}.$$

The loudness level is 65 phon.

When the second source is on, the total intensity is $2I$, so that the intensity level in dB is

$$10\log_{10}\left(\frac{2I}{10^{-12}}\right) = 10\log_{10}2 + 65 = 68\text{ dB}$$

The increase in the intensity level is 3 dB.

Problem 10. *The permissible intensity level of sound is* 60 dB *above the threshold. Taking the reference level for intensity and pressure as* $10^{-12}\,\mathrm{W/m^2}$ *and* 2×10^{-5} Pa, *respectively, determine the absolute intensity and the acoustic pressure.*

Solution. If I is the absolute intensity of sound, we have

$$60 = 10\log_{10}\left(\frac{I}{10^{-12}}\right).$$

So,

$$I = 10^6 \times 10^{-12} = 10^{-6}\,\mathrm{W/m^2}$$
$$= 1\mu\,\mathrm{W/m^2}.$$

Since

$$\mathrm{SPL} = 20\log_{10}\left(\frac{p}{p_0}\right),$$

where p is the absolute acoustic pressure and $p_0 = 2 \times 10^{-5}$ Pa, we have

$$60 = 20\log_{10}\left(\frac{p}{2\times 10^{-5}}\right)$$

$$\text{or,}\quad p = 10^3 \times 2\times 10^{-5}\,\mathrm{N/m^2} = 0.02\,\mathrm{Pa}.$$

Problem 11. *The sound pressure levels of two sources are* 20 dB *and* 40 dB, *respectively. What is the sound pressure level when the two sources emit sound simultaneously?*

Solution. Let I_1 and I_2 be the intensities of the two sources, and I_0 be the reference intensity. Then $20 = 10\log_{10}(I_1/I_0)$ and $40 = 10\log_{10}(I_2/I_0)$, giving $I_1/I_0 = 10^2$ and $I_2/I_0 = 10^4$. The intensity level of the two sources operating simultaneously is

$$\begin{aligned}10\log_{10}\left(\frac{I_1+I_2}{I_0}\right) &= 10\log_{10}(10^2+10^4)\\ &= 10\log_{10}(10^2\times 101)\\ &= 20 + 10\log_{10}101\\ &= 40.0043\,\mathrm{dB}.\end{aligned}$$

Problem 12. *A gas mixture enclosed in a vessel of volume V contains* 1 *gram mole of a gas A with $\gamma(=C_p/C_v) = 5/3$ and another gas B with $\gamma = 7/5$ at an absolute temperature T. The gram molecular weight of the gas A is* 4 *and that of the gas B is* 32. *The gases A and B are taken to be ideal and they do not react with each other. The gas mixture obeys the equation $PV^{19/13}$ = constant in an adiabatic process.* (i) *Find the speed of sound in the gas mixture at $T = 300$ K.* (ii) *If the temperature rises by* 1 K *from* 300 K, *what is the percentage change in the speed of sound in the gas mixture?*

Solution. The speed of sound in a gas is $c = \sqrt{\gamma P/\rho}$. For 1 mole of an ideal gas we have $PV = RT$ and $\rho = M/V$, where M is the molar mass. Hence $c = \sqrt{\gamma RT/M}$, where $\gamma = C_p/C_v$.

For the given gas mixture, $\gamma = 19/13$. Since $C_v = R/(\gamma - 1)$, we have

$$(C_v)_A = \frac{R}{5/3-1} = \frac{3}{2}R \quad\text{and}\quad (C_v)_B = \frac{R}{7/5-1} = \frac{5}{2}R.$$

Also

$$(C_v)_{\mathrm{mixture}} = \frac{R}{19/13-1} = \frac{13}{6}R.$$

For a temperature change δT, the sum of the energy change of the gases A and B must be equal to the total energy change of the gas mixture, i.e., if n_A and n_B are the number of moles of A and B in the mixture, we have

$$n_A(C_v)_A\delta T + n_B(C_v)_B\delta T = (n_A + n_B)(C_v)_{\text{mixture}}\delta T.$$

With $n_A = 1$, this gives

$$1 \times \frac{3}{2}R + n_B\frac{5}{2}R = (1 + n_B)\frac{13}{6}R$$

whence

$$n_B = 2.$$

(i) The speed of sound in the gas mixture is $c = \sqrt{\gamma RT/M}$, where $M = \frac{1\times4+2\times32}{3} = \frac{68}{3}$ gm/mole. So,

$$c = \left[\frac{\frac{19}{13} \times 8.31 \times 300}{\frac{68}{3} \times 10^{-3}}\right]^{1/2} = 400.9\ \text{m/s}.$$

(ii) Since $c^2 = \frac{\gamma RT}{M}$, we obtain $\frac{dC}{C} = \frac{dT}{2T}$. Therefore, the percentage change of the speed of sound is

$$\frac{dC}{C} \times 100 = \frac{dT}{2T} \times 100 = \frac{1}{2 \times 300} \times 100 = 0.167\%.$$

Problem 13. *A plane simple harmonic sound wave of amplitude* 0.001 mm *and frequency* 650 Hz *is propagated in a perfect gas of density* 1.29 kg/m^3 *and presure* 100 kPa. *If* $C_p/C_v = \gamma = 1.41$, *calculate the acoustic pressure amplitude of the wave.*

Solution. We have for the particle displacement

$$\xi = 10^{-6}\cos\frac{2\pi}{\lambda}(ct - x)\ \text{m},$$

where $c = \sqrt{\frac{K}{\rho_0}} = \sqrt{\frac{\gamma P}{\rho}} = \left[\frac{1.41\times10^5}{1.29}\right]^{1/2} = 330.61\ \text{m/s},$

$K = \gamma P = 1.41 \times 10^5$ Pa

and $\lambda = \frac{c}{\nu} = \frac{330.61}{650} = 0.51$ m.

The acoustic pressure is

$$p = -K\frac{\partial\xi}{\partial x} = 1.41 \times 10^5 \times 10^{-6} \times \frac{2\pi}{\lambda}\sin\frac{2\pi}{\lambda}(ct - x).$$

The acoustic pressure amplitude is

$$1.41 \times 10^{-1} \times \frac{2\pi}{\lambda} = \frac{1.41 \times 2 \times 3.1416}{10 \times 0.51} = 1.74\ \text{Pa}.$$

Problem 14. *A sound source describes a circular orbit of radius* 3 m *with an angular velocity* 10 rad/s. *A sound detector located far away from the source performs a simple harmonic motion along the line BD (Fig. 5.16) with an amplitude* $BC = CD = 6\,\text{m}$. *The frequency of oscillation of the detector is* $5/\pi$ per second. *The source is at the point A when the detector is at the point B. If the source emits a continuous sound wave of frequency* 340 Hz, *determine the maximum and the minimum frequencies recorded by the detector. (Velocity of sound* = 330 m/s*).*

Solution. The velocity of the sound source is $v_s = r\omega = 3 \times 10 = 30\,\text{m/s}$. The time period of rotation of the source is

$$T = \frac{2\pi}{\omega} = \frac{2\pi}{10} = \frac{\pi}{5}\,\text{s}.$$

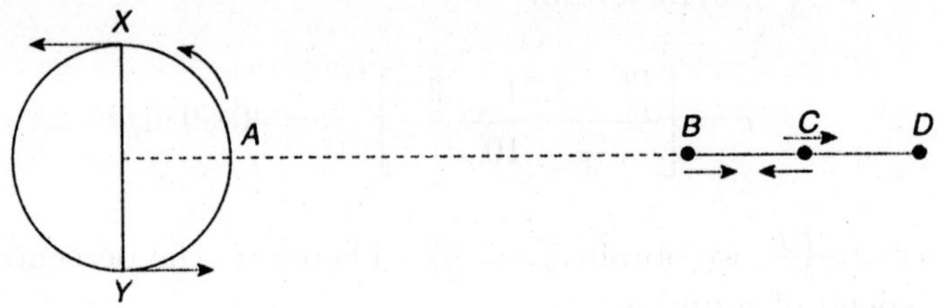

Fig. 5.16 Figure for Problem 14

Clearly the time period of the circular motion and of the simple harmonic motion are the same. When the source is at X, the detector is at C, and they are receding from each other. Then the apparent frequency is a minimum. The minimum frequency detected by the recorder is

$$n'_{\text{min}} = \frac{c - v_0}{c + v_s} n = \frac{330 - 60}{330 + 30} \times 340 = 255\ \text{Hz},$$

since the velocity of the detector at C is

$$v_0 = a\omega = 6 \times 10 = 60\ \text{m/s}.$$

When the source is at Y, the detector is again at C and they are now approaching each other. Then the apparent frequency recorded by the detector is a maximum and is given by

$$n'_{\text{max}} = \frac{c + v_0}{c - v_s} n = \frac{330 + 60}{330 - 30} \times 340 = 442\ \text{Hz}.$$

Problem 15. *A tuning fork producing sound of frequency* 440 Hz *is approaching a fixed wall with a velocity* 4 m/s. *How many beats* per second *can an observer detect if* (i) *the tuning fork is between the observer and the* *wall,* (ii) *the observer is between the tuning fork and the wall? (velocity of sound* = 332 m/s). [C.U. 2003]

Solution. The sound reflected by the wall is produced by the image of the tuning fork. The image is moving towards the observer with a velocity the same as that of the tuning fork, i.e., 4 m/s. Hence the apparent frequency of the reflected sound is

$$n_r = \frac{cn}{c - v_s} = \frac{332 \times 440}{332 - 4} = 445 \text{ Hz}.$$

(i) Here the tuning fork is moving away from the observer and so the apparent frequency of the tuning fork received by the stationary observer is

$$n_0 = \frac{cn}{c + v_s} = \frac{332 \times 440}{332 + 4} = 434 \text{ Hz}.$$

So the number of beats detected by the observer is $n_r - n_0 = 11$ Hz.

(ii) Here the tuning fork approaches the observer and so the apparent frequency of the fork received by the stationary observer is

$$n_0' = \frac{cn}{c - v_s} = n_r = 445 \text{ Hz}.$$

Since the two frequencies are equal, the number of beats will be zero.

REVIEW QUESTIONS

1. A plane progressive harmonic wave propagating along the positive x-direction is represented by

$$\xi = A \sin\left(\omega t - \frac{\omega x}{c} + \alpha\right)$$

where the notations have their usual significance. For the wave calculate the instantaneous kinetic and potential energies per unit volume. Show that the instantaneous energy density is not a constant, but its average over a complete period is a constant. How is the average energy density related to the intensity? [C.U. 1997]

2. (a) Define phase velocity and group velocity. [C.U. 2009]

(b) A wave group is formed by the superposition of two waves of equal amplitudes but of slightly different frequencies and wavelengths. Show that if c_g is the group velocity and c is the phase velocity, then

(i) $c_g = c - \lambda \frac{dc}{d\lambda}$ [C.U. 1997]

(ii) $c = \omega/k$ and $c_g = \frac{d\omega}{dk}$. [C.U. 1988]

3. What is Doppler effect? Derive an expression for the change in the frequency of a sound wave when the source and the observer are both in motion in a stationary medium. [C.U. 1996, 2003]

4. (a) Derive an expression for the intensity of a plane progressive harmonic sound wave. [C.U. 1993]

   (b) Define 'bel', 'decibel' and 'phon'. [C.U. 2004]

   (c) What do you mean by the 'characteristic acoustic impedance of a medium'? Define 'rayls' and 'mks rayls'.

5. Write the expression for the particle displacement for a plane progressive harmonic wave and also for a plane stationary harmonic wave. What is the essential difference between these two expressions? Give physical examples of such waves. [C.U. 1991]

6. Obtain an expression for the velocity of a plane longitudinal wave in a fluid medium. Write the assumptions you make. [C.U. 1990, '91, 2005, '08]

7. Consider a waveform $f(x,t) = Ae^{-B(x-vt)^2}$. Draw the rough sketches of $f(x,t)$ versus x at $t = 0$ and $t = 1\,\text{s}$ for $A = 1.0\,\text{m}$, $B = 1.0\,\text{m}^{-2}$ and $v = 2.0\,\text{m/s}$. [C.U. 1991]

8. What is meant by a plane progressive wave? Write the wave function for a plane harmonic progressive wave propagating with a velocity 350 m/s along a direction whose direction cosines are l, m and n with respect to x-, y-, and z- axes, respectively. [cf. C.U. 1990]

9. Let c be the speed of sound in a gas medium of $\gamma = 1.4$. Show that the rms speed v of the gas molecules is $v = (3/1.4)^{1/2}c$. [C.U. 1990, 2008]

10. ψ satisfies the wave equation $\frac{\partial^2\psi}{\partial t^2} = c^2\frac{\partial^2\psi}{\partial x^2}$. Show that $\frac{\partial\psi}{\partial t}$ and $\frac{\partial\psi}{\partial x}$ satisfy the same differential equation. If the above wave equation represents one dimensional longitudinal waves in a continuous homogeneous medium and ψ is the particle displacement from the equilibrium position, show that the condensation s and the particle velocity u are related by $\frac{\partial s}{\partial t} = -\frac{\partial u}{\partial x}$. Show also that the acoustic pressure satisfies the same wave equation as ψ. [C.U. 1989]

11. Explain why a monochromatic wave train of finite length leads to the concept of a wave group. [C.U. 1984]

12. Show that the function $f(ct - x)$ represents a plane progressive wave propagating in the positive x-direction whereas $f(ct + x)$ represents one propagating in the negative x-direction with a speed c.

13. What do you mean by acoustic pressure and acoustic intensity? Show that the acoustic intensity for a plane progressive harmonic wave is the product of the rms acoustic pressure and the rms particle velocity. Explain why the intensity is independent of the position coordinate along the direction of propagation. **[cf. C.U. 2006]**

14. A plane progressive longitudinal wave propagates along the positive x direction in a fluid medium. (i) Show that the pressure of the wave at a point x is $p = -K\frac{\partial \xi}{\partial x}$, where K is the bulk modulus of the medium and ξ is the displacement at a point x at time t. (ii) Hence set up a differential equation for the wave. (iii) Write a solution of the wave equation. (iv) Draw ξ-x curve and using the pressure expression show the region of compression in the graph. **[C.U. 1998]**

15. Determine the velocity of longitudinal waves in a thin solid rod in terms of Young's modulus and the density of the material of the rod. Is the formula valid for an extended solid medium?

[C.U. 2007; cf. 2010]

16. Calculate the energy of the stationary waves formed by the superposition of two plane harmonic waves of the same amplitude and frequency, propagating in opposite directions. Show that there is no transmission of energy along the line of propagation of the component waves.

17. Two plane progressive harmonic waves of the same frequency and unequal amplitudes, propagating in opposite directions along the x-axis, superimpose. Show that the resultant wave consists of a stationary wave and a progressive wave moving along the x-axis.

[C.U. 2004]

18. What is a shock wave? Mention its characteristics. Show that a supersonic flow velocity becomes subsonic on passing through a shock front.

19. Derive the differential equation for waves in three dimensions in terms of the acoustic pressure. Show that the same equation is obeyed by the condensation and the incremental density.

20. What are spherical waves? Write the differential wave equation for spherically symmetric waves in terms of the acoustic pressure. Discuss the nature of the solutions for spherical waves.

21. Assuming that the excess pressure varies sinusoidally, derive an expression for the specific acoustic impedance Z_s at a point in a divergent spherically symmetric acoustic wave. What is the significance of the complex nature of Z_s? How is the nature of Z_s altered for large distances from the origin?

22. Derive an expression for the intensity at a point in a divergent spherical wave in terms of the effective value of the harmonically varying acoustic pressure. Hence show that the power from the source of the wave is $4\pi r^2 p_{\text{rms}}^2/(\rho_0 c)$, where the symbols are standard.

23. Define group velocity and phase velocity. Derive a relationship between them. **[C.U. 2005]**

24. (a) Distinguish between progressive and stationary waves. **[C.U. 2005, '09]**

(b) From the equation of a wave, how can you judge whether the wave is progressive or stationary?

(c) Show that the average energy density contained in a stationary wave is twice that for a progressive wave. **[C.U. 2009]**

PROBLEMS

1. Two sound waves of equal frequency, one in air and the other in water have the same pressure amplitudes. What is the ratio of the intensity in water to that in air? Given, $\rho_0 c$ in air = 4.28 and in water = 1.50×10^5 CGS units. **[C.U. 1984]**

[**Ans.** 0.0000285 : 1]

2. A locomotive continuously sounds a whistle of frequency 200 Hz while approaching a tunnel. The tunnel reflects the sound so that the driver of the locomotive hears 5 beats per second. If the speed of sound is 350 m/s, find the speed of the locomotive.

[**Ans.** 4.32 m/s]

3. A source emitting a note of frequency 1 kHz, moves with a velocity 20 m/s towards a stationary reflector. What is the frequency of the reflected wave as measured by an observer moving with the source? (Speed of sound in air = 350 m/s) **[C.U. 1999]**

[**Ans.** 1.121 kHz]

4. What is the intensity of a 60 dB sound if the standard intensity is 10^{-12} W/m^2? **[C.U. 1991]**

[**Ans.** 10^{-6} W/m^2]

5. Calculate the maximum particle displacement in a plane progressive harmonic sound wave of wavelength 50 cm if the *rms* pressure is 1 dyne/cm^2. The bulk modulus of the medium is 1.4×10^{-6} dyne/cm^2.

[**Ans.** 7.96×10^{-6} cm]

6. A source of sound emits energy equal in all directions. If the total power radiated is 1W, calculate the intensity level in dB at a distance of 1m from the source, the audibility threshold being $10^{-12}\,\text{W/m}^2$.

   [**Ans.** 109 dB]

7. The intensity level of a source is n dB. If N such sources emit sound simultaneously, show that the intensity level is $(n + 10\log_{10} N)$ dB.

8. Calculate the intensity of a sound which is 12 dB lower than a sound of intensity $0.01\mu\,\text{W/m}^2$.

   [**Ans.** $6.31 \times 10^{-10}\,\text{W/m}^2$]

9. A sound of intensity I_1 has the level p dB. Another sound of intensity I_2 has the level q dB. Show that the intensity level of $(I_1 + I_2)$ is $10\log_{10}(10^{p/10} + 10^{q/10})$ dB.

10. The velocity of sound in air is 332.5 m/s and the density of air is 1.29 kg/m^3 under standard conditions. For a note of frequency 1 kHz, the intensity of sound is $3 \times 10^{-6}\,\text{W/m}^2$. Find the acoustic pressure and the amplitude of the wave.

    [**Ans.** 0.0359 Pa, 18.8 nm]

11. For gravity waves in a liquid, the phase velocity is given by $v_p = c\sqrt{\lambda}$, where c is a constant and λ is the wavelength. Show that the group velocity is half the phase velocity. [**Burd. U. 1998**]

12. Two trains approach each other in adjacent parallel tracks, the engine of one of them sounding continuously its whistle of frequency 500 Hz. Find the drop of frequency of the sound heard by a passenger in the other train as the whistling engine crosses him. The velocity of each train is 75 km/hr and the velocity of sound in air is 340 m/s.

    [**Ans.** 122.8 Hz]

13. A simple harmonic wave of amplitude 8 cm traverses a line of particles along the positive x-axis. At a given instant of time, the displacement of a particle at 10 cm from the origin is $4\sqrt{3}$ cm and that for a particle at 25 cm from the origin is 4 cm. Calculate the wavelength and the rms pressure, if the bulk modulus is 1.4×10^6 dyne/cm^2.

    [**Ans.** 180 cm; 2.79×10^5 dyne/cm^2] [**cf. C.U. 1996**]

14. A train approaching a hill with a speed of 40 km/hour sounds a whistle of frequency 580 Hz when it is at a distance of 1 km from the hill. Wind blows with a speed of 40 km/hour in the direction of motion of the train. Determine (i) the frequency of the whistle heard by a

listener on the hill. (ii) the frequency of the echo heard by the driver of the train. (velocity of sound in air = 1200 km/hour)

[**Ans.** 599.3 Hz; 620 Hz]

15. A vibrating tuning fork of frequency 350 Hz moves away from a stationary observer. Another vibrating tuning fork of the same frequency moves towards the stationary observer at the same speed. The observer hears 4 beats per second. If the velocity of sound in air is 340 m/s, find the speed of each fork with respect to the observer.

[**Ans.** 1.94 m/s]

16. The engine pulling a train sounds a whistle whose frequency appears to drop by 200 Hz to an observer standing on the platform as the train passes him at a speed of 110 km/hour. Find the frequency of the whistle, if (i) the air of still; (ii) the wind blows opposite to the direction of motion of the train at a speed of 10 m/s. (velocity of sound in still air = 332 m/s).

[**Ans.** 1079 Hz; 1072 Hz]

17. A progressive harmonic wave is represented by $y(x,t) = a\sin(0.5x - 10t)$ in SI units. Obtain the wave velocity. **[C.U. 2008]**

[**Ans.** 20 m/s]

18. The phase velocity of waves in a medium is given by $v_p = c\Big/\left(a + \frac{b}{\lambda^2}\right)$, where a, b, and c are constants, and λ is the wavelength. Prove that the group velocity is $v_g = v_p - \dfrac{2bv_p^2}{c\lambda^2}$.

What is the long-wavelength value of v_g?

[**Ans.** v_p]

Chapter 6

Vibration of Strings

6.1 Introduction

Following Lord Rayleigh, an ideal string is defined as "a perfectly uniform and perfectly flexible filament of solid matter stretched between two fixed points." Actual strings closely approximate but do not exactly meet the ideal conditions. A practical string is neither perfectly uniform nor perfectly flexible. The nonuniform density along the length of the string implies that the tension along the string is not constant everywhere. Practical strings have also some rigidity which increases with increasing diameter-to-length ratio. In the ideal string, the tension is the only restoring force. In an actual string, additional elastic forces appear due to the bending of the string, so that the total restoring force increases, thereby enhancing the frequency of vibration. During the vibration of an actual string, the points between which the string is stretched, are not perfectly at rest. The yielding of the supports increases the effective length of the string, thus affecting the pitch of the emitted note.

A string serves as a vibrator and also as a medium for wave motion. The study of vibration of strings is important from both theoretical and practical points of view. Stringed instruments like piano, guitar, sitar, violin, etc. play an important role in the world of music. They generate musical sound by the transverse vibration of stretched strings. A point of a string under tension is displaced from its rest position by plucking, striking, or bowing. The point vibrates and waves travel along the string on either side. The reflected waves from the fixed supports superimpose on the incident waves, producing stationary wave patterns. The vibration of the string gives out a musical note of certain pitch and quality.

6.2 Equation of Transverse Vibration of a Stretched String

Let a string stretched between two fixed supports lie at rest along the x-axis. When a portion of the string is drawn along the y-axis, i.e., perpendicular to its length such that the amplitude of vibration is small, the tension of the string can be taken to be a constant in time as transverse waves proceed along the string. Suppose that ab is the stretched string in its position of rest, and $AB(=\delta l)$ is a small segment of the string in the displaced position (Fig. 6.1). The string being perfectly flexible, the tension T can be taken to be the same along the length of the string, acting tangentially at every point. If θ is the angle made by the tangent at A with ab, the transverse component of the tension will be $T_v = T\sin\theta$. Since θ is small, $\sin\theta \simeq \tan\theta = \frac{\partial y}{\partial x}$. Hence $T_v = T\frac{\partial y}{\partial x}$. The vertical component of the tension at B is $T\frac{\partial}{\partial x}\left(y + \frac{\partial y}{\partial x}\delta x\right)$. So, the resultant force on the element AB along the y-direction is

$$T\frac{\partial}{\partial x}\left(y + \frac{\partial y}{\partial x}\delta x\right) - T\frac{\partial y}{\partial x} = T\frac{\partial^2 y}{\partial x^2}\delta x.$$

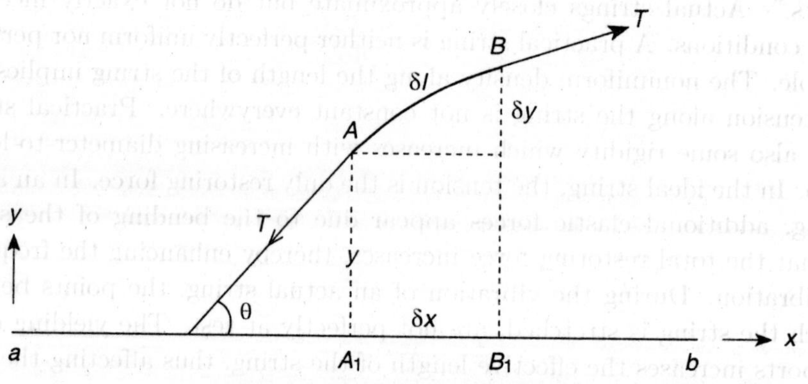

Fig. 6.1 A small portion of the displaced string under tension

From Newton's second law of motion, this force must be equated to the product of the mass of the element AB and its acceleration. If m is the mass per unit length of the string, the mass of the element AB is $m\delta x$ and its acceleration is $\frac{\partial^2 y}{\partial t^2}$. So,

$$T\frac{\partial^2 y}{\partial x^2}\delta x = m\delta x\frac{\partial^2 y}{\partial t^2}$$

$$\text{or,}\quad \frac{\partial^2 y}{\partial t^2} = \frac{T}{m}\frac{\partial^2 y}{\partial x^2} = c^2\frac{\partial^2 y}{\partial x^2}, \tag{6.1}$$

where

$$c = \sqrt{\frac{T}{m}}. \tag{6.2}$$

Equation (6.1) gives the differential equation for transverse wave motion along the stretched string, the wave velocity being c, given by Eq. (6.2). The general solution of the differential equation [Eq. (6.1)] is

$$y = f_1(ct - x) + f_2(ct + x). \tag{6.3}$$

The function f_1 represents a plane transverse wave propagating along the positive x-direction with the velocity c, and the function f_2 represents a plane transverse wave moving in the negative x-direction with the same velocity.

A general solution of the wave equation [Eq. (6.1)] can be obtained by the *method of separation of variables*. Since y is a function of both x and t, we write

$$y = F_1(x)F_2(t), \tag{6.4}$$

where $F_1(x)$ is a function of the position co-ordinate x alone and $F_2(t)$ is a function of time t alone. Substituting for y into Eq. (6.1) gives

$$F_1\frac{d^2F_2}{dt^2} = c^2F_2\frac{d^2F_1}{dx^2}.$$

Dividing through by F_1F_2 yields

$$\frac{1}{F_2}\frac{d^2F_2}{dt^2} = \frac{c^2}{F_1}\frac{d^2F_1}{dx^2}. \tag{6.5}$$

Since the left-hand side of Eq. (6.5) is a function of t only and the right-hand side is a function of x only, and x and t are independent variables, the two sides must be a constant, called the *separation constant.* As y must be finite and cannot increase or decrease indefinitely with x or t, the separation constant must be negative, say $-\omega^2$. Therefore, Eq. (6.5) gives

$$\frac{d^2F_2}{dt^2} + \omega^2F_2 = 0 \tag{6.6}$$

and $$\frac{d^2F_1}{dx^2} + \frac{\omega^2}{c^2}F_1 = 0. \tag{6.7}$$

Equations (6.6) and (6.7) are standard equations for simple harmonic motions. So, the general solutions of these equations can be written respectively as

$$F_2 = C_1\cos\omega t + C_2\sin\omega t \tag{6.8}$$

and $$F_1 = D_1\cos\frac{\omega x}{c} + D_2\sin\frac{\omega x}{c}, \tag{6.9}$$

where C_1, C_2, D_1 and D_2 are constants to be determined from the initial or the boundary conditions.

Thus the solution of the wave equation for the stretched string is

$$y = F_1F_2 = (C_1\cos\omega t + C_2\sin\omega t)\left(D_1\cos\frac{\omega x}{c} + D_2\sin\frac{\omega x}{c}\right). \tag{6.10}$$

6.3 Eigenfunctions and Eigenfrequencies

Suppose that a stretched string of length l, fixed rigidly at the two ends $x = 0$ and $x = l$, execute transverse vibrations. The transverse waves set up in the string move towards the ends and are reflected at the rigid supports. The superposition of these waves produces stationary waves with nodes at the two ends. Thus the boundary conditions are : (i) at $x = 0$, $y = 0$ for all values of t, and (ii) at $x = l$, $y = 0$ for all values of t.

The displacement y of the string as a function of x and t is given by Eq. (6.10). Using the boundary condition (i) in Eq. (6.10), we obtain

$$D_1(C_1 \cos \omega t + C_2 \sin \omega t) = 0,$$

so that $D_1 = 0$. The boundary condition (ii), used in Eq. (6.10), gives

$$D_2 \sin \frac{\omega l}{c}(C_1 \cos \omega t + C_2 \sin \omega t) = 0.$$

$$\text{or,} \quad \frac{\omega l}{c} = n\pi, \quad (n = 1, 2, 3, \text{ etc.}). \tag{6.11}$$

Thus only the *modes of vibrations* with angular frequency $\frac{\pi c}{l}$ and multiples thereof are permissible. We exclude the possibilities $D_2 = 0$ or $n = 0$ for then the string does not move at all.

The solution is thus written as

$$y_n = D_{2n} \sin \frac{\omega_n x}{c}(C_{1n} \cos \omega_n t + C_{2n} \sin \omega_n t),$$

the subscript n being used for the nth mode of vibration with angular frequency

$$\omega_n = \frac{nc\pi}{l}. \tag{6.12}$$

Including the product $D_{2n}C_{1n}$ in a single constant a_n and the product $D_{2n}C_{2n}$ in another single constant b_n, we can write

$$y_n = \sin \frac{n\pi x}{l}\left(a_n \cos \frac{n\pi ct}{l} + b_n \sin \frac{n\pi ct}{l}\right). \tag{6.13}$$

Since there is a value of y_n for each integral value of n, and the differential equation is linear and homogeneous, the general solution is given by the sum of the y_n's by the principle of superposition. Thus, the general solution is

$$y = \sum_{n=1}^{\infty} \sin \frac{n\pi x}{l}\left(a_n \cos \frac{n\pi ct}{l} + b_n \sin \frac{n\pi ct}{l}\right) \tag{6.14}$$

$$= \sum_{n=1}^{\infty} C_n \sin \frac{n\pi x}{l} \cos\left(\frac{n\pi ct}{l} - \theta_n\right), \tag{6.15}$$

where $C_n^2 = a_n^2 + b_n^2$ and $\tan \theta_n = \frac{b_n}{a_n}$.

In the nth mode, the string vibrates with the angular frequency $n\pi c/l$, the amplitude of vibration at x being $C_n \sin\frac{n\pi x}{l}$. The frequency for the nth mode of vibration is

$$f_n = \frac{\omega_n}{2\pi} = \frac{nc}{2l} = \frac{n}{2l}\sqrt{\frac{T}{m}}. \tag{6.16}$$

The *fundamental frequency* occurs for $n = 1$. It is given by

$$f_1 = \frac{1}{2l}\sqrt{\frac{T}{m}}. \tag{6.17}$$

The fundamental wavelength is $\lambda_1 = c/f_1 = 2l$. The higher harmonics have frequencies which are integral multiples of f_1 and wavelengths which are correspondingly shorter.

The frequencies ω_n given by Eq. (6.12) are called *characteristic frequencies* or *eigenfrequencies*. The space part of the displacement y_n of the stretched string fixed at both ends is $\sin\frac{n\pi x}{l}$ [see Eq. (6.13)], which is the sine series of the Fourier expansion. The functions $\sin\frac{n\pi x}{l}$ for different integral values of n are referred to as the *eigenfunctions*.

If an operator $\hat{l}$ acting on a function $f(x)$ gives $l_s f(x)$ where l_s is a constant, then $f(x)$ is said to be the eigenfunction of $\hat{l}$ whose eigenvalue is l_s. In our case, the eigenfunction is $\sin\frac{n\pi x}{l}$. Equation (6.7) shows that the operator $\frac{d^2}{dx^2}$ acts on $\sin\frac{n\pi x}{l}$. Then

$$\frac{d^2}{dx^2}\left(\sin\frac{n\pi x}{l}\right) = -\left(\frac{n\pi}{l}\right)^2 \sin\frac{n\pi x}{l},$$

so that the *eigenvalue* of the operator $\frac{d^2}{dx^2}$ is $-\frac{n^2\pi^2}{l^2}$.
The eigenfunctions of a stretched string are *orthogonal,* because they satisfy the *orthogonality condition*:

$$\int_0^l \sin\frac{n\pi x}{l}\sin\frac{n'\pi x}{l}dx = 0 \qquad (n \neq n').$$

Observations

(1) Equation (6.17) gives the following three laws (called *Mersenne's laws*) for the transverse vibration of strings:

(a) *Law of length*: The frequency of vibration is inversely proportional to the vibrating length when the tension and the mass per unit length remain constant. That is, $f_1 \propto 1/l$ when T and m are constants.

(b) *Law of tension*: The frequency of vibration is directly proportional to the square root of the tension when the length and the mass per unit length of the string remain the same. That is, $f_1 \propto \sqrt{T}$ when l and m are constants.

(c) *Law of mass*: The frequency of vibration is inversely proportional to the square root of the mass per unit length when the length and the tension of the string do not vary. That is, $f_1 \propto 1/\sqrt{m}$ when T and l are constants.

For strings of circular cross section, the law of mass can be broken up into two sublaws:

(i) *Law of diameter*: The frequency of vibration of a string varies inversely as the diameter of the string when the length, the tension, and the density of the material of the string remain unaltered. That is, $f_1 \propto 1/d$ when l, T and ρ are constants. Here d is the diameter of the string and ρ is the density of its material.

(ii) *Law of density*: The frequency of vibration of a string varies inversely as the square root of the density of the material of the string when the length, the tension, and the diameter of the string remain constant. That is, $f_1 \propto 1/\sqrt{\rho}$ when l, T and d are constants.

The above two sublaws follow from the relationship $m = \frac{1}{4}\pi d^2 \rho$.

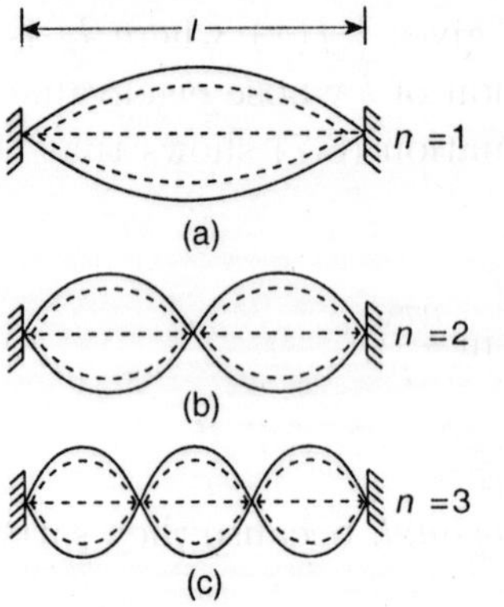

Fig. 6.2 Modes of vibration of a stretched string

(2) Equation (6.13) shows that for the nth mode, the displacement of the string is a maximum at $\sin\frac{n\pi x}{l} = 1$ or $\frac{n\pi x}{l} = \frac{\pi}{2}$ or $x = \frac{l}{2n}$. Here an *antinode* is formed. The displacement is zero at a *node* where $\sin\frac{n\pi x}{l} = 0$ or $\frac{n\pi x}{l} = \pi$ or $x = \frac{l}{n}$. Thus antinodes appear midway between nodes. Nodes always appear at $x = 0$ and $x = l$. In the nth mode of vibration, the string generates the nth harmonic and vibrates in n segments (or loops). The adjacent segments have opposite phases, i.e., when one segment moves up, the adjacent segment moves down.

In the fundamental mode ($n = 1$), the string vibrates with a frequency f_1 as a whole [Fig. 6.2(a)]. In the next mode ($n = 2$), the string vibrates in two segments [Fig. 6.2(b)], and generates the second harmonic, the frequency of vibration being $2f_1$. In the third mode ($n = 3$), the string vibrates in three segments with a frequency $3f_1$, emitting the third harmonic; and so on. In each mode, *the wavelength is two times the length occupied by a loop.*

6.4 Melde's Experiment

The resonant vibrations of stretched strings are demonstrated in Melde's experiment [Fig. 6.3(a) and (b)]. One end of a light thin wire W is connected to the extremity of one prong of an electrically maintained large tuning fork F. The string passes over a smooth pulley P and carries a light scale pan S at the other end. Suitable weights are placed on the scale pan to stretch the string. In the arrangement of Fig. 6.3(a), the fork vibrates perpendicular to the length of the string. This arrangement is therefore called the *transverse arrangement.* In Fig. 6.3(b), the fork vibrates parallel to the length of the string; this arrangement is the *longitudinal arrangement.*

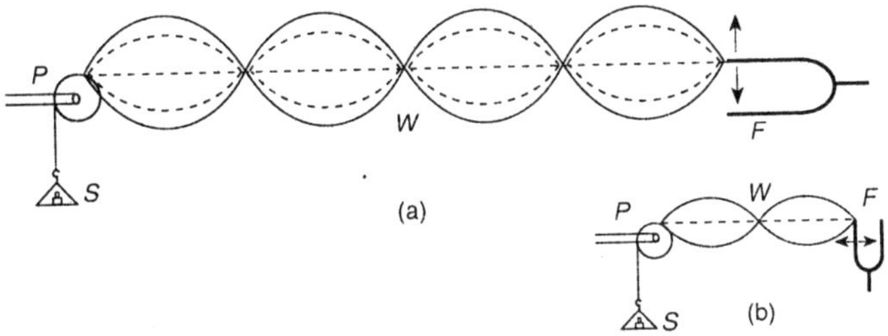

Fig. 6.3 Melde's experiment: (a) transverse arrangement; (b) longitudinal arrangement

(a) *Transverse arrangement*: Suppose that the load stretching the string produces a tension T in it. When the fork is excited, let the string be set into resonant stationary vibrations, the number of loops in which the string vibrates being n.

If f is the frequency of the fork, for resonance with the nth harmonic of the string, we have [see Eq. (6.16)]

$$f = \frac{n}{2l}\sqrt{\frac{T}{m}}.$$

Here l is the length of the string between the fork and the pulley. The quantities f, l and m being constants, we have $n\sqrt{T}$ = constant or n^2T = constant. Thus the load stretching the string is inversely proportional to the square of the number of loops in which the string vibrates. For example, if the string vibrates in one segment with a load of 1 kg, it vibrates in two segments with a load of 0.25 kg, and in three segments with a load of (1/9) kg., and so on.

It may be noted that at the fork end, the string has a slight transverse displacement. Hence the effective length of the wire differs slightly from its true length.

(b) *Longitudinal arrangement*: In this configuration [Fig. 6.3 (b)], the string is the tightest when the vibrating prong is farthest from the pulley. When the prong is nearest to the pulley, the string sags most [Fig. 6.4(a)]. As the prong swings away from this position to the other end, the string is stretched most [Fig. 6.4(b)]. When the prong moves towards

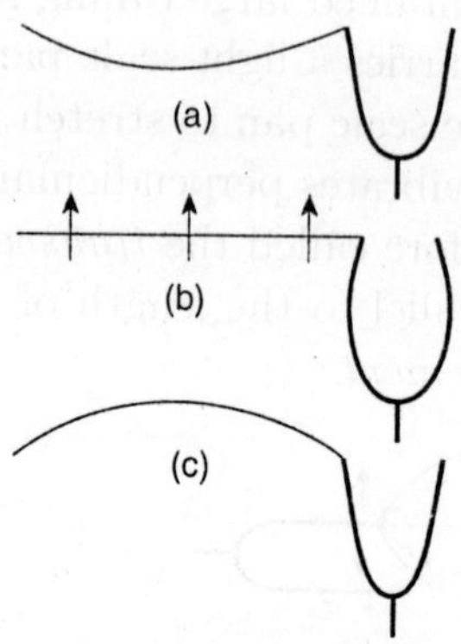

Fig. 6.4 Vibrations of the string in the longitudinal arrangement

its position nearest to the pulley, the string does not sag; rather it moves upwards due to inertia [Fig. 6.4(c)]. When the prong is nearest to the pulley again, the string attains its maximum displacement above the horizontal. Thus in one full oscillation of the fork, the string performs a half oscillation. Hence, the frequency of the string is half that of the fork.

If the string vibrates in n loops, and f is the frequency of the fork, we have

$$\frac{f}{2} = \frac{n}{2l}\sqrt{\frac{T}{m}}.$$

Obviusly, as in the case of the transverse arrangement, we obtain n^2T = constant. But the tension now required for a particular number of loops is one-fourth that needed in the transverse arrangement.

Equation (6.16) and hence the laws of transverse vibrating of stretched strings can be verified by Melde's experiment. A tuning fork of known frequency f is brought into resonance with the string by varying the load on the scale pan. The resonance is recongnized by the maximum amplitude of vibration of the string. If the string vibrates in n loops at resonance, the frequency of vibration of the string is

$$f_s = \frac{n}{2l}\sqrt{\frac{T}{m}}.$$

The tension T is given by $T = Mg$ where M is the weight placed on the scale pan and g is the local acceleration due to gravity. Knowing l, m and T and counting n, f_s can be calculated. In the transverse arrangement f_s is found to be equal to f_s whereas in the longitudinal arrangement, f_s is found to be $f/2$. This agreement between f_s and f establishes Eq. (6.16) and hence the laws of vibration of stretched strings.

6.5 Energy of a Vibrating String

For a stretched vibrating string fixed at the two ends, the transverse displacement y at any point x at time t is given by Eq. (6.15). Putting

$$\xi_n(t) = C_n \cos\left(\frac{n\pi ct}{l} - \theta_n\right), \tag{6.18}$$

we write Eq. (6.15) as

$$y = \sum_{n=1}^{\infty} \xi_n(t) \sin\frac{n\pi x}{l}. \tag{6.19}$$

Hence

$$\dot{y} = \frac{\partial y}{\partial t} = \sum_{n=1}^{\infty} \dot{\xi}_n \sin\frac{n\pi x}{l}. \tag{6.20}$$

The kinetic energy of the string at any time t is

$$\begin{aligned} E_k &= \frac{1}{2}\int_0^l m\dot{y}^2 dx = \frac{m}{2}\int_0^l \left(\sum_{n=1}^{\infty} \dot{\xi}_n \sin\frac{n\pi x}{l}\right)^2 dx \\ &= \frac{m}{2}\int_0^l \left(\sum \dot{\xi}_n^2 \sin^2\frac{n\pi x}{l} + 2\sum_{n\neq m} \dot{\xi}_n\dot{\xi}_m \sin\frac{n\pi x}{l}\sin\frac{m\pi x}{l}\right) dx. \end{aligned}$$

By the orthogonality property,

$$\int_0^l \sin\frac{n\pi x}{l}\sin\frac{m\pi x}{l} dx = 0, \quad \text{when } n \neq m.$$

When $n = m$, we have

$$\int_0^l \sin\frac{n\pi x}{l}\sin\frac{m\pi x}{l} dx = \frac{l}{2}.$$

Therefore,

$$E_k = \frac{m}{2}\cdot\frac{l}{2}\sum_{n=1}^{\infty}\dot{\xi}_n^2 = \frac{\pi^2c^2M}{4l^2}\sum_{n=1}^{\infty} n^2C_n^2\sin^2\left(\frac{n\pi ct}{l} - \theta_n\right), \tag{6.21}$$

where $M(= ml)$ is the mass of the whole string.

The potential energy of an element dx of the string is the work done in stretching this element to its displaced position where its length is $ds = \sqrt{dy^2 + dx^2} = dx\sqrt{1 + \left(\frac{\partial y}{\partial x}\right)^2}$. Since the tension T remains constant, the work done in question is

$$dE_p = T(ds - dx) = T\left[\sqrt{1 + \left(\frac{\partial y}{\partial x}\right)^2} - 1\right] dx.$$

As $\frac{\partial y}{\partial x} \ll 1$, we have $\sqrt{1 + \left(\frac{\partial y}{\partial x}\right)^2} \simeq 1 + \frac{1}{2}\left(\frac{\partial y}{\partial x}\right)^2$.

Hence

$$dE_p = \frac{T}{2}\left(\frac{\partial y}{\partial x}\right)^2 dx. \tag{6.22}$$

Equation (6.19) yields

$$\frac{\partial y}{\partial x} = \frac{\pi}{l}\sum_{n=1}^{\infty} n\xi_n(t)\cos\frac{n\pi x}{l}.$$

Substituting this value of $\frac{\partial y}{\partial x}$ in Eq. (6.22) and integrating from 0 to l, we have for the potential energy of the string at time t

$$E_p = \frac{\pi^2 T}{2l^2} \int_0^l \left(\sum_{n=1}^{\infty} n\xi_n(t) \cos \frac{n\pi x}{l} \right)^2 dx$$

$$= \frac{\pi^2 T}{2l^2} \int_0^l \left(\sum n^2 \xi_n^2 \cos^2 \frac{n\pi x}{l} + 2 \sum_{n \neq m} nm\xi_n\xi_m \cos \frac{n\pi x}{l} \cos \frac{m\pi x}{l} \right).$$

Now,

$$\int_0^l \cos^2 \frac{n\pi x}{l} dx = \frac{l}{2}$$

$$\text{and} \quad \int_0^l \cos \frac{n\pi x}{l} \cos \frac{m\pi x}{l} dx = 0, \quad \text{for } n \neq m.$$

So,

$$E_p = \frac{\pi^2 T}{4l} \sum_{n=1}^{\infty} n^2 C_n^2 \cos^2 \left(\frac{n\pi ct}{l} - \theta_n \right)$$

$$= \frac{\pi^2 c^2 M}{4l^2} \sum_{n=1}^{\infty} n^2 C_n^2 \cos^2 \left(\frac{n\pi ct}{l} - \theta_n \right), \tag{6.23}$$

where Eq. (6.2) has been used. Hence the total energy of the vibrating string is

$$E = E_k + E_p = \frac{\pi^2 c^2 M}{4l^2} \sum_{n=1}^{\infty} n^2 C_n^2 = \frac{M}{4} \sum_{n=1}^{\infty} \omega_n^2 C_n^2, \tag{6.24}$$

since $\omega_n = \frac{nc\pi}{l}$ [see Eq. (6.12)]. Thus the total energy of the string is a constant independent of x and t. Equation (6.24) shows that the energy of vibration for a particular mode n is proportional to the square of the eigenfrequency and to the square of the amplitude for that mode of vibration. Note that

$$E_k = \frac{M}{4} \sum \dot{\xi}_n^2 \quad \text{and} \quad E_p = \frac{M}{4} \sum \omega_n^2 \xi_n^2,$$

so that

$$E = \frac{M}{4} \left(\sum \dot{\xi}_n^2 + \sum \omega_n^2 \xi_n^2 \right).$$

Comparing this relationship with Eq. (2.103) shows that the string can be regarded as an assembly of an infinite number of strongly coupled particle oscillators. Thus the string has an infinite number of modes of vibration. For mode n, the quantity ξ_n is referred to as the *normal coordinate* and ω_n as the *normal frequency*.

Since the average value of $\sin^2\left(\frac{n\pi ct}{l}-\theta_n\right)$ or $\cos^2\left(\frac{n\pi ct}{l}-\theta_n\right)$ over a complete period is $\frac{1}{2}$, we find from Eqs. (6.21), (6.23) and (6.24) that

$$\bar{E}_k = \bar{E}_p = \frac{\pi^2 c^2 M}{8l^2}\sum_{n=1}^{\infty} n^2 C_n^2 = \frac{1}{2}E,$$

where $\bar{E}_k$ and $\bar{E}_p$ are the time averages of the kinetic and the potential energies, respectively. Thus, the time-averaged kinetic energy is equal to the time-averaged potential energy, and each is equal to half the total energy of the vibrating string.

6.6 Plucked String

Suppose that a string of length l stretched between two fixed supports at the ends be pulled at a point distant a from the end $x = 0$ through a small transverse distance b and then released. The string is said to be *plucked* at $x = a$, the resulting vibration being that due to *plucking*. The distance b is so small that the tension can assumed to be constant. The initial configuration of the string is given by (see Fig. 6.5)

$$\begin{aligned} y_0 &= \frac{b}{a}x \qquad (0 \le x \le a) \\ &= \frac{b}{l-a}(l-x) \qquad (a \le x \le l). \end{aligned} \tag{6.25}$$

The initial velocity of the string is zero everywhere, i.e.,

$$\dot{y}_0 = 0. \tag{6.26}$$

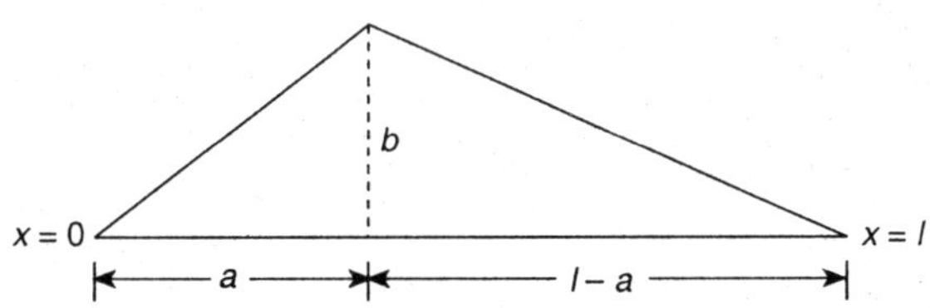

Fig. 6.5 Initial configuration of a plucked string

The general expression for the displacement $y(x,t)$ of the string at the point x at time t is given by Eq. (6.14), viz.

$$y = \sum_{n=1}^{\infty} \sin\frac{n\pi x}{l}\left(a_n \cos\frac{n\pi ct}{l} + b_n \sin\frac{n\pi ct}{l}\right). \tag{6.27}$$

The velocity at the same point x at time t is

$$\dot{y}(x,t) = \frac{\pi c}{l}\sum_{n=1}^{\infty} n \sin\frac{n\pi x}{l}\left(-a_n \sin\frac{n\pi ct}{l} + b_n \cos\frac{n\pi ct}{l}\right). \tag{6.28}$$

Putting $t = 0$, we have

$$y(x,0) = y_0 = \sum_{n=1}^{\infty} a_n \sin \frac{n\pi x}{l}, \tag{6.29}$$

and
$$\dot{y}(x,0) = 0 = \sum_{n=1}^{\infty} nb_n \sin \frac{n\pi x}{l}. \tag{6.30}$$

For Eq. (6.30) to hold true, we must have $b_n = 0$. Hence Eq. (6.27) reduces to

$$y(x,t) = \sum_{n=1}^{\infty} a_n \sin \frac{n\pi x}{l} \cos \frac{n\pi ct}{l}. \tag{6.31}$$

Equation (6.29) shows that y_0 is simply a sine series of Fourier expansion. Multiplying the two sides of Eq. (6.29) by $\sin \frac{m\pi x}{l}$ where m is an integer, viz. $m = 1, 2, 3, \cdots$, and integrating from $x = 0$ to $x = l$, we obtain

$$\int_0^l y_0 \sin \frac{m\pi x}{l} dx = \int_0^l \sum_{n=1}^{\infty} a_n \sin \frac{n\pi x}{l} \sin \frac{m\pi x}{l} dx.$$

The integrals on the right-hand side are zero when $n \neq m$. When $n = m$, we get

$$\int_0^l y_0 \sin \frac{n\pi x}{l} dx = a_n \int_0^l \sin^2 \frac{n\pi x}{l} dx = \frac{a_n l}{2},$$

whence

$$a_n = \frac{2}{l} \int_0^l y_0 \sin \frac{n\pi x}{l} dx. \tag{6.32}$$

Noting that y_0 is given by Eq. (6.25), we break up the interval 0 to l into two parts, one from 0 to a,and the other from a to l. Substituting for y_0 from Eq. (6.25) into Eq. (6.32) gives

$$a_n = \frac{2}{l} \left[\int_0^a \frac{b}{a} x \sin \frac{n\pi x}{l} dx + \int_a^l \frac{b}{l-a} (l-x) \sin \frac{n\pi x}{l} dx \right].$$

Both the integrals appearing in this equation can be evaluated upon integration by parts. Putting $k = \frac{n\pi}{l}$ for brevity, the first integral is

$$\int_0^a x \sin kx \, dx = \left[x \left(-\frac{\cos kx}{k} \right) \right]_0^a + \int_0^a \frac{\cos kx}{k} dx = -\frac{a \cos ka}{k} + \frac{\sin ka}{k^2}$$

$$= \frac{l^2}{n^2\pi^2} \sin \frac{n\pi a}{l} - \frac{al}{n\pi} \cos \frac{n\pi a}{l}.$$

The second integral is

$$\int_a^l (l-x)\sin kx\,dx = \left[(l-x)\left(\frac{-\cos kx}{k}\right)\right]_a^l - \int_a^l \frac{\cos kx}{k}dx$$
$$= (l-a)\frac{\cos ka}{k} + \frac{\sin ka}{k^2}$$
$$= \frac{(l-a)l}{n\pi}\cos\frac{n\pi a}{l} + \frac{l^2}{n^2\pi^2}\sin\frac{n\pi a}{l}.$$

Substituting the values of these integrals, we get

$$a_n = \frac{2}{l}\left[\frac{bl^2}{an^2\pi^2}\sin\frac{n\pi a}{l} - \frac{bl}{n\pi}\cos\frac{n\pi a}{l} + \frac{bl}{n\pi}\cos\frac{n\pi a}{l} + \frac{bl^2}{(l-a)n^2\pi^2}\sin\frac{n\pi a}{l}\right]$$
$$= \frac{2bl^2}{a(l-a)n^2\pi^2}\sin\frac{n\pi a}{l}. \tag{6.33}$$

Hence Eq. (6.31) yields

$$y(x,t) = \frac{2bl^2}{a(l-a)\pi^2}\sum_{n=1}^{\infty}\frac{1}{n^2}\sin\frac{n\pi a}{l}\sin\frac{n\pi x}{l}\cos\frac{n\pi ct}{l}.$$
$$= \frac{2bl^2}{a(l-a)\pi^2}\left[\sin\frac{\pi a}{l}\sin\frac{\pi x}{l}\cos\frac{\pi ct}{l} + \frac{1}{4}\sin\frac{2\pi a}{l}\sin\frac{2\pi x}{l}\cos\frac{2\pi ct}{l} + \frac{1}{9}\sin\frac{3\pi a}{l}\sin\frac{3\pi x}{l}\cos\frac{3\pi ct}{l} + \cdots\right]. \tag{6.34}$$

The successive terms on the right-hand side of Eq. (6.34) give respectively the displacements for the fundamental, second harmonic, third harmonic, etc. For the nth mode, the displacement is

$$y_n(x,t) = \frac{2bl^2}{a(l-a)\pi^2}\frac{1}{n^2}\sin\frac{n\pi a}{l}\sin\frac{n\pi x}{l}\cos\frac{n\pi ct}{l}. \tag{6.35}$$

Note that the coefficients a_n vary inversely as n^2. Therefore, the amplitudes of the higher harmonics fall off rapidly.

Let $a = l/p$, where p is an integer. Then, $n\pi a/l = n\pi/p$. If n is an integral multiple of p, i.e., if $n = p, 2p, 3p$, etc., then $\sin\frac{n\pi a}{l} = 0$. For $p = 2$, i.e., if the string is plucked at the midpoint, n is an even number, so that all the even harmonics will vanish. All these harmonics have a node at $a = l/2$. Hence follows the result:

For a plucked string, all the harmonics having a node at the point of plucking are absent.

This is referred to as *Young's law* or *Young-Helmholtz law.*

Mandoline, harp, guitar, banjo, etc. are plucked stringed instruments of the western variety, while tanpura, vina, sarod, sitar, etc. are such instruments of the Indian variety.

Observations

(i) Equation (6.35) shows that the maximum amplitude of vibration for the fundamental is $A_1 = \frac{2bl^2}{a(l-a)\pi^2} \sin\frac{\pi a}{l}$, whereas that for the third harmonic is $A_3 = \frac{2bl^2}{9a(l-a)\pi^2} \left|\sin\frac{3\pi a}{l}\right|$. If I_1 and I_3 are the intensities of the fundamental and the third harmonic, respectively, then

$$\frac{I_1}{I_3} = \left(\frac{A_1}{A_3}\right)^2 = 81\frac{\sin^2 \pi a/l}{\sin^2 3\pi a/l}. \tag{6.36}$$

If the string is plucked at the midpoint, i.e., if $a = l/2$, then $I_1/I_3 = 81$. Thus, the fundamental is 81 times as intense as the third harmonic.

(ii) Note that the vibration of the string can be stopped by touching the point of plucking after the plucked string is set into vibrations. For, the components which have a node at the point of plucking are not excited at all according to Young's law. The other components which do not have a node at the point of plucking are stopped by damping resulting from touching. Consequently, all vibrations will cease to exist.

(iii) If the string is plucked at, say, $a = l/3$ and touched lightly at another point, say, $x = l/4$, then $\sin(n\pi x/l) = \sin(n\pi/4) = 0$ for $n = 4, 8, 12, 16$, etc. These harmonics have nodes at $l/4$; so they are not affected by touching at $l/4$. Of these harmonics, the 12th, 24th, etc. which are multiples of 3 and are not excited, due to Young-Helmholtz law, by plucking at $l/3$, will be absent. Hence the harmonics which will be present are 4th, 8th, 16th, 20th, 28th, etc. The remaining components which are excited will be damped and finally eliminated by touching at $l/4$.

(iv) If the string is plucked at a single point, as assumed in the theory, by a sharp object, the emitted note is very brilliant due to the presence of a large number of modes. The situation exists in the mandoline which is plucked by a thin celluloid sheet termed the plectrum. But if the string is plucked over a finite length by a round object like finger, a less number of modes are excited so that the sound is softer. The quality of sound given out by a harp is inferior to that given out by a mandoline because the former is plucked by fingers. In the extreme case, if the initial configuration of a string is a sine wave so that $y_0 = A\sin\frac{\pi x}{l}$, then only the fundamental is excited.

(v) If the bridges supporting the wire at the ends are not fixed but yield slightly, then the effective vibrating length l increases. Hence the frequencies of all the components decrease by the same factor.

(vi) In the plucked stringed instruments of Indian origin, such as, sitar, tanpura, etc. the surface of the supporting bridge is almost flat and the string makes a small acute angle with it. This makes the vibrations very rich in harmonics, the amplitudes of the higher modes being much greater than those for mandoline.

(vii) If the string is thick, then apart from tension, additional elastic forces due to bending appear. So, the pitch of all the components increases, but the increase is greater for the higher frequencies. Thus, the frequencies no longer form a harmonic series. The stiffness of the wire thus reduces the melody of its vibration.

6.7 Struck String

If a string stretched between two fixed supports at its ends is struck at a point by a hammer, a sudden impulse is imparted to the point of striking but the velocity of all other points on the string remains initially zero. Thus, the initial conditions of a struck string differ from those of a plucked string. A plucked string has an initial displacement and no initial velocity whereas a struck string has no initial displacement but an initial velocity at the region of striking. The initial conditions of a plucked string are thus *static*, while those of a struck string are *dynamic*.

In the analysis of a struck string, the duration of contact between the striking hammer and the string is assumed to be so small that the impact can be considered to have ceased before the disturbance finds time to move over to other regions of the length. The initial velocity is thus confined to a very short length, but otherwise the motion is free. Waves move in opposite directions from the point of excitation and are reflected at the fixed ends. Thus standing waves are produced.

Let l be the length of the stretched string supported at $x = 0$ and $x = l$. It is struck at an infinitesimally small region from $x = a$ to $x = a + dx$. Thus, this region attains the velocity u at time $t = 0$.

The displacement of the string at the point x at time t is given, in general, by Eq. (6.14), viz.

$$y = \sum_{n=1}^{\infty} \sin\frac{n\pi x}{l}\left(a_n \cos\frac{n\pi ct}{l} + b_n \sin\frac{n\pi ct}{l}\right). \tag{6.37}$$

Putting the initial conditions that $y = 0$ everywhere at $t = 0$ we obtain $a_n = 0$.

Hence

$$y = \sum_{n=1}^{\infty} b_n \sin\frac{n\pi x}{l} \sin\frac{n\pi ct}{l}. \tag{6.38}$$

The velocity at any point is

$$\dot{y} = \frac{\partial y}{\partial t} = \frac{\pi c}{l}\sum_{n=1}^{\infty} nb_n \sin\frac{n\pi x}{l}\cos\frac{n\pi ct}{l}. \tag{6.39}$$

The initial velocity at any point x is

$$\dot{y}_0 = \dot{y}(x,0) = \frac{\pi c}{l}\sum_{n=1}^{\infty} nb_n \sin\frac{n\pi x}{l}. \tag{6.40}$$

The coefficient b_n is found by multiplying Eq. (6.41) through $\sin\frac{m\pi x}{l}$, where m is an integer, and integrating from $x = 0$ to $x = l$. All the integrals on the right-hand side disappear when $m \neq n$. When $m = n$, we have

$$\int_0^l \dot{y}_0 \sin\frac{n\pi x}{l}dx = \frac{\pi c}{l} nb_n \int_0^l \sin^2\frac{n\pi x}{l}dx = \frac{\pi cnb_n}{2}. \tag{6.41}$$

Note that $\dot{y}_0 = 0$ for all values of x except for the region from $x = a$ to $x = a + dx$, where $\dot{y}_0 = u$. Hence Eq. (6.41) yields

$$\frac{\pi cnb_n}{2} = \int_a^{a+dx} u\sin\frac{n\pi a}{l}dx = A\sin\frac{n\pi a}{l},$$

$$\text{where} \quad A = \int_a^{a+dx} u\,dx.$$

Therefore,

$$b_n = \frac{2A}{n\pi c}\sin\frac{n\pi a}{l}. \tag{6.42}$$

Substituting for b_n from Eq. (6.42) into Eq. (6.38), we obtain

$$y(x,t) = \frac{2A}{\pi c}\sum_{n=1}^{\infty}\frac{1}{n}\sin\frac{n\pi a}{l}\sin\frac{n\pi x}{l}\sin\frac{n\pi ct}{l} \tag{6.43}$$

$$= \frac{2A}{\pi c}\left[\sin\frac{\pi a}{l}\sin\frac{\pi x}{l}\sin\frac{\pi ct}{l} + \frac{1}{2}\sin\frac{2\pi a}{l}\sin\frac{2\pi x}{l}\sin\frac{2\pi ct}{l} + \frac{1}{3}\sin\frac{3\pi a}{l}\sin\frac{3\pi x}{l}\sin\frac{3\pi ct}{l} + \cdots\right]. \tag{6.44}$$

Note that the coefficients b_n are inversely proportional to n. Hence the amplitudes of the higher harmonics fall off with increasing n, but not as rapidly as for a plucked string (where the coefficients decrease as $1/n^2$).

Let $a = l/p$, where p is an integer. So, $n\pi a/l = n\pi/p$. If $n = p, 2p, 3p$ etc. then $\sin\frac{n\pi a}{l} = 0$. When the string is struck at the middle, $p = 2$, and n is an even integer. All these harmonics have a node at $a = l/2$. Hence, as in the case of plucked string, we have *Young's law* or *Young-Helmholtz law* which states that *all the harmonics having a node at the point of striking are absent.*

Piano is the only example of a stringed struck instrument. The wires for high-pitched notes in a piano are short and thin, subjected to a high tension. The strings for low-pitched notes are loaded to increase the linear density. The strings are struck at approximately (1/7)th of the length, so that the seventh harmonic which tends to produce dissonance with other notes in a chord is eliminated.

Observations

(i) Equation (6.44) shows that the amplitude of the fundamental for a struck string is $\frac{2A}{\pi c}\sin\frac{\pi a}{l}$, whereas the same for the third harmonic is $\frac{2A}{3\pi c}\left|\sin\frac{3\pi a}{l}\right|$. As the intensity is proportional to the square of the amplitude, the ratio of the intensity I_1 of the fundamental to the intensity I_3 of the third harmonic is

$$\frac{I_1}{I_3} = 9\frac{\sin^2\frac{\pi a}{l}}{\sin^2\frac{3\pi a}{l}}. \tag{6.45}$$

When the string is struck at the midpoint, then $a = l/2$ and $I_1/I_3 = 9$. Thus the third harmonic is much stronger here than for a plucked string.

(ii) As in the case of a plucked string, the vibrations of a struck string can be stopped by touching the region of striking. The components having a node at the point of striking are not generated in accordance with Young's law. The other components not having a node at the point of striking are damped out by touching, so that all vibrations disappear.

(iii) If the string is struck at $a = l/3$ and then touched at $x = l/4$, we have the same result as for the plucked string [see observation (iii) in Sec. 6.6].

(iv) Young's law is not strictly obeyed in musical instruments due to the reflection from the belly of the instrument acting as a resonator. The vibration of a struck string is affected by the finite time and area of contact between the hammer and the string.

6.8 Bowed String

A string stretched between two supports can be set into vibrations by drawing a resined bow of horse-hair across it. Such a string is referred to as a *bowed string*. Violin, viola, cello, etc. are bowed western musical instruments, whereas esraj, sarangi, dilruba, etc. are the Indian counterparts.

The bow moving across the string drags the string along with it due to the static friction between them. The restoring force of the string thus increases and when it exceeds the frictional force, the string slips towards its initial position and overshoots it due to inertia. The bow which is always moving keeping contact with the string, siezes the string once again and drags it forward. The process is repeated rapidly. Figure 6.3 shows the variation of the displacement y with time t for a point on the string. A two-step straight zigzag is obtained as the point is carried forward by the bow, and then slips. During slipping, the dynamic friction operates. As the static friction is greater than the dynamic friction, the work done on the string exceeds that done by it. The difference of the two works is expended in maintaining the vibration.

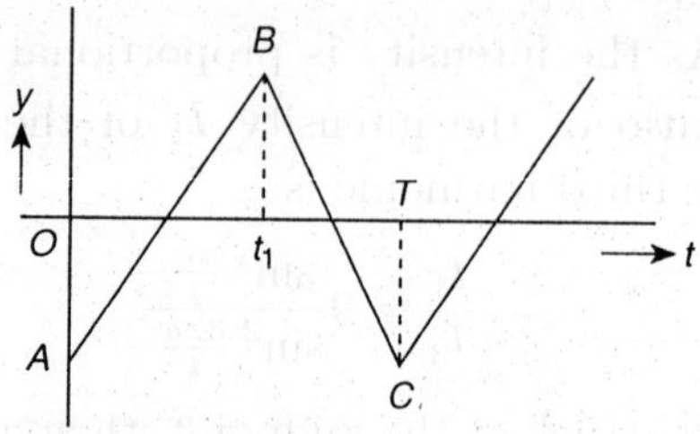

Fig. 6.6 Time-displacement curve of a point on a bowed string

Let T be the periodic time of vibration (Fig. 6.6). In the time interval 0 to t_1, the point on the string moves forward with a uniform velocity v_1 given by the slope of the straight line AB. In the time interval t_1 to T, the point travels with the uniform velocity $-v_2$ given by the slope of the straight line BC.

The general expression for the transverse displacement at a point x at time t is obtained from Eq. (6.14):

$$y = \sum_{n=1}^{\infty} \sin \frac{n\pi x}{l} \left(a_n \cos \frac{n\pi ct}{l} + b_n \sin \frac{n\pi ct}{l} \right). \tag{6.46}$$

The instantaneous velocity at the point is

$$\dot{y} = \sum_{n=1}^{\infty} \frac{n\pi c}{l} \sin \frac{n\pi x}{l} \left(-a_n \sin \frac{n\pi ct}{l} + b_n \cos \frac{n\pi ct}{l} \right). \tag{6.47}$$

To evaluate a_n, we multiply the two sides of Eq. (6.47) by $\sin \frac{m\pi ct}{l}$, where m is an integer, and integrate from $t = 0$ to $t = T$. All the integrals on the right-hand side are zero except the one for $m = n$. Hence

$$\int_0^T \dot{y} \sin \frac{n\pi ct}{l} dt = -\frac{n\pi c}{l} \sin \frac{n\pi x}{l} a_n \int_0^T \sin^2 \frac{n\pi ct}{l} dt$$
$$= -\frac{a_n T}{2} \frac{n\pi c}{l} \sin \frac{n\pi x}{l} = -a_n n\pi \sin \frac{n\pi x}{l}, \qquad (6.48)$$

since $\pi c/l = \omega$ and $\omega T = 2\pi$.

From $t = 0$ to $t = t_1$, we have $\dot{y} = v_1$ and from $t = t_1$ to $t = T$, $\dot{y} = -v_2$. Putting these values in Eq. (6.48) yields

$$\begin{aligned} -a_n \sin \frac{n\pi x}{l} &= \frac{1}{n\pi}\left[\int_0^{t_1} v_1 \sin \frac{n\pi ct}{l} dt + \int_{t_1}^T (-v_2) \sin \frac{n\pi ct}{l} dt\right] \\ &= \frac{1}{n\pi}\left[\frac{v_1 l}{n\pi c}\left(1 - \cos \frac{n\pi ct_1}{l}\right) + \frac{v_2 l}{n\pi c}\left(1 - \cos \frac{n\pi ct_1}{l}\right)\right] \\ &= \frac{(v_1 + v_2)l}{n^2\pi^2 c}\left(1 - \cos \frac{n\pi ct_1}{l}\right) \\ &= \frac{2l(v_1 + v_2)}{n^2\pi^2 c} \sin^2 \frac{n\pi ct_1}{2l} \\ &= \frac{T(v_1 + v_2)}{n^2\pi^2} \sin^2 \frac{n\pi ct_1}{2l}. \end{aligned} \qquad (6.49)$$

Similarly,

$$b_n \sin \frac{n\pi x}{l} = \frac{T(v_1 + v_2)}{n^2\pi^2} \sin \frac{n\pi ct_1}{2l} \cos \frac{n\pi ct_1}{2l}. \qquad (6.50)$$

Using Eqs. (6.49) and (6.50) in Eq. (6.46), we write

$$\begin{aligned} y(x,t) &= \frac{T(v_1 + v_2)}{\pi^2} \sum_{n=1}^{\infty} \frac{1}{n^2}\left(-\sin^2 \frac{n\pi ct_1}{2l} \cos \frac{n\pi ct}{l}\right. \\ &\qquad \left. + \sin \frac{n\pi ct_1}{2l} \cos \frac{n\pi ct_1}{2l} \sin \frac{n\pi ct}{l}\right) \\ &= \frac{T(v_1 + v_2)}{\pi^2} \sum_{n=1}^{\infty} \frac{1}{n^2} \sin \frac{n\pi ct_1}{2l} \sin \frac{n\pi c}{l}\left(t - \frac{t_1}{2}\right). \end{aligned} \qquad (6.51)$$

Equation (6.46) can also be written as

$$y(x,t) = \sum_{n=1}^{\infty} C_n \sin \frac{n\pi x}{l} \sin \frac{n\pi c}{l}(t - \alpha), \qquad (6.52)$$

where α is a constant. Comparing Eqs. (6.51) and (6.52), we find

$$C_n \sin \frac{n\pi x}{l} = \frac{T(v_1 + v_2)}{n^2\pi^2} \sin \frac{n\pi ct_1}{2l} = \frac{T(v_1 + v_2)}{n^2\pi^2} \sin \frac{n\pi t_1}{T}. \qquad (6.53)$$

If $\frac{n\pi x}{l}$ is an integral multiple of π, then Eq. (6.52) shows that $y = 0$ for all values of time t. Since $v_1 + v_2 \neq 0$, Eq. (6.53) shows that in this case $\sin \frac{n\pi t_1}{T} = 0$. The two conditions are satisfied by putting

$$\frac{t_1}{T} = \frac{1}{n} = \frac{x}{l}. \tag{6.54}$$

At the midpoint $x = l/2$, we have $t_1/T = 1/2$. Thus, at the midpoint equal times are required for the forward and backward motion.

C_1 is the amplitude at the midpoint for the transverse stationary wave at the fundamental frequency. Then the amplitude of the wave at a point x is $C_1 \sin \frac{\pi x}{l}$. Equation (6.53) shows that the amplitude at x is $\frac{T(v_1+v_2)}{\pi^2} \sin \frac{\pi x}{l}$. Thus $\frac{T(v_1+v_2)}{\pi^2}$ is a constant, or $v_1 + v_2$, the sum of the forward and the backward velocities, is a constant for all values of x.

If A is the amplitude of vibration at the midpoint, then

$$v_1 + v_2 = 2v_1 = 2\frac{2A}{T/2} = \frac{8A}{T}. \tag{6.55}$$

Substituting for $v_1 + v_2$ from Eq. (6.55) into Eq. (6.51) gives

$$y = \frac{8A}{\pi^2} \sum_{n=1}^{\infty} \frac{1}{n^2} \sin \frac{n\pi x}{l} \sin n\omega \left(t - \frac{t_1}{2}\right), \tag{6.56}$$

where we have used Eq. (6.54) and the relationship $T = \frac{2\pi}{\omega} = \frac{2l}{c}$.

We find from Eq. (6.56) that if $t = t_1/2$, then $y = 0$ for all values of x. This shows that all points on the string pass through their equilibrium positions simultaneously.

The theory developed above follows Helmholtz. It shows that the *action of the moving bow* is to supply the energy of the periodic vibration of the string. However, it is not a case of forced vibration. The period of vibration of the string is not determined by the bow, so that the vibration of the string can be considered to be *free*.

In practice, the quality of the tone emitted from the bowed string is somewhat determined by the speed of the bow, the location of the point of bowing, and the pressure exerted by the bow on the string. The tone structure can be effectively modified by the bow pressure. The higher harmonics become relatively prominent if the point of bowing is closer to the bridge. The amplitude of vibration of the string, and consequently the intensity of the emitted sound, depend on the bow hairs in contact with the string and on the speed of the bow. An increase of the speed of the bow enhances the intensity. Thus, the bow speed can effectively control the sound intensity.

Wolf note

The quality of the sound from the stringed musical instruments differs from that expected from the vibration of the string alone. The forced vibrations and resonances in the body of the instrument modify the relative prominance of the harmonics. In instruments of the violin family, a particular note cannot be sometimes smoothly elicited. If the pitch of the note emitted from the string coincides with the fundamental or with a predominant harmonic of the wooden structure or the air cavity, a resonance occurs. The energy of the reinforced note is provided by the larger pressure needed to maintain the amplitude of vibration of the string. In this case, an undesirable *howling effect* is produced. Hence the note is referred to as the *Wolf note.*

Professor C. V. Raman argued that a greater bow pressure is required for the fundamental than for the octave when the string is bowed near one end. At the Wolf's pitch, the string begins to vibrate in its fundamental mode, but the resonant vibration of the body of the instrument absorbs so much energy from the bow that the fundamental cannot be maintained. Hence the vibration changes to a mode with the octave predominating. When the vibration of the body of the instrument dies out, the fundamental reappears. Such cyclic changes between the fundamental and the octave are typical in violin-type instruments.

6.9 Some Comments on Stringed Musical Instruments

Stringed musical instruments are plucked, struck, or bowed types. Examples of different types and some of their features have already been discussed in the previous sections. In all stringed instruments, musical notes are generated by the vibration of stretched strings. The wires have different lengths, masses and tensions. The ends of the strings are fixed to pegs or screws mounted on a board. As the strings vibrate, their pull on the pegs changes, so that the pegs communicate the vibrations of the strings to the board. The forced vibrations of the board generate sound waves in the surrounding air. The strings themselves are poor radiators of sound energy. The vibrations of the board and the adjoining air are responsible for the effective radiation of the sound energy. The resonant vibrations of the board magnify some frequency components relative to others.

Notes are tuned by changing the tension of the string with the aid of the pegs to which one end of the string is fixed. As the instrument is played, the frequency of the note can be altered by damping the string at suitable points.

In plucked and bowed type instruments, the amplitude of a partial is inversely proportional to the square of the number of the partial in the harmonic series, whereas in struck instruments the same varies inversely as the number of the partial. It is thus expected that the sound from a struck type instrument will be richer in harmonics.

6.10 Solved Problems

Problem 1. *The wind is blowing over some suspension cable on a bridge, causing a force* $F = 0.02 \sin 21\pi t$ N. *Is resonance possible (at fundamental or at some harmonic) if the tension in the cable is* 49000 N, *the cable has a mass of* 10 kg/m, *and it is* 10 m *long?* **[C.U. 1997]**

Solution. The fundamental frequency of vibration of the cable is

$$f_1 = \frac{1}{2l}\sqrt{\frac{T}{m}}$$

Here T = tension = 49000 N, m = mass per unit length = 10 kg/m, and l = cable length = 10 m. So,

$$f_1 = \frac{1}{2 \times 10}\sqrt{\frac{49000}{10}} = 3.5 \text{ Hz}.$$

The impressed force has the frequency $f = \frac{21\pi}{2\pi} = 10.5$ Hz. Since $f/f_1 = 3$, resonance is possible at the third harmonic frequency.

Problem 2. *A string of length l fixed at both ends has a fundamental frequency f_0. The string is plucked at a point $l/4$ from one end and immediately afterwards an observer touches the string lightly at its midpoint. What is the lowest frequency excited in this case? What will be the frequency of the next harmonic?* **[C.U. 1988]**

Solution. Since the point of plucking is $l/4$, by Young's law, the harmonics having a node at $l/4$ will be absent. Thus, the frequencies $4f_0$, $8f_0$ etc. are absent. Again, since the string is touched at the midpoint, the frequencies not having a node at $l/2$ will be eliminated. Thus the frequencies $f_0, 3f_0, 5f_0, 7f_0$ etc. will be absent. Therefore, the lowest excited frequency is $2f_0$. The next excited harmonic is $6f_0$.

Problem 3. *A stretched string of length l and fixed at the ends is plucked at the midpoint through a distance h. Show that the displacement at a point x on the string at time t is*

$$y(x,t) = \frac{8h}{\pi^2}\left(\sin\frac{\pi x}{l}\cos\frac{\pi ct}{l} - \frac{1}{9}\sin\frac{3\pi x}{l}\cos\frac{3\pi ct}{l} + \frac{1}{25}\sin\frac{5\pi x}{l}\cos\frac{5\pi ct}{l} - \cdots\right).$$

Solution. The general expression for the displacement for a plucked string is

$$y(x,t)=\frac{2bl^2}{a(l-a)\pi^2}\sum_{n=1}^{\infty}\frac{1}{n^2}\sin\frac{n\pi a}{l}\sin\frac{n\pi x}{l}\cos\frac{n\pi ct}{l}.$$

Here $a=\frac{l}{2}$ and $b=h$; hence $\frac{2bl^2}{a(l-a)\pi^2}=\frac{8h}{\pi^2}$. Also, $\sin\frac{n\pi a}{l}=\sin\frac{n\pi}{2}$. (since $a=l/2$). So, $n=2,4,6$ etc. will be absent in the series. Therefore,

$$y(x,t)=\frac{8h}{\pi^2}\bigg(\sin\frac{\pi x}{l}\cos\frac{\pi ct}{l}-\frac{1}{9}\sin\frac{3\pi x}{l}\cos\frac{3\pi ct}{l}$$
$$+\frac{1}{25}\sin\frac{5\pi x}{l}\cos\frac{5\pi ct}{l}-\cdots\bigg),\ \text{proved.}$$

Problem 4. *A stretched string of length l is plucked through b at a distance $l/3$ from one end. Find the point on the string where the amplitude of the second harmonic is a maximum. What is the maximum amplitude?*

Solution. The displacement for the second harmonic ($n=2$) at a point x at time t when $a=l/3$ is

$$y_2(x,t)=\frac{9b}{\pi^2}\times\frac{1}{2^2}\sin\frac{2\pi}{3}\sin\frac{2\pi x}{l}\cos\frac{2\pi ct}{l}.$$

Hence the amplitude of the second harmonic at x is

$$A_2(x)=\frac{9b}{4\pi^2}\frac{\sqrt{3}}{2}\left|\sin\frac{2\pi x}{l}\right|.$$

The maximum amplitude occurs at points where $\left|\sin\frac{2\pi x}{l}\right|=1$. That is, $\frac{2\pi x}{l}=\frac{\pi}{2}$ or, $\frac{2\pi x}{l}=\frac{3\pi}{2}$. So, $x=\frac{l}{4}$ or, $x=\frac{3l}{4}$. The maximum amplitude is

$$A_{2m}=\frac{9\sqrt{3}b}{8\pi^2}.$$

Problem 5. *A stretched string of length l supported at its ends is pulled perpendicular to its length by a distance h at its midpoint and then released. Find the amplitude of the fundamental at a distance $l/4$ from one end.*

Solution. From Problem 3, we have for the amplitude of the fundamental at x

$$A(x)=\frac{8h}{\pi^2}\sin\frac{\pi x}{l}.$$

At $x=l/4$, we obtain

$$A=\frac{8h}{\pi^2}\times\frac{1}{\sqrt{2}}=\frac{4\sqrt{2}h}{\pi^2}.$$

Problem 6. *A stretched string of length l, fixed at its ends, is plucked by a distance 'b' at a point distant 'a' from one of its ends. Find the energy of the sth harmonic.* **[C.U. 1996]**

Solution. The displacement at a point x on the string at time t for the sth harmonic is

$$y_s(x,t) = A_s \sin\frac{s\pi x}{l}\cos\frac{s\pi ct}{l}, \quad \text{where } A_s = \frac{2bl^2}{a(l-a)\pi^2 s^2}\sin\frac{s\pi a}{l}.$$

If m is the mass per unit length of the string, the kinetic energy of an elemental length dx of the string for the sth harmonic is

$$dE_k = \frac{1}{2}m\left(\frac{\partial y_s}{\partial t}\right)^2 dx = \frac{1}{2}m\frac{s^2\pi^2c^2}{l^2}A_s^2\sin^2\frac{s\pi x}{l}\sin^2\frac{s\pi ct}{l}dx.$$

The vibration being simple harmonic, the total energy of the element dx is its maximum kinetic energy. Thus

$$dE = dE_{k(\max)} = \frac{ms^2\pi^2c^2}{2l^2}A_s^2\sin^2\frac{s\pi x}{l}dx.$$

The total energy of the entire string for the sth harmonic is obtained by integrating the above expression over $x = 0$ to $x = l$. So,

$$E = \frac{ms^2\pi^2c^2}{2l^2}A_s^2\int_0^l \sin^2\frac{s\pi x}{l}dx$$
$$= \frac{ms^2\pi^2c^2}{4l}A_s^2 = \frac{M\omega_s^2}{4}A_s^2$$

where $M(= ml)$ is the mass of the string and $\omega_s\ (= \frac{s\pi c}{l})$ is the angular frequency of the sth harmonic.

Problem 7. *A uniform string of length l is stretched between its fixed ends $x = 0$ and $x = l$. Obtain an expression for the transverse displacement $y(x,t)$ of the string when it is struck at the centre so that the velocity varies linearly from zero at the ends to v_0 at the centre.* **[Burd. U. 1995]**

Solution. With the initial condition that $y = 0$ everywhere at $t = 0$, we have

$$y(x,t) = \sum_{n=1}^{\infty} b_n \sin\frac{n\pi x}{l}\sin\frac{n\pi ct}{l}.$$

The velocity at any point is

$$\dot{y} = \frac{\partial y}{\partial t} = \frac{\pi c}{l}\sum_{n=1}^{\infty} nb_n \sin\frac{n\pi x}{l}\cos\frac{n\pi ct}{l}.$$

The initial velocity at any point x is

$$\dot{y}_0 = \dot{y}(x, 0) = \frac{\pi c}{l} \sum_{n=1}^{\infty} n b_n \sin \frac{n\pi x}{l}$$

where

$$\frac{\pi c n b_n}{2} = \int_0^l \dot{y}_0 \sin \frac{n\pi x}{l} dx.$$

By the given condition,

$$\begin{aligned} \dot{y}_0 &= \frac{v_0}{(l/2)} x \qquad \text{for} \quad 0 < x < \frac{l}{2} \\ &= \frac{v_0 (l - x)}{l/2} \qquad \text{for} \quad \frac{l}{2} < x < l. \end{aligned}$$

So,

$$\begin{aligned} \frac{\pi c n b_n}{2} &= \int_0^{l/2} \frac{v_0 x}{l/2} \sin \frac{n\pi x}{l} dx + \int_{l/2}^{l} \frac{v_0 (l-x)}{l/2} \sin \frac{n\pi x}{l} dx \\ &= \frac{2v_0}{l} \left[\int_0^{l/2} x \sin \frac{n\pi x}{l} dx + \int_{l/2}^{l} (l - x) \sin \frac{n\pi x}{l} dx \right] \\ &= \frac{2v_0}{l} \left[\left(\frac{l^2}{n^2\pi^2} \sin \frac{n\pi}{2} - \frac{l^2}{2n\pi} \cos \frac{n\pi}{2} \right) \right. \\ &\qquad \left. + \left(\frac{l^2}{2n\pi} \cos \frac{n\pi}{2} + \frac{l^2}{n^2\pi^2} \sin \frac{n\pi}{2} \right) \right] \\ &= \frac{4v_0 l}{n^2\pi^2} \sin \frac{n\pi}{2}. \\ \text{or,} \quad b_n &= \frac{8v_0 l}{n^3\pi^3 c} \sin \frac{n\pi}{2}. \end{aligned}$$

Hence

$$\begin{aligned} y(x, t) &= \frac{8v_0 l}{\pi^3 c} \sum_{n=1}^{\infty} \frac{1}{n^3} \sin \frac{n\pi}{2} \sin \frac{n\pi x}{l} \sin \frac{n\pi ct}{l} \\ &= \frac{8v_0 l}{\pi^3 c} \left[\sin \frac{\pi x}{l} \sin \frac{\pi ct}{l} - \frac{1}{27} \sin \frac{3\pi x}{l} \sin \frac{3\pi ct}{l} \right. \\ &\qquad \left. + \frac{1}{125} \sin \frac{5\pi x}{l} \sin \frac{5\pi ct}{l} - \cdots \right]. \end{aligned}$$

Problem 8. *For a struck string of length l, the striking point is at a distance of $l/3$ from the end $x = 0$. Calculate the amplitude of the fundamental at $x = l/4$ and $x = l/2$.*

Solution. The displacement is

$$y(x,t) = \frac{2A}{\pi c}\sum_{n=1}^{\infty}\frac{1}{n}\sin\frac{n\pi a}{l}\sin\frac{n\pi x}{l}\sin\frac{n\pi ct}{l}.$$

Here $a = l/3$, so that the amplitude of the fundamental is

$$A_1 = \frac{2A}{\pi c}\sin\frac{\pi}{3}\sin\frac{\pi x}{l}.$$

At $x = l/4$, $A_1 = \frac{2A}{\pi c}\times\frac{\sqrt{3}}{2}\times\frac{1}{\sqrt{2}} = \sqrt{\frac{3}{2}}\frac{A}{\pi c}$.
At $x = l/2$, $A_1 = \frac{2A}{\pi c}\times\frac{\sqrt{3}}{2}\times 1 = \sqrt{3}\frac{A}{\pi c}$.

Problem 9. *A string has a linear mass density* 0.25 kg/m *and is stretched along the x-axis with a tension* 25 N. *One end of the string* ($x = 0$) *is given a sinusoidal motion with a frequency* 5 Hz *and amplitude* 0.1 m. *At time* $t = 0$, *the end has zero displacement and it is moving in the positive y-direction.*

(i) *Find the wave speed in the string and the amplitude.* (ii) *Write a wave equation for the wave.* (iii) *Find the position of the point at* $x = 0.25$ m *at* $t = 0.1$ s. (iv) *Find the slope of the string at* $x = 0.25$ m *at* $t = 0.1$ s

[C.U. 1992]

Solution. (i) The wave speed is $c = \sqrt{\frac{T}{m}} = \sqrt{\frac{25}{0.25}} = 10$ m/s.
The amplitude is 0.1 m.
(ii) The wave equation is $\frac{\partial^2 y}{\partial t^2} = c^2\frac{\partial^2 y}{\partial x^2}$.
(iii) The wave is represented by

$$y = 0.1\sin 2\pi f\left(t - \frac{x}{c}\right)\text{ m} = 0.1\sin 10\pi\left(t - \frac{x}{10}\right)\text{ m}.$$

At $x = 0.25$ m and $t = 0.1$s, we have

$$y = 0.1\sin 10\pi(0.1 - 0.025)\text{ m} = 0.1\sin 0.975\pi\text{ m}.$$

(iv) The slope of the string is

$$\dot{y} = \frac{\partial y}{\partial x} = -0.1\pi\cos 10\pi\left(t - \frac{x}{10}\right).$$

At $x = 0.25$ m and $t = 0.1$s, we have

$$\dot{y} = -0.1\pi\cos 10\pi(0.1 - 0.025) = -0.1\pi\cos 0.975\pi.$$

Problem 10. *Considering two simple harmonic plane waves moving in the opposite directions along a stretched string of length l fixed at both ends, show that the vibrating string can be regarded as a stationary wave.*

Solution. The general solution of the differential wave equation for the string, viz., $\frac{\partial^2 y}{\partial t^2} = c^2\frac{\partial^2 y}{\partial x^2}$ has the form $y = f_1(ct-x)+f_2(ct+x)$. Here f_1 and f_2 are two plane waves moving in the positive and negative x-axis, respectively. In the simplest case, f_1 and f_2 are simple harmonic with angular frequency ω and amplitudes y_{10} and y_{20}, respectively. These two waves are reflected at the ends and form stationary waves on superposition. We write

$$y = y_{10}\sin(\omega t - kx) + y_{20}\sin(\omega t + kx). \quad \text{(i)}$$

The boundary conditions are (a) at $x = 0$, $y = 0$ for all t and (b) at $x = l$, $y = 0$ for all t. The first condition gives $y_{10} = -y_{20} = -\frac{C}{2}$, say. The second boundary condition gives $\sin(\omega t + kl) - \sin(\omega t - kl) = 0$ or, $\cos\omega t \sin kl = 0$, or, $\sin kl = 0$

$$\text{or,} \qquad k = \frac{n\pi}{l} \quad (n \text{ is an integer}). \quad \text{(ii)}$$

The angular frequency ω is

$$\omega = kc = \frac{n\pi c}{l}. \quad \text{(iii)}$$

Thus, k and ω are determined by n. The constant C also depends, in general, on n. Hence the solution for a particular value of n is

$$\begin{aligned} y_n &= \frac{C_n}{2}\left[\sin(\omega_n t + k_n x) - \sin(\omega_n t - k_n x)\right] \\ &= C_n \cos\omega_n t \sin k_n x \\ &= C_n \cos\frac{n\pi ct}{l}\sin\frac{n\pi x}{l} \quad \text{(iv)} \end{aligned}$$

Thus the vibrating string can be regarded as a stationary wave formed by the superposition of two oppositely moving waves $\frac{C_n}{2}\sin(\omega_n t + k_n x)$ and $-\frac{C_n}{2}\sin(\omega_n t - k_n x)$. The amplitude changes with x and attains a maximum at an antinode where $\sin\frac{n\pi x}{l} = 1$ or, $\frac{n\pi x}{l} = \frac{\pi}{2}$ or, $x = \frac{l}{2n}$. The amplitude is zero at a node where $\sin\frac{n\pi x}{l} = 0$, or $\frac{n\pi x}{l} = \pi$ or, $x = \frac{l}{n}$.

Nodes always appear at $x = 0$ and $x = l$. In the nth mode of vibration, the string vibrates in n segments (or loops). The adjacent segments have opposite phase of vibration, i.e., when one segment goes up, the adjacent segment goes down.

Problem 11. *In the transverse arrangement of Melde's experiment a copper wire of diameter* 0.2 mm *resonates with a tuning fork of frequency* 64 Hz. *If the length of each loop is* 25 cm, *find the tension in* kg wt. *Given: density of copper* = 8.93×10^3 kg/m^3.

Solution. If the tension is M kg wt$= Mg$ newton, the velocity of the wave is $c = \sqrt{Mg/m} = \sqrt{Mg/(\pi d^2\rho/4)}$. Since the length of each loop is 25 cm, the wavelength is $\lambda = 2 \times 25 = 50$ cm $= 0.5$ m. So,

$$c = f\lambda = \sqrt{4Mg/\pi d^2\rho}, \quad \text{or, ;} M = \frac{\pi d^2 \rho f^2 \lambda^2}{4g}$$

$$= \frac{3.1416 \times (0.2 \times 10^{-3})^2 \times (8.93 \times 10^3) \times 64^2 \times 0.5^2}{4 \times 9.8} = 0.0293 \text{ kg wt.}$$

Problem 12. *In the transverse arrangement of Melde's experiment, the string vibrates in 3 loops when the tension is 0.32 kg wt. Determine the tension when the string vibrates in 2 loops in the longitudinal arrangement.*

Solution. In the transverse arrangement we have $f = \frac{n}{2l}\sqrt{\frac{T}{m}}$, where f is the frequency of the fork, l is the effective length of the string, n is the number of loops, T is the tension, and m is the mass per unit length of the string. Here $n = 3$, and $T = 0.32g$, where g is the local acceleration due to gravity. So,

$$f = \frac{3}{2l}\sqrt{\frac{0.32g}{m}}. \tag{i}$$

In the longitudinal arrangement we have $\frac{f}{2} = \frac{n}{2l}\sqrt{\frac{T}{m}}$. Here $n = 2$, so that

$$f = \frac{2}{l}\sqrt{\frac{T}{m}} \tag{ii}$$

From (i) and (ii), we get

$$\frac{3}{2l}\sqrt{\frac{0.32g}{m}} = \frac{2}{l}\sqrt{\frac{T}{m}}$$

whence $T = \frac{9}{16} \times 0.32g = 0.18$ kg wt.

Problem 13. *A string of length l is fixed at both ends under tension T. When plucked at the point $x = a$ to a transverse distance b (Fig. 6.5) and then released from rest, it performs small transverse oscillations.* (i) *Show that the initial potential energy of the string is $\frac{mc^2b^2l}{2a(l-a)}$.* (ii) *Show that the total energy at any instant of time is equal to the initial potential energy of the string.*

Solution. (i) At $t = 0$, the initial configuration of the string is given by Eq. (6.25). The potential energy of the string can be obtained by integrating Eq. (6.22) :

$$E_p = \int_0^l \frac{T}{2}\left(\frac{\partial y}{\partial x}\right)^2 dx$$

Using Eq. (6.25) we have for the initial potential energy

$$E_{po} = \frac{T}{2}\left[\int_0^a \left(\frac{b}{a}\right)^2 dx + \int_a^l \left(\frac{-b}{l-a}\right)^2 dx\right]$$

$$= \frac{mc^2}{2}\left[\frac{b^2}{a} + \frac{b^2}{l-a}\right] \qquad (\text{since } c = \sqrt{T/m})$$

$$= \frac{mc^2b^2l}{2a(l-a)}, \text{ proved.}$$

(ii) The kinetic energy of the string at any instant of time is

$$E_k = \frac{m}{2}\int_0^l \left(\frac{\partial y}{\partial t}\right)^2 dx.$$

So the total instantaneous energy of the string is

$$E = E_k + E_p = \frac{m}{2}\int_0^l \left(\frac{\partial y}{\partial t}\right)^2 dx + \frac{mc^2}{2}\int_0^l \left(\frac{\partial y}{\partial x}\right)^2 dx.$$

The displacement for the nth harmonic is given by Eq. (6.35). Hence the instantaneous deflection of the string is

$$y(x,t) = \sum_n y_n(x,t) = \sum_n A_n \sin\frac{n\pi x}{l}\cos\frac{n\pi ct}{l}, \tag{i}$$

where $A_n = \dfrac{2bl^2}{\pi^2 n^2 a(l-a)}\sin\dfrac{n\pi a}{l}$. So,

$$E_k = \frac{m}{2}\int_0^l \left(\frac{\partial y}{\partial t}\right)^2 dx = \sum_n \frac{n^2\pi^2c^2A_n^2 m}{4l}\sin^2\frac{n\pi ct}{l},$$

since the cross product terms do not contribute and $\displaystyle\int_0^l \sin^2\frac{n\pi x}{l}dx = \frac{l}{2}$.

Also,

$$E_p = \frac{mc^2}{2}\int_0^l \left(\frac{\partial y}{\partial x}\right)^2 dx = \sum_n \frac{n^2\pi^2c^2A_n^2 m}{4l}\cos^2\frac{n\pi ct}{l},$$

since again the cross product terms do not contribute and $\displaystyle\int_0^l \cos^2\frac{n\pi x}{l}dx = \frac{l}{2}$. Hence

$$E = E_k + E_p = \sum_n \frac{n^2\pi^2c^2A_n^2 m}{4l} = \frac{mc^2b^2l^3}{\pi^2a^2(l-a)^2}\sum_n \frac{1}{n^2}\sin^2\frac{n\pi a}{l}. \tag{ii}$$

Putting $t = 0$ and $x = a$ in Eq. (i), we obtain

$$y(a,0) = b = \sum A_n \sin\frac{n\pi a}{l} = \frac{2bl^2}{\pi^2 a(l-a)}\sum_n \frac{1}{n^2}\sin^2\frac{n\pi a}{l}$$

$$\text{or,}\quad \sum_n \frac{1}{n^2}\sin^2\frac{n\pi a}{l} = \frac{\pi^2 a(l-a)}{2l^2}.$$

With this result, Eq. (ii) gives

$$E = \frac{mc^2b^2l}{2a(l-a)} = E_{po},\quad \text{proved.}$$

REVIEW QUESTIONS

1. (a) Obtain the differential equation for the transverse vibration of a stretched string. **[cf. C.U. 2002, '04]**

   (b) Solve the equation by the method of separation of variables.

2. Show that the fundamental frequency of the transverse vibration of a uniform string of length l and mass per unit length m under a tension T is $\frac{1}{2l}\sqrt{\frac{T}{m}}$.

3. Define eigenfunctions, eigenvalues and eigenfrequencies for the transverse vibration of a stretched string.

4. Show that the energy of vibration of a stretched string for a particular mode varies as the square of the eigenfrequency and the square of the amplitude for that mode. **[cf. C.U. 1992]**

5. Show from energy considerations that a vibrating string can be considered to be an assembly of an infinite number of strongly coupled particle oscillators. Obtain the normal co-ordinate for the nth mode of vibration of the string.

6. Briefly discuss the formation of stationary waves for the transverse vibration of a string under tension, and fixed at the two ends. **[C.U. 1997, 2004]**

7. Describe Melde's experiment. How far can the laws of vibration of a string be verified in this experiment?

8. For a struck string, the displacement at a point x at time t can be represented by a superposition of harmonics, denoted by s, as follows:

$$y(x,t) = A \sum_{s=1,2,3,\cdots} \frac{1}{s} \sin\frac{s\pi a}{l} \sin\frac{s\pi x}{l} \sin\frac{s\pi ct}{l},$$

where l is the length of the string, 'a' is the point of striking, A is a constant, and c is the velocity of the transverse wave through the string.

(i) Find the nodal points of the sth harmonic. (ii) Show that if the point of striking coincides with any of the nodal points of the sth harmonic, it will be absent from the vibration. **[C.U. 1997]**

9. Obtain the amplitudes of the different frequencies of a uniformly stretched string with two ends fixed when it is struck over a small region at a distance h from one end. Assume that the region moves with the instantaneous velocity v at time $t = 0$. **[cf. C.U. 1999, 2008]**

10. A string of a length l is fixed at the two ends and is plucked at a distance 'a' from one end through a distance h. Find the expressions for the displacement and velocity at any point x on the string at time t.

State Young's law and show that the vibration of plucked string obeys the law.

What modes of vibration will be absent if the string is touched (i) at the plucked point and (ii) at one of the nodal points of the sth allowed mode of vibration? **[C.U. 1984, '91]**

11. Obtain an expression for the transverse displacement of a bowed string stretched between two fixed points at the ends. Show that the sum of the forward and backward velocity of the string is a constant at all points. What is a Wolf note?

12. Explain why in a stringed struck musical instrument, the string is struck at approximately (1/7)th of the length. **[cf. C.U. 2002]**

13. For stationary waves produced in a stretched string performing transverse vibrations, the displacement at a point x at time t is given by

$$y(x,t) = \sum_{s=1}^{\infty} \frac{A}{s} \sin\frac{s\pi a}{l} \sin\frac{s\pi x}{l} \sin\frac{s\pi ct}{l}$$

where l is the length of the string, $x = a$ is the point of excitation, $s = 1, 2, 3, \cdots$, A is a constant, and c is the velocity of the transverse wave along the string.

(i) Find the initial displacement at a point x.

(ii) Calculate the frequency and the wavelength of the fundamental tone. Represent this mode of vibration diagrammatically.

(iii) The string is excited at $x = l/3$ and touched at $x = l/4$. Calculate the harmonics present. **[cf. C.U. 2002]**

14. For stationary waves in a stretched transversely vibrating string, the displacement at a point x at a time t is

$$y(x,t) = \sum_{s=1}^{\infty} \frac{A}{s^2} \sin\frac{s\pi a}{l} \sin\frac{s\pi x}{l} \cos\frac{s\pi ct}{l}$$

where $x = a$ is the point of excitation and c is the velocity of propagation of the wave along the string. The other symbols are standard.

(i) Find the initial displacement of the fifth harmonic.

(ii) Calculate the frequency and the wavelength of the second harmonic ($a \neq l/2$).

(iii) The string is excited at $x = a = l/3$. Find the first few harmonics that are present.

(iv) What happens if the string after excitation be touched at the same point $x = a$? **[C.U. 2004]**

15. (a) Mention two dissimilarities and one similarity between the vibrations of a plucked string and of a struck string.

(b) Give examples of plucked and struck stringed instruments.

16. (a) Compare the intensities of the fundamental and the third harmonic if a stretched string fixed at the two ends, is plucked at the midpoint.

(b) What happens if a string is plucked over a finite length instead of being plucked at a single point?

(c) What will be the effect on the emitted note from a plucked string, if (i) the bridges supporting the string at the ends yield, and (ii) the string is made thicker?

17. For a stretched string of length l, fixed rigidly at the two ends, the transverse displacement at a point x at time t, is

$$y(x,t) = \sum_{n=1}^{\infty} \sin\frac{n\pi x}{l}\left(a_n \cos\frac{n\pi ct}{l} + b_n \sin\frac{n\pi ct}{l}\right)$$

with usual notations. For a plucked string, obtain the expressions for a_n and b_n.

If the above string is plucked at $l/3$ and touched at $l/4$, what will happen? **[C.U. 2009]**

PROBLEMS

1. A uniform flexible string of length L is stretched between its ends. The string is plucked with an amplitude h at a distance $l/3$ from one end and then released. Investigate the subsequent motion of the string.

   [**Ans.** $y(x,t) = \frac{9h}{\pi^2}\sum_{n=1}^{\infty} \frac{1}{n^2} \sin\frac{n\pi}{3} \sin\frac{n\pi x}{l} \cos\frac{n\pi ct}{l}$] **[Burd. U. 1994]**

2. The initial displacement of a stretched string of length L connected to two fixed supports at the ends is zero. An initial impulse is given to the string so that its velocity is

$$\begin{aligned} v_0(x) &= \alpha x, & 0 < x < \frac{L}{4} \\ &= \alpha\left(\frac{L}{2} - x\right), & \frac{L}{4} < x < \frac{L}{2} \\ &= 0 & \frac{L}{2} < x < L, \end{aligned}$$

   where α is a constant. Find the harmonics that are absent from the vibration of the string.

   [**Ans.** 4th, 8th, 12th etc. harmonics are absent]

3. In the transverse arrangement of Melde's experiment, 4m length of the wire vibrates in eight segments for a load of 0.3 kg. If the mass of this length of the wire is 1.5×10^{-3} kg, determine the frequency of the fork.

   [**Ans.** 88.5 Hz]

4. In Melde's experiment, the string vibrates in 5 loops when the tension is T. If T is increased by 0.05 kg, the number of loops becomes 4. Calculate T.

   [**Ans.** 0.089 kg]

5. A string of length l and of mass per unit length m is under a tension T. The string is struck at $l/5$ and released. What harmonics are excited? Sketch the fundamental and the second harmonic.

   Calculate the frequency of the fundamental tone, given $l = 50$ cm, $m = 0.0092$ g/cm, $T = 6$ kg wt., $g = 9.8$ m/s^2. **[C.U. 2006]**

   [**Ans.** 1st, 2nd, 3rd, 4th, 6th, 7th, 8th, 9th, 11th, etc; 252.8 Hz]

6. A string of length 50 cm and mass 0.5 g is stretched by 9 kg weight. Calculate the velocity of transverse vibration of the string. **[C.U. 2009]**

   [**Ans.** 297 m/s]

Chapter 7

Vibrations of Bars, Membranes, and Air Columns

7.1 Introduction

The vibrations of bars, membranes, and air columns produce stationary waves due to the superposition of oppositely moving waves which are reflected from the boundaries. Both longitudinal and transverse waves can be produced in bars whereas only longitudinal waves are possible in air columns. The vibrations of membranes and plates are transverse. Stretched membranes are the two-dimensional analogues of stretched strings.

7.2 Longitudinal Vibration of a Bar

The propagation of longitudinal waves in a solid bar has been discussed already in Sec. 5.10. The one dimensional differential wave equation was found to be

$$\frac{\partial^2 \xi}{\partial t^2} = c^2 \frac{\partial^2 \xi}{\partial x^2}, \tag{7.1}$$

where ξ is the displacement at the point x at time t, and

$$c = \sqrt{\frac{Y}{\rho}} \tag{7.2}$$

is the wave velocity. Here Y is Young's modulus of the material of the bar of density ρ. As in the case of a stretched string, the wave equation for the bar can be solved by the method of separation of variables. Thus writing $\xi = F_1(x)F_2(t)$, and denoting the separation constant by $-\omega^2$, we obtain as in Sec. 6.2,

$$F_1 = D_1 \cos\frac{\omega x}{c} + D_2 \sin\frac{\omega x}{c} \tag{7.3}$$

$$\text{and} \quad F_2 = C_1 \cos\omega t + C_2 \sin\omega t, \tag{7.4}$$

where D_1, D_2 and C_1, C_2 are arbitrary constants determined by the boundary and initial conditions. Hence we can write

$$\xi = (C_1 \cos\omega t + C_2 \sin\omega t)\left(D_1 \cos\frac{\omega x}{c} + D_2 \sin\frac{\omega x}{c}\right). \tag{7.5}$$

Case 1. Bar free to move at both ends

We consider that the bar of length L rests lightly on two knife edges so that both ends of the bar at $x = 0$ and $x = L$ can move freely. Here the stress $F/\alpha = -Y\frac{\partial\xi}{\partial x}$ or the strain is zero at each end. So the boundary conditions are:

(i) At $x = 0$, $\frac{\partial\xi}{\partial x} = 0$ for all values of t.

(ii) At $x = L$, $\frac{\partial\xi}{\partial x} = 0$ for all values of time t.

Equation (7.5) gives

$$\frac{\partial\xi}{\partial x} = (C_1 \cos\omega t + C_2 \sin\omega t)\left(-\frac{D_1\omega}{c}\sin\frac{\omega x}{c} + \frac{D_2\omega}{c}\cos\frac{\omega x}{c}\right). \tag{7.6}$$

Using the boundary condition (i) in Eq. (7.6) we obtain $D_2 = 0$. Hence

$$\frac{\partial\xi}{\partial x} = (C_1 \cos\omega t + C_2 \sin\omega t)\left(-\frac{D_1\omega}{c}\sin\frac{\omega x}{c}\right). \tag{7.7}$$

Using the boundary condition (ii) in Eq. (7.7) gives

$$\sin\frac{\omega L}{c} = 0$$

$$\text{or,} \quad \frac{\omega L}{c} = n\pi \quad \text{(where } n \text{ is an integer).} \tag{7.8}$$

Each value of n gives a solution ξ_n:

$$\begin{aligned} \xi_n &= (C_{1n} \cos\omega_n t + C_{2n} \sin\omega_n t) D_{1n} \cos\frac{n\pi x}{L} \\ &= \cos\frac{n\pi x}{L}(a_n \cos\omega_n t + b_n \sin\omega_n t) \end{aligned} \tag{7.9}$$

where a_n and b_n are new constants. As the wave equation is linear, the general solution is obtained by the linear superposition of the ξ_n's. Thus

$$\xi = \sum_{n=1}^{\infty} \cos\frac{n\pi x}{L}(a_n \cos\omega_n t + b_n \sin\omega_n t). \tag{7.10}$$

Equation (7.8) gives the *characteristic frequencies*

$$\omega_n = \frac{n\pi c}{L} \tag{7.11}$$

$$\text{or,} \quad f_n = \frac{\omega_n}{2\pi} = \frac{nc}{2L} = \frac{n}{2L}\sqrt{\frac{Y}{\rho}}. \tag{7.12}$$

The wavelength for the nth mode is

$$\lambda_n = \frac{c}{f_n} = \frac{2L}{n}. \tag{7.13}$$

Here *all the harmonics are present* and the vibrational modes are similar to those for a stretched string. The fundamental frequency and wave length are obtained from Eqs. (7.12) and (7.13) respectively by putting $n = 1$.

Note that, if the bar which is free at both ends, is clamped at the middle (as in the case of *Kundt's tube*), then $\xi = 0$ at $x = L/2$. Equation (7.10) then gives $\cos\frac{n\pi x}{L} = 0$ at $x = L/2$. So, $n = 1, 3, 5, \cdots$, i.e., only the *odd* harmonics are present. the frequencies being given by Eq. (7.12).

Case 2. Bar fixed at one end and free to move at the other end

Here the boundary conditions are

(i) at $x = 0$, $\xi = 0$ for all values of t, the bar being rigidly fixed at the end $x = 0$,

(ii) at $x = L$, $\frac{\partial\xi}{\partial x} = 0$, for all values of t, the bar being free at the end $x = L$.

Using the boundary condition (i) in Eq. (7.5), we get $D_1 = 0$. Hence

$$\xi = (C_1 \cos\omega t + C_2 \sin\omega t) D_2 \sin\frac{\omega x}{c}$$

$$\text{and} \quad \frac{\partial\xi}{\partial x} = (C_1 \cos\omega t + C_2 \sin\omega t)\frac{D_2\omega}{c}\cos\frac{\omega x}{c}. \tag{7.14}$$

Using the boundary condition (ii) in Eq. (7.14), we obtain

$$\cos\frac{\omega L}{c} = 0$$

$$\text{or,} \quad \frac{\omega L}{c} = (2n-1)\frac{\pi}{2} \quad (n = 1, 2, 3, \cdots). \tag{7.15}$$

Thus the characteristic frequencies are

$$\omega_n = (2n-1)\frac{\pi c}{2L}$$

$$\text{or,} \quad f_n = \frac{\omega_n}{2\pi} = (2n-1)\frac{c}{4L} = \frac{2n-1}{4L}\sqrt{\frac{Y}{\rho}}. \tag{7.16}$$

The fundamental frequency ($n = 1$) is $f_1 = \frac{c}{4L}$, and the corresponding wavelength is $\lambda_1 = 4L$. The next higher harmonic ($n = 2$) has the frequency $3c/4L = 3f_1$. Clearly, only the *odd harmonics* are present. The general solution for ξ is

$$\xi = \sum_{n=1}^{\infty} \sin \frac{(2n-1)\pi x}{2L}(a_n \cos \omega_n t + b_n \sin \omega_n t). \tag{7.17}$$

Case 3. Bar fixed at both ends

This situation is similar to that of a stretched string fixed at both ends. The boundary conditions are the same as in the case of strings, viz., $\xi = 0$ at $x = 0$ and at $x = L$. Thus we can proceed along the lines developed for strings to obtain the solution, given below:

$$\xi = \sum_{n=1}^{\infty} \sin \frac{n\pi x}{L}\left(a_n \cos \frac{n\pi ct}{L} + b_n \sin \frac{n\pi ct}{L}\right).$$

Here *all the harmonics* will be present.

7.3 Vibrating Bar as a Stationary Wave

The wave equation for a bar gives a plane wave solution. These plane waves move along the bar and are reflected from the ends. If the reflection is total, the incident and the reflected waves, on superposition, produce antinodes where the displacement is a maximum and nodes where the displacement is zero. If the end of the bar is free, an antinode occurs there, while if the end is fixed, a node appears there. If the reflection is partial, the displacement at the nodes will be a minimum but not zero.

The reflection is total when both ends of the bar are either free or fixed. The general solution of the wave equation [Eq. (7.1)] has the form $\xi = f_1(ct-x)+f_2(ct+x)$, where f_1 and f_2 represent plane waves propagating in the positive and the negative x-axis, respectively. In the simplest case, f_1 and f_2 are simple harmonic with amplitudes ξ_{10} and ξ_{20}, respectively. Hence we write

$$\xi = \xi_{10} \sin(\omega t - kx) + \xi_{20} \sin(\omega t + kx). \tag{7.18}$$

When the ends of the bar are rigidly fixed, the boundary conditions are (i) at $x = 0$, $\xi = 0$ for all values of t, and (ii) at $x = L$, $\xi = 0$ for all values of t. The first condition gives $\xi_{10} + \xi_{20} = 0$, whence $\xi_{10} = -\xi_{20} = -\frac{C}{2}$, say. The second boundary condition gives

$$-\frac{C}{2}\left[\sin(\omega t - kL) - \sin(\omega t + kL)\right] = 0$$

$$\text{or,} \quad C \cos \omega t \sin kL = 0 \quad \text{or,} \quad \sin kL = 0$$

$$\text{or,} \quad kL = n\pi, \quad \text{(where } n \text{ is an integer).} \tag{7.19}$$

The angular frequency ω is given by

$$\omega = kc = \frac{n\pi c}{L}. \tag{7.20}$$

Note that both k and ω depend on n. The constant C is also generally determined by n. So, the solution for a particular integral value of n can be written as

$$\begin{aligned} \xi_n &= \frac{C_n}{2}\left[\sin(\omega_n t + k_n x) - \sin(\omega_n t - k_n x)\right] \\ &= C_n \cos\omega_n t \sin k_n x. \end{aligned} \tag{7.21}$$

The above analysis applies exactly to strings as well.

The amplitude changes with x and attains a maximum at an antinode where $\sin k_n x = 1$, or $\frac{n\pi x}{L} = \frac{\pi}{2}$ or, $x = \frac{L}{2n}$. The amplitude is a minimum at a node where $\sin k_n x = 0$ or, $\frac{n\pi x}{L} = \pi$ or, $x = \frac{L}{n}$. Clearly, in the nth mode the bar vibrates in n segments. The neighbouring segments have opposite phase of vibration.

When one end of the bar is fixed and the other end is free, the above analysis has to be modified. Let the end $x = 0$ be fixed and the end $x = L$ be free. The forward wave is represented by $\xi_1 = \xi_{10}\sin(\omega t - kx)$ and the backward reflected wave by $\xi_2 = \xi_{20}\sin(\omega t + kx)$. At $x = 0$, we have $\xi_{10} = -\xi_{20} = -\frac{C}{2}$, say. Hence

$$\begin{aligned} \xi &= \xi_1 + \xi_2 = \frac{C}{2}\left[\sin(\omega t + kx) - \sin(\omega t - kx)\right] \\ &= C \sin kx \cos\omega t. \end{aligned} \tag{7.22}$$

Therefore,

$$\frac{\partial \xi}{\partial x} = Ck \cos kx \cos\omega t. \tag{7.23}$$

At the free end $x = L$, the stress $-Y\frac{\partial\xi}{\partial x}$ vanishes for all t. Hence

$$\cos kL = 0$$

$$\text{or,} \quad k = \frac{(2n-1)\pi}{2L}, \tag{7.24}$$

$$\text{and} \quad \omega = kc = \frac{(2n-1)\pi c}{2L}. \tag{7.25}$$

k, ω and also C, in general, depend on n. Hence the solution for a particular n is

$$\xi_n = C_n \sin\frac{(2n-1)\pi x}{2L}\cos\frac{(2n-1)\pi ct}{2L}. \tag{7.26}$$

The general solution is formed by the superposition of the solutions for different values of n. So,

$$\xi = \sum_{n=1}^{\infty} C_n \sin \frac{(2n-1)\pi x}{2L} \cos \frac{(2n-1)\pi ct}{2L}. \tag{7.27}$$

The characteristic frequencies are

$$f_n = \frac{\omega_n}{2\pi} = \frac{(2n-1)c}{4L},$$

agreeing with Eq. (7.16).

7.4 Transverse Vibration of Rods and Bars

When thin rods and bars execute transverse vibrations, the strains are elongation and contraction on the two sides of the neutral axis. The forces coming into play are thus primarily those appearing in the bending of a beam. When a force F is applied perpendicular to the neutral axis to bend a bar, the bending moment M is

$$M = YAK^2 \frac{\partial^2 y}{\partial x^2}, \tag{7.28}$$

where Y is Young's modulus of the material of the bar, A is the cross sectional area of the bar, and K is the radius of gyration of the cross sectional area about the neutral plane. The x-axis is taken along the neutral axis, and y is the depression at the point x. Furthermore, we have

$$F = -\frac{\partial M}{\partial x}. \tag{7.29}$$

Let PQ be a thin slab of the beam of width δx (Fig. 7.1). The faces of the slab are perpendicular to the neutral axis. If the force at P is F, that at Q is $F + \frac{\partial F}{\partial x}\delta x$. Thus, the unbalanced force is $\frac{\partial F}{\partial x}\delta x$, and this must equal the product of the mass and the acceleration of the slab PQ. Therefore,

$$\frac{\partial F}{\partial x}\delta x = (A\rho\delta x)\frac{\partial^2 y}{\partial t^2}, \tag{7.30}$$

where ρ is the material density.

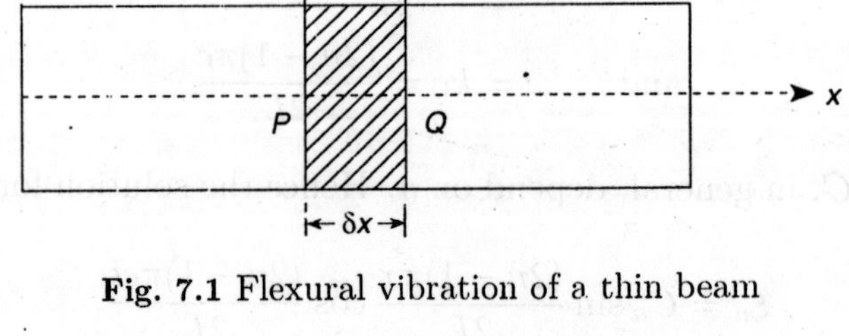

Fig. 7.1 Flexural vibration of a thin beam

Equation (7.29) gives

$$\frac{\partial F}{\partial x} = -\frac{\partial^2 M}{\partial x^2} = -YAK^2\frac{\partial^4 y}{\partial x^4}, \tag{7.31}$$

on using Eq. (7.28). Substituting for $\frac{\partial F}{\partial x}$ from Eq. (7.31) into Eq. (7.30) yields

$$-YAK^2\frac{\partial^4 y}{\partial x^4} = A\rho\frac{\partial^2 y}{\partial t^2}$$

$$\text{or,}\quad \frac{\partial^4 y}{\partial x^4} + \left(\frac{\rho}{YK^2}\right)\frac{\partial^2 y}{\partial t^2} = 0. \tag{7.32}$$

Equation (7.32) gives the differential wave equation for the flexural waves in a thin rod or bar.

Seeking plane wave solutions of Eq. (7.32) we write

$$y = ae^{j(\omega t - kx)} \tag{7.33}$$

where a is a constant. Then

$$\frac{\partial^4 y}{\partial x^4} = k^4 y \quad \text{and} \quad \frac{\partial^2 y}{\partial t^2} = -\omega^2 y.$$

Putting these values in Eq. (7.32) gives

$$k^4 y - \left(\frac{\rho}{YK^2}\right)\omega^2 y = 0. \tag{7.34}$$

Since the wave velocity c is given by $c = \omega/k$, we have

$$k = \frac{\omega}{c}. \tag{7.35}$$

Using Eq. (7.35) in Eq. (7.34) we obtain

$$c^4 = \omega^2\frac{YK^2}{\rho} \quad \text{or,} \quad c = \sqrt{\omega c_0 K}, \tag{7.36}$$

where $c_0 (= \sqrt{Y/\rho})$ is the velocity of longitudinal waves in the beam. Since c, given by Eq. (7.36), is a function of the angular frequency ω, the medium is *dispersive*. If a number of frequencies are present, the velocity of propagation is given by the group velocity

$$c_g = \frac{\partial \omega}{\partial k} = 2c_0 K\frac{\omega}{c} \tag{7.37}$$

7.5 Simple Harmonic Transverse Vibrations of Bars

Suppose that the bar vibrates simple harmonically with an amplitude $\phi(x)$, where ϕ is a function of x. Then $y = \phi(x)\cos\omega t$. Substituting for y in the wave equation [Eq. (7.32)] gives

$$\frac{d^4\phi(x)}{dx^4} = \frac{\rho\omega^2}{YK^2}\phi(x) = \alpha^4\phi(x), \tag{7.38}$$

$$\text{where} \quad \alpha^4 = \frac{\rho\omega^2}{YK^2}. \tag{7.39}$$

Let $\phi(x) = Ae^{mx}$, then Eq. (7.38) gives $m^4 = \alpha^4$, so that $m = \alpha, -\alpha, j\alpha, -j\alpha$. Hence the solution to the wave equation is

$$y = (A_1e^{\alpha x} + A_2e^{-\alpha x} + A_3e^{j\alpha x} + A_4e^{-j\alpha x})\cos\omega t. \tag{7.40}$$

Noting that $e^{\pm\alpha x} = \cosh\alpha x \pm \sinh\alpha x$ and $e^{\pm j\alpha x} = \cos\alpha x \pm j\sin\alpha x$ we have on taking the real part of the complex solution

$$y = (a\cosh\alpha x + b\sinh\alpha x + c\cos\alpha x + d\sin\alpha x)\cos\omega t, \tag{7.41}$$

where $a, b, c,$ and d are arbitrary constants to be determined from the boundary conditions.

Case 1. Bar fixed at one end and free at the other

Let the bar be clamped at the end $x = 0$ and free at the end $x = L$. The boundary conditions are (i) at $x = 0$, $y = 0$ and $\frac{\partial y}{\partial x} = 0$; (ii) at $x = L$, the bending moment and the shearing force vanish because there is no bar beyond the free end to produce a moment or a shearing stress. The vanishing bending moment gives $\frac{\partial^2 y}{\partial x^2} = 0$ and the vanishing shearing stress gives $\frac{\partial^3 y}{\partial x^3} = 0$. The last equation follows from the fact that at the free end $F = -\frac{\partial M}{\partial x} = 0$, where M is given by Eq. (7.28). The conditions in (i) give $a + c = 0$ and $b + d = 0$. Hence Eq. (7.41) reduces to

$$y = (a\cosh\alpha x - a\cos\alpha x + b\sinh\alpha x - b\sin\alpha x)\cos\omega t.$$

Hence

$$\left.\frac{\partial^2 y}{\partial x^2}\right|_{x=L} = \alpha^2\left[a\cosh\alpha L + a\cos\alpha L + b\sinh\alpha L + b\sin\alpha L\right]\cos\omega t = 0,$$

so that

$$a(\cosh\alpha L + \cos\alpha L) + b(\sinh\alpha L + \sin\alpha L) = 0. \tag{7.42}$$

Similarly, $\left.\frac{\partial^3 y}{\partial x^3}\right|_{x=L} = 0$ yields

$$a(\sinh\alpha L - \sin\alpha L) + b(\cosh\alpha L + \cos\alpha L) = 0. \tag{7.43}$$

For nonvanishing a and b, the determinant of the Eqs. (7.42) and (7.43) vanish. Thus

$$\begin{vmatrix} \cosh \alpha L + \cos \alpha L & \sinh \alpha L + \sin \alpha L \\ \sinh \alpha L - \sin \alpha L & \cosh \alpha L + \cos \alpha L \end{vmatrix} = 0.$$

Simplifying we obtain

$$\sec \alpha L = -\cosh \alpha L. \tag{7.44}$$

The solution of the transcendental Eq. (7.44) is obtained by plotting the two sides of the equation against αL and locating the point of intersection of the two curves. The solutions are $\alpha L = 1.88, 4.69, 7.85$ etc.
Equation (7.39) gives $\omega = \alpha^2 K\sqrt{Y/\rho}$. Hence the frequency is

$$f = \frac{\omega}{2\pi} = \frac{\alpha^2 K}{2\pi}\sqrt{\frac{Y}{\rho}}. \tag{7.45}$$

Thus, the frequency of the fundamental is

$$f_1 = \frac{K}{2\pi}\sqrt{\frac{Y}{\rho}} \times \frac{(1.88)^2}{L^2}. \tag{7.46}$$

The frequencies of the overtones are

$$f_2 = \frac{K}{2\pi}\sqrt{\frac{Y}{\rho}}\frac{(4.69)^2}{L^2} \tag{7.47}$$

$$f_3 = \frac{K}{2\pi}\sqrt{\frac{Y}{\rho}}\frac{(7.85)^2}{L^2}, \tag{7.48}$$

and so on. Thus, $f_1 : f_2 : f_3 = 1 : 6.27 \cdot 17.55$. This shows that the partials do not form a harmonic series unlike the transverse vibration of strings. Therefore, the sound from the vibrating bar is *nonmelodious.*

Case 2. Bar free at both ends

Here both $\frac{\partial^2 y}{\partial x^2}$ and $\frac{\partial^3 y}{\partial x^3}$ are zero at $x = 0$ and $x = L$, which are the free ends. We obtain from Eq. (7.41)

$$\frac{\partial^2 y}{\partial x^2} = \alpha^2\left[a\cosh \alpha x + b\sinh \alpha x - a\cos \alpha x - d\sin \alpha x\right]\cos \omega t$$

$$\text{and}\quad \frac{\partial^3 y}{dx^3} = \alpha^3\left[a\sinh \alpha x + b\cosh \alpha x + c\sin \alpha x - d\cos \alpha x\right]\cos \omega t.$$

The condition $\frac{\partial^2 y}{\partial x^2} = 0$ at $x = 0$ gives $a = c$, and the condition $\frac{\partial^3 y}{\partial x^3} = 0$ at $x = 0$ gives $b = d$. From the conditions $\frac{\partial^2 y}{\partial x^2} = 0$ and $\frac{\partial^3 y}{\partial x^3} = 0$ at $x = L$ we obtain $a(\cosh \alpha L - \cos \alpha L) = b(\sin \alpha L - \sinh \alpha L)$ and $a(\sinh \alpha L + \sin \alpha L) =$

$-b(\cosh \alpha L - \cos \alpha L)$. The nonzero a and b demands that the determinant of these equations is zero. This leads to the equation

$$\sec \alpha L = \cosh \alpha L. \tag{7.49}$$

The solution of this transcendental equation gives $\alpha L = 4.73,\ 7.85,\ 10.996$ etc. So the fundamental frequency is

$$f_1 = \frac{K}{2\pi}\sqrt{\frac{Y}{\rho}} \times \frac{(4.73)^2}{L^2}. \tag{7.50}$$

The overtones are

$$f_2 = \frac{K}{2\pi}\sqrt{\frac{Y}{\rho}}\frac{(7.85)^2}{L^2} \tag{7.51}$$

$$f_3 = \frac{K}{2\pi}\sqrt{\frac{Y}{\rho}}\frac{(10.996)^2}{L^2} \tag{7.52}$$

etc. So $f_1 : f_2 : f_3 = 1 : 2.76 : 5.40$. Again the partials *do not* give a harmonic series.

7.6 Tuning Fork

When a bar, free at both ends, sounding its fundamental, is gradually bent into the form of a U, it is found that the two nodes approach the bend (Fig. 7.2). This enhances the length of each outer portion of the vibrator,

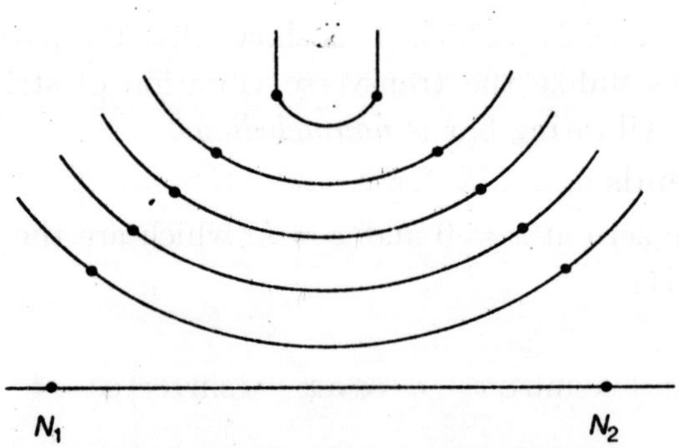

Fig. 7.2 Formation of a tuning fork

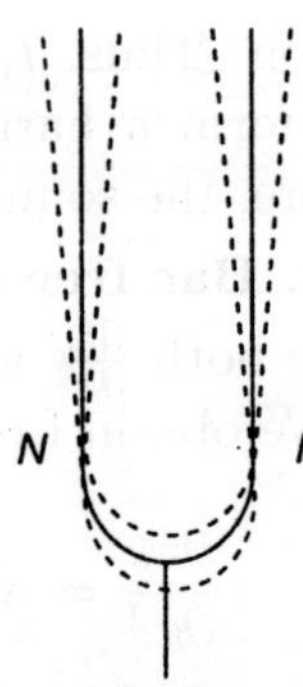

Fig. 7.3 A vibrating tuning fork

thus decreasing the frequency. The amplitude of vibration at the centre is also reduced. When the arms are parallel, we obtain a tuning fork. The original vibrational mode being the same, the two arms of the fork move inwards or outwards simultaneously. When a stem is attached at the centre, the amplitude of vibration at the antinode produced there is reduced further, and the nodes approach still closer. The outward movement of the arms raises the stem whereas their inward movement lowers the stem (Fig. 7.3). Consequently, the transverse vibration of the prongs produces the longitudinal vibration of the stem. If the stem is pressed on a surface, the vibrations of the stem cause forced transverse vibration of the surface.

A fork, when struck mildly, emits the fundamental tone only. Each prong behaves approximately as a straight bar fixed at one end and free at the other end. The frequency of the fundamental is given by Eq. (7.46). If the fork is struck vigorously, overtones whose frequencies are nearly 6 and 17 times the fundamental frequency are produced. But they are very feeble and die out quickly. Thus only the pure fundamental tone persists for a long time. For a rectangular bar, the radius of gyration about the neutral surface is $K = d/\sqrt{12}$, where d is the thickness of the prong. For steel, $\sqrt{Y/\rho} = 5.24 \times 10^5$ cm/s. Hence Eq. (7.46) gives

$$\begin{aligned} f_1 &= \frac{d}{2\pi\sqrt{12}} \times 5.24 \times 10^5 \times \frac{(1.88)^2}{L^2} \\ &= 8.46 \times 10^4 \frac{d}{L^2} \text{ Hz}, \end{aligned} \tag{7.53}$$

where d and L are in cm. Clearly, a fork with thick and short prongs will give out a high-frequency tone. On the contrary, a fork having thin and long prongs will emit a low-frequency tone.

When the temperature increases, the length of the prong increases whereas the density and Young's modulus decrease. The change of Y with temperature primarily determines the variation of frequency of the tuning fork with temperature, so that the frequency decreases slowly with increasing temperature. Good tuning forks are made of elinvar (a nickel steel alloy) for which the frequency is quite stable with respect to temperature variations.

The vibration of a tuning fork can be maintained over any desired period of time by feeding energy into it from an electromagnet driven by a battery (for low frequencies $\sim$100 Hz) or by an electronic oscillator (for high frequencies $\sim$1 kHz).

Tuning bar : At frequencies above 4 kHz and below 20 kHz, *tuning bars* made of a light alloy are often used instead of tuning forks. Such bars are clamped at the middle and struck on the end with a rubber hammer. They then execute longitudinal vibrations over a much longer time than high-frequency forks.

7.7 Vibration of Membranes

A membrane is a thin stretched skin-like material extended in two dimensions. The vibrations of a membrane find practical applications in musical instruments like tabla, mridanga, drum, etc. To derive the equation of motion of a vibratory membrane, we consider a stretched membrane of uniform density and thickness. It lies in the xy-plane when undisturbed. The tension is taken to be constant for a small displacement in the z-direction.

Figure 7.4 shows an element of a stretched vibratory membrane. Since the transverse displacement is different at different points on the membrane, at any instant of time the element of the membrane $ABCD$ is curved. $A'B'C'D'$ is the displaced position of the membrane when the displacement along the z-direction is dz. In the configuration $ABCD$, the inside pressure exceeds the outside pressure. So, for the displacement dz, the work done by the excess pressure p is

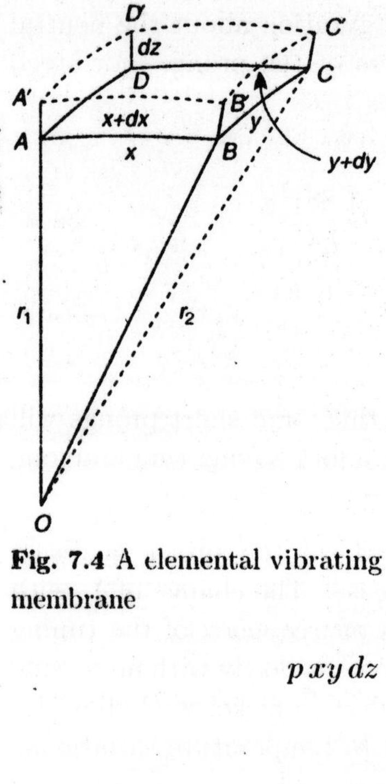

Fig. 7.4 A elemental vibrating membrane

$$dW = p\,xy\,dz,$$

where x and y are respectively the lengths of the sides AB and BC. In the displaced configuration $A'B'C'D'$, the area increases by $d(xy)$, so that the increase in potential energy due to stretching against the tension T acting tangentially per unit length of the membrane, is $Td(xy)$. Since the work done by the excess pressure equals the increase in potential energy due to stretching, we have

$$\begin{aligned} p\,xy\,dz &= Td(xy) = T(x\,dy + y\,dx) \\ &= T\,xy\left(\frac{dy}{y} + \frac{dx}{x}\right) \end{aligned}$$

$$\text{or,}\quad p\,dz = T\left(\frac{dy}{y} + \frac{dx}{x}\right). \tag{7.54}$$

Let r_1 and r_2 be the radii of curvatures of the arcs AB and BC, respectively. Then, from Fig. 7.4 we get

$$\frac{x + dx}{r_1 + dz} = \frac{x}{r_1} = \frac{(x + dx) - x}{(r_1 + dz) - r_1} = \frac{dx}{dz}$$

$$\text{or,}\quad dx = \frac{x}{r_1}dz.$$

Similarly,

$$dy = \frac{y}{r_2} dz.$$

Substituting the above values of dx and dy into Eq. (7.54) gives

$$p = T\left(\frac{1}{r_1} + \frac{1}{r_2}\right). \tag{7.55}$$

If ρ is the mass per unit area of the membrane, we have $p =$ mass per unit area $\times$ acceleration $= \rho\frac{\partial^2 z}{\partial t^2}$. Also, $\frac{1}{r_1} = \frac{\partial^2 z}{\partial x^2}$ and $\frac{1}{r_2} = \frac{\partial^2 z}{\partial y^2}$. Substituting these values in Eq. (7.55) we obtain

$$\frac{\partial^2 z}{\partial t^2} = \frac{T}{\rho}\left(\frac{\partial^2 z}{\partial x^2} + \frac{\partial^2 z}{\partial y^2}\right)$$

$$\text{or,} \quad \frac{\partial^2 z}{\partial t^2} = c^2\left(\frac{\partial^2 z}{\partial x^2} + \frac{\partial^2 z}{\partial y^2}\right), \tag{7.56}$$

where

$$c = \sqrt{\frac{T}{\rho}}. \tag{7.57}$$

Equation (7.56) is the differential equation for the motion of the membrane. We solve this equation by the method of separation of variables. Let the solution

$$z = X(x)Y(y)T(t), \tag{7.58}$$

where X, Y, T are respectively the functions of x, y and t alone. Putting Eq. (7.58) into Eq. (7.56), we get

$$\frac{1}{X}\frac{d^2X}{dx^2} + \frac{1}{Y}\frac{d^2Y}{dy^2} = \frac{1}{c^2T}\frac{d^2T}{dt^2}. \tag{7.59}$$

The left-hand side of Eq. (7.59) is a function of x and y, whereas the right-hand side is a function of t. So, each side is equal to a constant, say $-\omega^2/c^2$, which is independent of x, y, and t. The constant $-\omega^2/c^2$ is called the *separation constant.* Then we obtain

$$\frac{1}{X}\frac{d^2X}{dx^2} + \frac{1}{Y}\frac{d^2Y}{dy^2} = -\frac{\omega^2}{c^2} \tag{7.60}$$

$$\text{and} \quad \frac{d^2T}{dt^2} = -\omega^2 T. \tag{7.61}$$

$$T = A\sin(\omega t + \alpha). \tag{7.62}$$

Multiplying Eq. (7.60) through by XYT gives

$$\frac{\partial^2 z}{\partial x^2} + \frac{\partial^2 z}{\partial y^2} + \frac{\omega^2}{c^2}z = 0, \tag{7.63}$$

which is referred to as *Helmholtz equation.*

Equation (7.60) is rewritten as

$$\frac{1}{X}\frac{d^2X}{dx^2}+\frac{\omega^2}{c^2}=-\frac{1}{Y}\frac{d^2Y}{dy^2}=\phi^2, \tag{7.64}$$

where ϕ^2 is another separation constant independent of x, y. Then

$$\frac{d^2Y}{dy^2}+\phi^2 Y=0 \tag{7.65}$$

$$\text{and}\qquad \frac{d^2X}{dx^2}+\left(\frac{\omega^2}{c^2}-\phi^2\right)X=0. \tag{7.66}$$

Equations (7.65) and (7.66) assume the solutions

$$Y = B\sin(\phi y+\beta) \tag{7.67}$$

$$\text{and}\quad X = C\sin\left[\sqrt{\frac{\omega^2}{c^2}-\phi^2}\; x+\gamma\right], \tag{7.68}$$

respectively.

7.8 Rectangular Membrane

Consider a rectangular membrane whose sides along x and y have the lengths a and b, respectively. Then the boundary conditions are: $z = 0$ at $x = 0$ or $x = a$, and $y = 0$ or $y = b$, for all values of time t. The solution is $z = XYT$, where X, Y, and T are given by Eqs. (7.68), (7.67) and (7.62), respectively.

Since $z = 0$ at $x = 0$ and $x = a$ for all values of y and t, we have $X = 0$ at $x = 0$ and $x = a$. Hence Eq. (7.68) gives $C\sin\gamma = 0$ and $C\sin\left[\sqrt{\frac{\omega^2}{c^2}-\phi^2}\,a+\gamma\right]=0$. So, $\gamma = 0$ and $\sqrt{\frac{\omega^2}{c^2}-\phi^2}\,a = n\pi$, where n is an integer. For a particular value of n, the solution is

$$X_n = C_n \sin\frac{n\pi x}{a}. \tag{7.69}$$

The general solution is obtained by linear superposition of the solutions for different values of n. Thus

$$X=\sum_{n=1}^{\infty} C_n \sin\frac{n\pi x}{a}. \tag{7.70}$$

Again the conditions $z = 0$ at $y = 0$ and $y = b$ for all values of x and t imply that $Y = 0$ at $y = 0$ and $y = b$. Then Eq. (7.67) gives $B\sin\beta = 0$ and $B\sin(\phi b+\beta) = 0$, so that $\beta = 0$ and $\phi b = m\pi$, where m is an integer. The solution for a particular m is

$$Y_m = B_m \sin\frac{m\pi y}{b} \tag{7.71}$$

and the general solution is

$$Y = \sum_{m=1}^{\infty} B_m \sin \frac{m\pi y}{b}. \tag{7.72}$$

The complete solution for the displacement z is

$$z = \sum_{n=1}^{\infty} \sum_{m=1}^{\infty} AB_m C_n \sin \frac{n\pi x}{a} \sin \frac{m\pi y}{b} \sin(\omega_{nm} t + \alpha), \tag{7.73}$$

where

$$\omega_{nm}^2 = c^2 \left(\frac{n^2\pi^2}{a^2} + \frac{m^2\pi^2}{b^2} \right) = \pi^2 c^2 \left(\frac{n^2}{a^2} + \frac{m^2}{b^2} \right). \tag{7.74}$$

Equation (7.73) can be put in the form

$$z = \sum_{n=1}^{\infty} \sum_{m=1}^{\infty} (A_{nm} \cos \omega_{nm} t + B_{nm} \sin \omega_{nm} t) \sin \frac{n\pi x}{a} \sin \frac{m\pi y}{b} \tag{7.75}$$

where A_{nm} and B_{nm} are two new constants to be determined from the initial conditions.

Let at time $t = 0$ the displacement be $z = z_0$ and the velocity be $\dot{z} = \dot{z}_0$. Then Eq. (7.75) yields

$$z_0 = \sum_{n=1}^{\infty} \sum_{m=1}^{\infty} A_{nm} \sin \frac{n\pi x}{a} \sin \frac{m\pi y}{b}. \tag{7.76}$$

Multiplying through by $\sin \frac{n\pi x}{a}$ and integrating from $x = 0$ to $x = a$, we obtain

$$\int_0^a z_0 \sin \frac{n\pi x}{a} dx = \frac{a}{2} \sum_{m=1}^{\infty} A_{nm} \sin \frac{m\pi y}{b}$$

$$\text{or,} \quad \frac{2}{a} \int_0^a z_0 \sin \frac{n\pi x}{a} dx = \sum_{m=1}^{\infty} A_{nm} \sin \frac{m\pi y}{b}.$$

Again multiplying through by $\sin \frac{m\pi y}{b}$ and integrating from $y = 0$ to $y = b$, one gets

$$A_{nm} = \frac{4}{ab} \int_0^a \int_0^b z_0 \sin \frac{n\pi x}{a} \sin \frac{m\pi y}{b} dx\, dy. \tag{7.77}$$

Similarly,

$$B_{nm} = \frac{4}{ab\omega_{nm}} \int_0^a \int_0^b \dot{z}_0 \sin \frac{n\pi x}{a} \sin \frac{m\pi y}{b} dx\, dy. \tag{7.78}$$

In the one-dimensional case of a string, we have nodal points where the displacement is zero. In the two-dimensional membrane we have

correspondingly *nodal lines* where $z = 0$. These lines are defined by $\sin \frac{n\pi x}{a} = 0$ and $\sin \frac{m\pi y}{b} = 0$, i.e., by $x = \frac{a}{n}, \frac{2a}{n}, \ldots, \frac{(n-1)a}{n}$ and $y = \frac{b}{m}, \frac{2b}{m}, \ldots, \frac{(m-1)b}{m}$. The nodal lines $x = 0$ and $y = 0$ lie on the boundary. The other nodal lines are parallel to the co-ordinate axes. There are $(n-1)$ nodal lines parallel to the y-axis and $(m-1)$ nodal lines parallel to the x-axis. Hence the nodal system divides the rectangular membrane into nm equal parts.
The frequency f_{nm} of the nm mode is obtained from Eq. (7.74):

$$f_{nm} = \frac{\omega_{nm}}{2\pi} = \frac{c}{2}\left(\frac{n^2}{a^2} + \frac{m^2}{b^2}\right)^{1/2}. \tag{7.79}$$

For a *square membrane*, we have $a = b$, so that

$$f_{nm} = f_{mn} = \frac{c}{2a}(n^2 + m^2)^{1/2}. \tag{7.80}$$

The fundamental frequency is obtained when $n = m = 1$. Thus $f_{11} = \frac{\sqrt{2}c}{2a}$. The next higher tone frequency corresponds to $n = 1$ and $m = 2$ or vice-versa. Thus, $f_{12} = f_{21} = \frac{\sqrt{5}c}{2a}$. Other higher tones can be similarly obtained. Clearly, the partials do not belong to a harmonic series. Hence the sound is nonmelodious with the fundamental predominating.

7.9 Circular Membrane

For circular membranes one has to resort to polar co-ordinates so that the mathematical treatment is rather involved and will not be undertaken here. The nodal lines for circular membranes are concentric circles or diameters. The frequencies of the higher tones relative to the fundamental are 1.59, 2.14, 2.65, and so on. Thus these tones do not make a harmonic series. The relative frequencies are significantly changed by the pressure fluctuations inside the hollow chamber under the stretched circular membrane.

The fundamental frequency for a circular membrane of radius r and thickness t is given by

$$f = \frac{0.76}{r}\sqrt{\frac{T}{4dt}}, \tag{7.81}$$

where d is the material density of the membrane and T is the tension per unit length. Clearly, the frequency f can be altered by varying T. For a small t, i.e., for a thin membrane and for a small density d, the frequency f is high.

Most of the musical instruments like tabla, kettledrum etc. have stretched circular membranes. The characteristics of the tabla are markedly different from those of the drums. The membrane in the tabla is loaded at the centre and the density varies along the radius, so that a note having harmonic partials can be generated from it.

7.10 Vibration of Plates

The fundamental difference between the vibration of a membrane and that of a thin plate is that the restoring force in a membrane is tension whereas the same in a thin plate is stiffness, as in a bar. The mathematical analysis for the vibration of a plate is quite involved and will not be attempted here.

The transverse vibrations of a rectangular plate were studied experimentally by Chladni in 1787. Some sand is uniformly sprinkled on a plate supported at the centre and free at the rim. The plate is set into vibrations by bowing with a violin bow at a point on the rim, which would be theoretically an antinode. The plate is touched with a finger at another point which would be a node. As the plate vibrates, the sand is thrown off and is collected along the nodal lines. These figures of nodal lines are referred to as *Chladni's figures.* Chladni preferred glass plates in his experiment.

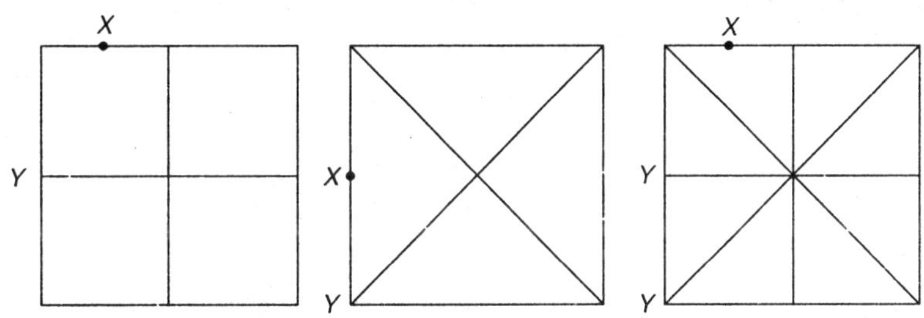

7.5 Some simple Chladni's figures

Some simple Chladni's figures are shown in Fig. 7.5. These are formed on a square plate 30 cm × 30 cm × 2 cm. The plate is fixed at the centre, bowed at the point X and touched at the point Y. The number of figures may be quite large; Chladni obtained 52 figures for a rectangular plate, 43 for a circular plate, and about 100 for plates of other shapes.

Chladni's figures originate from the superposition of two stationary wave patterns corresponding to the simultaneous modes of vibrations whose nodal lines are perpendicular.

7.11 Vibration of Air Columns

We shall discuss here the vibration of air columns in hollow cylinders of uniform cross section. If the cylindrical tube is closed at one end and open at the other, it is called a *closed pipe.* If the tube is open at both ends, it is known as an *open pipe.* Within the pipe there are plane waves, and at the open ends the waves move out spherically into the surrounding air. The reflections at the ends produce oppositely moving waves which superimpose on the incident wave to form stationary waves in the pipe. Instruments like organ pipes, woodwind, horn-loudspeakers etc. use tube shaped sound generating devices.

The simplifying assumptions made in the theory of vibration in pipes are as follows:

(i) Plane waves are produced so that the motion is uniform over a cross section.

(ii) The velocity and the pressure changes in the waves are so small that their squares can be neglected.

(iii) The vibrations are very rapid. Hence the process is adiabatic; enough time is not allowed for the exchange of heat between the adjacent layers to equalize the temperature.

(iv) Viscosity effects are negligible, since the diameter of the pipe is assumed to be much smaller than the wave length and the length of the tube.

(v) Rotatory motions and vortices are not set up in the pipe.

(vi) The walls of the pipe are perfectly rigid.

The longitudinal vibration of the air column satisfies the differential wave equation

$$\frac{\partial^2 \xi}{\partial t^2} = c^2 \frac{\partial^2 \xi}{\partial x^2}, \tag{7.82}$$

where $c(= \sqrt{K/\rho})$ is the wave velocity, K is the bulk modulus, and ρ is the density of the air. ξ is the displacement of a particle in the air column at the position x at time t. By the method of separation of variables, the solution of Eq. (7.82) can be written as

$$\xi = (C_1 \cos \omega t + C_2 \sin \omega t)\left(D_1 \cos \frac{\omega x}{c} + D_2 \sin \frac{\omega x}{c}\right), \tag{7.83}$$

where the constants C_1, C_2 and D_1, D_2 are determined by the boundary and the initial conditions.

Case 1. Pipe open at both ends

Both the ends of the pipe at $x = 0$ and $x = l$ are displacement antinodes here. Thus at $x = 0$ or at $x = l$, ξ is a maximum, i.e., $\frac{\partial \xi}{\partial x} = 0$.
The condition $\frac{\partial \xi}{\partial x} = 0$ at $x = 0$ for all values of t, used in Eq. (7.83), gives $D_2 = 0$. Hence

$$\xi = (a \cos \omega t + b \sin \omega t) \cos \omega \frac{x}{c}, \tag{7.84}$$

where a and b are new constants.

The condition $\frac{\partial \xi}{\partial x} = 0$ at $x = l$, used in Eq. (7.84) yields

$$\sin \frac{\omega l}{c} = 0,$$

$$\text{or,} \qquad \frac{\omega l}{c} = n\pi \quad (n = 1, 2, 3, \text{ etc.}). \tag{7.85}$$

So the general solution for ξ is

$$\xi = \sum_{n=1}^{\infty} \left(a_n \cos \frac{n\pi ct}{l} + b_n \sin \frac{n\pi ct}{l} \right) \cos \frac{n\pi x}{l}. \tag{7.86}$$

The nth harmonic frequency is

$$f_n = \frac{\omega_n}{2\pi} = \frac{nc}{2l} = \frac{n}{2l}\sqrt{\frac{K}{\rho}}. \tag{7.87}$$

The frequency of the fundamental tone is

$$f_1 = \frac{1}{2l}\sqrt{\frac{K}{\rho}} = \frac{c}{2l}. \tag{7.88}$$

The fundamental wave length is

$$\lambda = \frac{c}{f_1} = 2l. \tag{7.89}$$

A pulse of compression travelling from the mouth inside the tube is reflected from the other end with a reversal of phase, i.e., as a wave of rarefaction. The reflected wave travels back in the tube and is again reflected from the mouth with a phase reversal, i.e., as a pulse of compression. This explains that the wavelength for the fundamental is twice the length of the tube. The superposition of the incident and the reflected waves in the tube form stationary waves.

Case 2. Pipe closed at one end and open at the other

Here the boundary conditions are: (i) at the closed end ($x = 0$), the displacement $\xi = 0$; (ii) at the open end ($x = l$), the displacement ξ is a maximum, i.e., $\frac{\partial \xi}{\partial x} = 0$.

The first condition, used in Eq. (7.83), gives $D_1 = 0$. So

$$\xi = (a \cos \omega t + b \sin \omega t) \sin \frac{\omega x}{c}. \tag{7.90}$$

Using the second condition in Eq. (7.90), we have $\cos \frac{\omega l}{c} = 0$. That is,

$$\frac{\omega l}{c} = (2n + 1)\frac{\pi}{c} \qquad (n = 0, 1, 2, 3, \text{ etc.}). \tag{7.91}$$

Therefore,

$$\xi = \sum_{n=1}^{\infty} \left(a_n \cos \frac{(2n+1)\pi ct}{2l} + b_n \sin \frac{(2n+1)\pi ct}{2l} \right) \sin \frac{(2n+1)\pi x}{2l}. \quad (7.92)$$

The frequency of the nth harmonic is

$$f_n = \frac{\omega_n}{2\pi} = \frac{(2n+1)c}{4l} = \frac{2n+1}{4l}\sqrt{\frac{K}{\rho}}. \quad (7.93)$$

Note that only odd harmonics are present here in contrast with the pipe open at both ends where all the harmonics are present.
The fundamental frequency is

$$f_1 = \frac{1}{4l}\sqrt{\frac{K}{\rho}} = \frac{c}{4l}. \quad (7.94)$$

The corresponding wave length is

$$\lambda_1 = \frac{c}{f_1} = 4l. \quad (7.95)$$

This shows that the initial pulse at the open end must travel the length of the pipe four times before attaining the same phase. A pulse of compression starting from the open end travels in the tube and is reflected at the closed end without suffering a phase change. So the reflected wave is also a pulse of compression. It travels back and is reflected at the open end with a phase reversal. Thus, the reflected pulse is now one of rarefaction. This pulse again traverses the tube and is reflected at the closed end as a pulse of rarefaction, which, in turn, is reflected from the open end with a phase reversal to restore the initial pulse of compression. Thus the wave travels the length of the tube four times to repeat the cycle.

End correction: Antinodes have been assumed to occur at the open ends of the pipe. This assumption is not correct. Beyond the open end, the plane waves inside the tube produce spherical waves. As a result, at the open end, there is some excess pressure which falls to zero at a point outside. This distance, called the *end correction*, is estimated to be $0.6r$, where r is the radius of the pipe. The end correction needs to be added to the length of the tube at each open end. Hence the effective wave length of the fundamental from a pipe open at both ends is $\lambda_1 = 2(l + 2 \times 0.6r) = 2(l + 1.2r)$. The same for a pipe open at one end and closed at the other is $\lambda_1 = 4(l + 0.6r)$. So, the fundamental frequency of the former is a little less than twice that of the latter.

Humidity, temperature, nature of the gas etc. affect the speed of sound in the pipe, and hence the frequency of the emitted note.

7.12 Energy of a Vibrating Air Column

Putting $\omega_n = \frac{n\pi c}{l}$ and $k_n = \frac{n\pi}{l}$, the particle displacement $\xi(x,t)$ for an open pipe, given by Eq. (7.86), can be rewritten as

$$\begin{aligned}\xi &= \sum_{n=1}^{\infty} c_n \cos(\omega_n t - \theta_n) \cos k_n x \\ &= \sum_{n=1}^{\infty} s_n \cos k_n x, \end{aligned} \tag{7.96}$$

where $c_n = \sqrt{a_n^2 + b_n^2}$, $\tan\theta_n = b_n/a_n$, and

$$s_n = c_n \cos(\omega_n t - \theta_n).$$

The kinetic energy of an element of length dx of the air column is

$$dE_k = \frac{1}{2}(\rho\alpha dx)\left(\frac{\partial\xi}{\partial t}\right)^2$$

where ρ is the air density and α is the cross-sectional area of the air column. The total kinetic energy of the air column of length l is

$$\begin{aligned}E_k &= \frac{1}{2}\rho\alpha\int_0^l \left(\frac{\partial\xi}{\partial t}\right)^2 dx \\ &= \frac{1}{2}\rho\alpha\int_0^l \left(\sum_{n=1}^{\infty} \dot{s}_n \cos k_n x\right)^2 dx \\ &= \frac{1}{2}\rho\alpha\int_0^l \sum_{n=1}^{\infty}\sum_{m=1}^{\infty} \dot{s}_n \dot{s}_m \cos k_n x \cos k_m x \, dx.\end{aligned}$$

As

$$\begin{aligned}\int_0^l \cos k_n x \cos k_m x \, dx &= 0 \quad \text{for} \quad n \neq m \\ &= \frac{l}{2} \quad \text{for} \quad n = m.\end{aligned}$$

we get

$$E_k = \frac{1}{4}(\rho\alpha l)\sum_{n=1}^{\infty} \dot{s}_n^2 = \frac{M}{4}\sum_{n=1}^{\infty} \omega_n^2 c_n^2 \sin^2(\omega_n t - \theta_n), \tag{7.97}$$

where $M(= \rho\alpha l)$ is the mass of the vibrating air column.

The potential energy dE_p of the element of length dx of the air column is given by Eq. (5.59) where the excess pressure p is

$$p = -K\frac{\partial\xi}{\partial x},$$

K being the bulk modulus. Here $V_0 = \alpha dx$. Since the linear term in p does not contribute on the whole, the total potential energy of the air column is

$$E_p = \frac{1}{2}K\alpha\int_0^l \left(\frac{\partial \xi}{\partial x}\right)^2 dx.$$

Proceeding as in the case of E_k, we have

$$E_p = \frac{1}{4}K\alpha l \sum_{n=1}^{\infty} s_n^2 k_n^2 = \frac{M}{4}\sum_{n=1}^{\infty} \omega_n^2 c_n^2 \cos^2(\omega_n t - \theta_n), \tag{7.98}$$

since $c = \sqrt{K/\rho}$ and $\omega_n = ck_n$. So, the total energy of the vibrating air column is

$$E = E_k + E_p = \frac{M}{4}\sum_{n=1}^{\infty} \omega_n^2 c_n^2. \tag{7.99}$$

Thus the total energy is the sum of the energies of the different modes of vibration. In other words, the energy of the air column is distributed among the modes of vibration.

The same result is obtained for a pipe closed at one end.

7.13 Vortex Sound

Vortex motion produces tones like Aeolian tones, jet tones, and edge tones. We shall discuss these tones below.

(i) Aeolian tones: Such tones are produced when a fluid streams past a linear obstacle with an appreciable speed. Eddies appear behind the obstacle just like the case when a stream of water flows past a pile or rock. The eddies originate from the viscous forces between the surface of the obstacle and the fluid. The eddies or vortices on the opposite side of an obstacle rotate in opposite directions and exert a lateral thrust (Fig. 7.6). If the velocity of flow attains a certain value, the eddies grow in size, detach themselves from the obstacle and move downstream. The fluttering of a flag on a pole serves as an example of the alternating production of such vortices and exhibits the alternating thrusts.

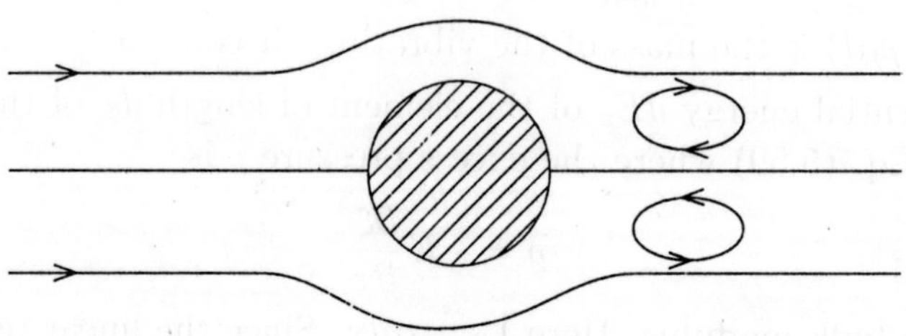

Fig. 7.6 Production of eddies behind an obstacle

As wind blows past a stationary wire, the alternating thrusts set the wire into vibration. If the frequency of alternation of the thrusts happens to coincide with one of the normal modes of transverse vibration of the wire, resonance occurs and the sound is magnified. Such sounds are known as *Aeolian tones.* The singing of telegraph or telephone wires in a wind, and the whistling of the wind through trees are Aeolian tones.

The frequency of alternating thrusts, f, does not depend on the length or the tension of the wire. It is given by $f = 0.185\, v/d$, where v is the velocity of the wind and d is the diameter of the wire. The greater the wind velocity and the smaller the wire diameter, the higher the pitch.

(ii) Jet tones: When a jet of air issues from a linear slit, a weak fluctuating tone having the features of a 'hiss' is generated. Such a tone is called a *jet tone.*

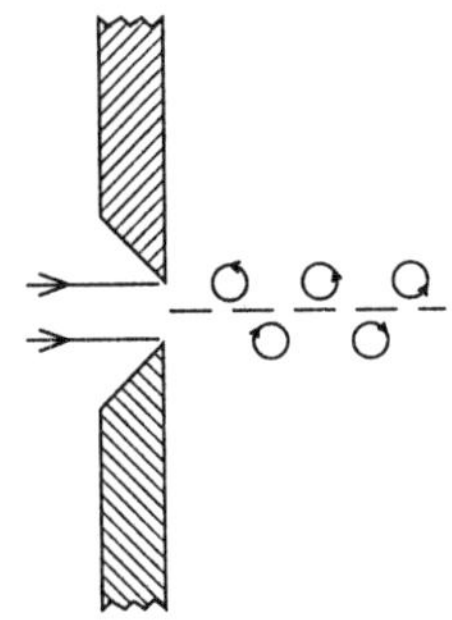

Fig. 7.7 Production of a jet tone

Let a jet of air issue from a rectangular slit (Fig. 7.7). There are surfaces of discontinuity between the issuing jet and the air at rest. The issuing air tends to curl outward into the stationary air and produces alternate vortices on the two sides of the jet. The pressure fluctuations in the stationary air generate the jet tone whose frequency is $0.055v/\omega$, where v is the velocity of the jet and ω is the width of the slit.

The reason for the weak hissing character of the jet tone is that unlike the solid wire in the Aeolian tone, there is no solid resonator here to take up the vibration.

(iii) Edge tones: If an edge such as a small-angle wedge is placed in front of the slit and parallel to it, a tone stronger and stabler than the jet tone, is produced. This is called the *edge tone.* The vortices for the edge tones do not originate at the orifice unlike the case with the jet tone. A wedge placed symmetrically with respect to the jet, divides the air stream into two parts. A slight disturbance may deflect the air stream; the reaction of the wedge then deflects it in the opposite direction. Thus a momentary compression is produced. If this cycle of events is periodically repeated, the resulting compressions propagate , thus generating the edge tone.

The frequency of the edge tone is proportional to the velocity of efflux of the air from the slit and depends on the distance of the wedge from the slit. There is a minimum distance d_m of the wedge from the slit for any given velocity of efflux. When the distance slightly exceeds d_m, a disturbance strikes the edge at the moment the following one emerges from the slit. A clear note is then produced, the space between the slit and the wedge serving as a resonator. As the wedge distance is increased from d_m, the pitch drops continuously till at the distance $2d_m$ it suddenly rises to an octave. This is explained by the fact that the original spacing of the disturbances reappears but with twice as many between the slit and the edge. With further increase of the wedge distance, the pitch again falls off followed by a sudden rise. The process goes on till the note merges with noise.

7.14 Organ Pipes

(i) Flue organ pipe: In this type of organ pipe stationary vibrations are set up in an air column in a pipe by an air blast directed on to a sharp edge. Such pipes are either wooden and rectangular in cross-section (as shown in Fig. 7.8), or metallic and circular in cross-section. One end of the pipe is open or closed by an adjustable piston P. The effective length of the air column can be changed for tuning purposes by adjusting the position of the piston. (Tuning in an open pipe is achieved by means of a short sleeve fitted at the open end. By sliding the sleeve up and down, the effective length can be changed). The pipe length depends on the desired note. The tube diameter does not affect the pitch so long as it is small compared to the wave length.

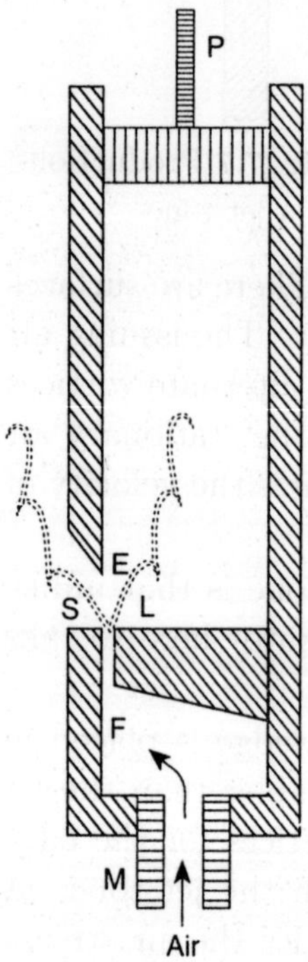

Fig. 7.8 Flue organ pipe

The mouth M of the pipe is connected to a windchest which is fed by bellows. The air blast from the chest enters the tube through the mouth piece M. It passes through a narrow slit or *flue* F (which accounts for the name of the organ), and leaves the pipe through the opening S. In the process, it impinges on the sharp edge E. Eddies produced on both sides of E generate edge tones. These tones are coupled to the natural frequencies of the air column in the pipe, of which S is an open end. When resonance in the air column is excited, the tube 'speaks', i.e., it emits a sustained characteristic note.

The frequency of the edge tone is determined by the distance between the slit and the edge, i.e., by the distance LE. It is also determined by the velocity of efflux of the air stream. The organ pipe is designed in such a way that at the normal blowing pressure, the pitch of the edge tone coincides with that of the fundamental of the pipe.

Effect of overblowing or underblowing the pipe: Let the blowing pressure be increased from the normal value which sets up the fundamental vibration in the pipe. When the pipe is removed, the frequency of the edge tone is proportional to the velocity of efflux, which increases with the blowing pressure. In the presence of the pipe, the edge tones are coupled to the vibrations of the pipe. As the period of the pipe is forced on that of the edge tone, the distance between the successive disturbances is changed so as to maintain the vibration of the pipe. This adjustment continues till the velocity of efflux causes the pitch of the pre edge tone to be equal to that of the next harmonic of the pipe. The vibrations of the pipe then jump to that pitch.

If the blowing pressure is reduced from the normal value, the pitch of the natural edge tone in the absence of the pipe is decreased. The presence of the pipe, however, forces its period on the edge tone as before, so that the vibrations of the pipe are maintained. When the blowing pressure is very low, the edge tone is hardly produced. But the pipe tends to impose one of its overtones on the eddies, so that the fundamental is feebly sounded.

Effect of the lateral dimension of the pipe: The end correction at the blown end is fairly large, about two to three times the radius of the pipe. The vibrating length is thus considerably longer than the geometrical length. The end correction is also determined by the frequency. The vibrations of the pipe, therefore, do not strictly form a harmonic series. For a narrow tube, the end correction is reduced and the departure from the harmonic series is consequently less.

(ii) Reed pipe: In the reed pipe, stationary vibrations in an air column are sustained by an air blast directed through a channel periodically closed by a flexible strip of metal or cane, called a *reed*. If the vibrating part of the reed is a little less than the aperture over which it vibrates, it is known as a *free reed*. However, if the vibrating part of the reed is slightly larger than the aperture, it is called a *beating reed*. Beating reeds are curved outwards and hence do not completely close the aperture when at rest.

A reed organ pipe is schematically represented in Fig. 7.9. An opening on the side of a small tube—the *shallot* (Sh)—serves as the aperture. To the top of the shallot is connected a conical pipe P. Air is blown into a chamber, known as the *boot* B. The reed R is often of the beating type, and it is tuned by moving the spring S.

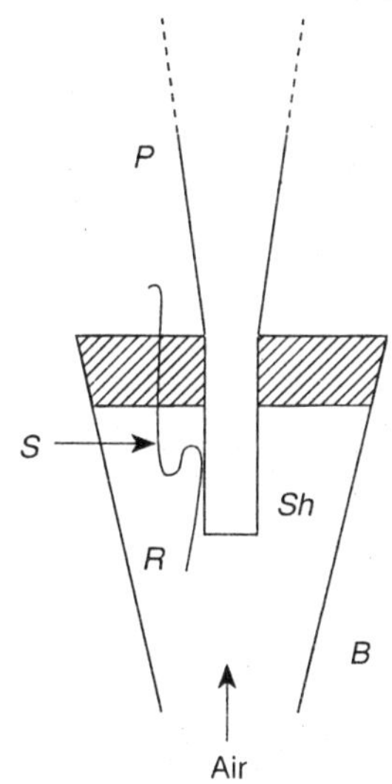

Fig. 7.9 Reed pipe.

As air rushes from the boot through the small opening between the reed and the shallot, the reed is set into vibration. The vibrating reed closes and opens the aperture periodically. Each blast of air entering the pipe makes a condensation move up the pipe. From the open end of the pipe, it is reflected as a rarefaction. When the rarefaction reaches the reed, the aperture is closed by the reed. The low pressure in the rarefaction assists the reed in the process. The rarefaction is now reflected without a phase change. From the open end it is returned as a condensation. As the condensation reaches the reed, it finds the aperture open. The process is repeated. Clearly, in one full cycle of vibration of the reed the pulse traverses four times the length of the pipe. Therefore, the emitted wave length is four times the length of the pipe. The coupling of the pipe and the reed causes the frequency to be less than the natural frequency of vibration.

7.15 Vibrations Maintained by Heat

When some part of a solid or a gas is periodically or intermittently heated, the expansion of the heated portion compresses the adjacent part of the solid or the gas. If phase conditions are favourable, a continuous vibration is set up. *Trevelyan's rocker* is an example of heat-maintained vibration of a solid. The *singing flame* is an example where vibrations are produced by the intermittent supply of heat to a portion of a gas.

(i) Trevelyan's rocker: Trevelyan's rocker consists of a copper or brass prism P with a rod terminating in a knob K (Fig. 7.10). One edge of the prism is grooved to produce two parallel ridges. A lead block B with a rounded and rough top surface is placed on the working table. The rocker is heated to a temperature below the melting point of lead and placed with the groove downwards on the cold lead block. The rocker is in a slanting possition with the knob K resting on the table. With a smart tap near the ridges, the prism goes on tilting about the ridges, exhibiting vibrations as long as the rocker is hot.

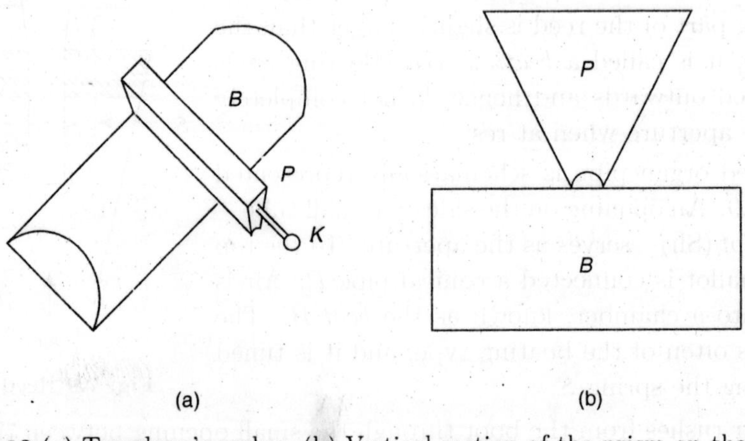

Fig. 7.10 (a) Trevelyan's rocker, (b) Vertical section of the prism on the block

The action is explained as follows. Consider that as the rocker vibrates, one ridge is in contact with the block and the other ridge is at a raised position. At the region of contact, heat is conducted from the prism to the block. The local heating of the block expands it slightly and produces a hum, which tilts the rocker in the other direction. Hence the other ridge touches the block, the contact at the previous ridge being broken. The process is repeated; the rocking motion on the two ridges takes place under the action of the torque produced by gravity, and the vibration is maintained by the flow of heat from the prism to the block.

For satisfactory results, the following conditions have to be met: (a) The ridges must be smooth and clean, whereas the lead surface must be roughened to avoid slipping. (b) The thermal conductivity of the two metals must differ considerably. Heat must be readily transferred from the hot body to the cold one, but this heat must stay in the vicinity of the region of contact to produce a local expansion and not readily conducted through the metal block. (c) There must be a small time lag between contact and expansion. (d) The metal block on which the rocker is placed must have a sizable temperature coefficient of expansion.

Mathematical treatment : Let m be the mass of the rocker, k its radius of gyration about the ridge, and $2l$ the distance between the ridges (Fig. 7.11). G is the centre of gravity of the rocker, through which its weight mg acts vertically downward, g being the local acceleration due to gravity. If θ is the angle of tilt of the rocker, the restoring torque due to the weight of the rocker is $-mgl\cos\theta \simeq -mgl$, since θ is a very small angle. The equation of motion is

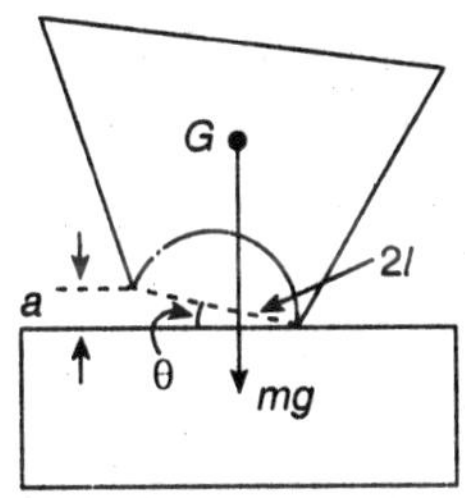

Fig. 7.11 Motion of the rocker

$$mk^2\frac{d^2\theta}{dt^2} = -mgl \tag{7.100}$$

At time $t = 0$, the rocker starts tilting with the maximum angular velocity, ω_m; and at $t = T/2$ (where T is the time interval between two successive contacts of the ridge with the block), the ridge attains the maximum height a and the angular velocity decreases to zero. The maximum tilt is $\theta_m = a/(2l)$. The mean angular velocity in the time interval $T/2$ is $\omega_m/2$. Thus, $\omega_m/2 = \theta_m/(T/2) = a/(lT)$, or

$$\omega_m = \frac{2a}{lT} \tag{7.101}$$

Integrating Eq. (7.100), we obtain

$$k^2\int_0^{T/2}\frac{d^2\theta}{dt^2}dt = -gl\int_0^{T/2}dt$$

$$\text{or,}\quad k^2(0-\omega_m) = -\frac{1}{2}glT$$

$$\text{or,}\quad \omega_m = \frac{glT}{2k^2} \tag{7.102}$$

Equations (7.101) and (7.102) give

$$T^2 = \frac{4ak^2}{gl^2} \tag{7.103}$$

If f is the frequency of contacts of the ridges with the block, we have

$$af^2 = \frac{a}{T^2} = \frac{gl^2}{4k^2}, \tag{7.104}$$

which is a constant for a given rocker. When the rocker is very hot, the height a is large and the frequency f is low since af^2 is a constant. As the rocker cools down, a decreases and hence the frequency increases.

(ii) Singing flame: The experimental arrangement for a singing flame is shown in Fig. 7.12. An upright tube P, about 10 cm long and having a pinhole at the top, is attached to a gas pipe S. The upper part of the tube P is inserted in a wider vertical tube T, about 30 cm. long. If the gas is lighted at the pinhole and when the flame is about 7 cm from the bottom of the tube, it starts to sing, emitting a characteristic note whose frequency is determined by the length of the tube T.

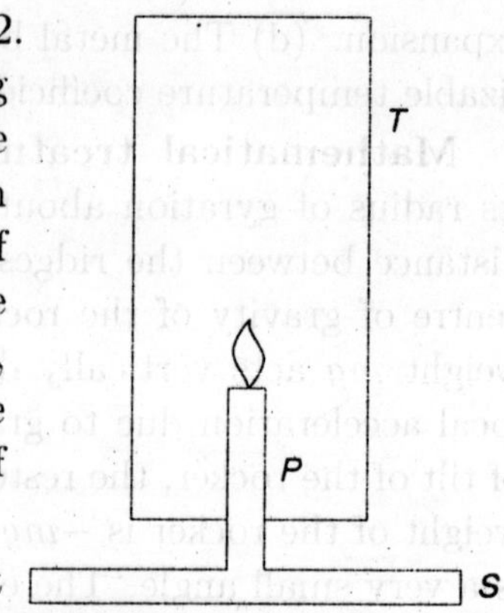

Fig. 7.12 Singing flame

The singing is sustained by the intermittent supply of heat by the burning jet. The heating exerts a force on the air in the tube at the right phase at a point of maximum pressure amplitude. When the jet burns steadily without singing, it looks like a band in a rotating mirror [Fig. 7.13(a)]. When the flame sings, it has a tooth-like appearance in the rotating mirror [Fig. 7.13(b)]. An analysis of this appearance shows that the singing flame burns intermittently with a period equal to that of the emitted note.

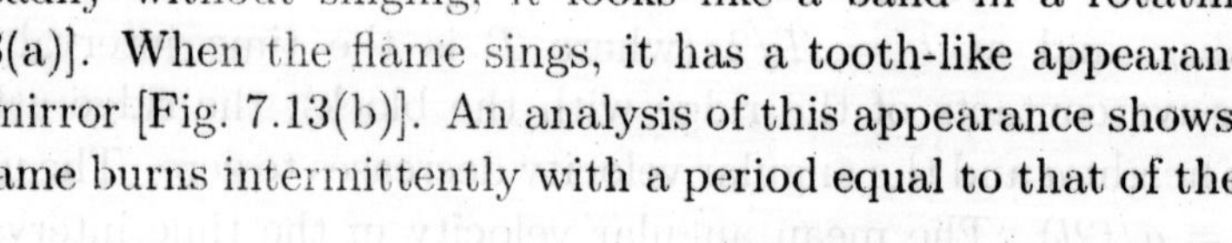

Fig. 7.13 Appearance of the jet in a rotating mirror when the flame (a) does not sing; (b) sings

The reflections at the boundaries of the tube produce stationary waves in the tube. Stationary waves are also formed in the gas supply tube giving a maximum pressure at the nodes. For the flame to sing efficiently, the following conditions are to be satisfied : (i) The flame should be at a node of the singing tube, where the pressure amplitude is a maximum. (ii) The jet should also lie at a node of the supply tube. (iii) The intermittent supply of heat, i.e., the driving force, must be in tune with the vibrations in the tube. The pressure changes in the singing tube must be synchronized to that in the supply tube.

The length of the tube should be slightly greater than $(n+1)\lambda/4$, where n is an integer and λ is the wavelength of the emitted note.

7.16 Solved Problems

Problem 1. *The second harmonic of the note emitted from a bar, free at both ends, coincides with the third harmonic of the note emitted from another bar, clamped at one end and free at the other. Find the ratio of the lengths of the two bars, assuming that both the bars execute longitudinal vibrations.*

Solution. For the bar free at both ends, the second harmonic has the frequency $f_2 = \frac{c}{l_1}$, where c is the velocity of sound and l_1 is the length of the bar. For the bar fixed at one end and free at the other, the third harmonic frequency is $f_3 = 3c/(4l_2)$ where l_2 is the length of the bar. So,

$$\frac{c}{l_1} = \frac{3c}{4l_2} \quad \text{or,} \quad \frac{l_1}{l_2} = \frac{4}{3}.$$

Problem 2. *The longitudinal vibrations of a bar, free at both ends, are initiated by striking the free end $x = l$, so that at time $t = 0$, the velocity is zero except over a small length dl at $x = l$. Find an expression for the displacement ξ.*

Solution. ξ is given by Eq. (7.10), viz.,

$$\xi = \sum_{n=1}^{\infty} \cos\frac{n\pi x}{l}(a_n \cos\omega_n t + b_n \sin\omega_n t).$$

Since at $t = 0$, $\xi = 0$ for all values of x, we have $a_n = 0$. Hence

$$\xi = \sum_{n=1}^{\infty} b_n \cos\frac{n\pi x}{l} \sin\omega_n t.$$

At time $t = 0$, the velocity is

$$\dot{\xi}_0 = \sum_{n=1}^{\infty} \omega_n b_n \cos\frac{n\pi x}{l}$$
$$= \frac{\pi c}{l}\sum_{n=1}^{\infty} n b_n \cos\frac{n\pi x}{l}.$$

Hence,

$$\int_0^l \dot{\xi}_0 \cos\frac{n\pi x}{l}dx = \frac{\pi c\, n b_n l}{l\ 2}$$

$$\text{or,}\quad b_n = \frac{2}{\pi cn}\int_{l-dl}^{l}\dot{\xi}_0 \cos\frac{n\pi x}{l}dx = \frac{2(-1)^n}{\pi cn}\int_{l-dl}^{l}\dot{\xi}_0 dx = (-1)^n\frac{2P}{\pi cn},$$

$$\text{where}\quad P = \int_{l-dl}^{l}\dot{\xi}_0 dx.$$

Thus,

$$\xi = \frac{2P}{\pi c}\sum_{n=1}^{\infty}\frac{(-1)^n}{n}\cos\frac{n\pi x}{l}\sin\frac{n\pi ct}{l}.$$

Problem 3. *The longitudinal vibrations of a rod, free at both ends and clamped at the middle, are initiated by striking the free end at $x = l$, as in Problem 2. Find an expression for the displacement ξ. What is the fundamental frequency?*

Solution. In addition to the boundary conditions for the bar free at both ends, the additional condition here is $\xi = 0$ at $x = l/2$. Hence the expression for ξ obtained in Problem 2 yields $\cos\frac{n\pi x}{l} = 0$ at $x = l/2$. Hence $n = 1, 3, 5$, etc., i.e., only odd harmonics are present. So,

$$\xi = -\frac{2P}{\pi c}\left[\cos\frac{\pi x}{l}\sin\frac{\pi ct}{l} + \frac{1}{3}\cos\frac{3\pi x}{l}\sin\frac{3\pi ct}{l} + \frac{1}{5}\cos\frac{5\pi x}{l}\sin\frac{5\pi ct}{l} + \cdots\right].$$

The fundamental frequency is $f_1 = \frac{c}{2l} = \frac{1}{2l}\sqrt{\frac{Y}{\rho}}$.

Problem 4. *A square membrane of side* 15 cm *is under a uniform tension of* 2×10^4 N/m. *If the surface density of the membrane is* 0.8 kg/m^2, *calculate the speed of sound in the membrane, and the fundamental and the next higher tone frequencies.*

Solution. The speed of sound is

$$c = \sqrt{\frac{T}{\rho}} = \sqrt{\frac{2\times 10^4}{0.8}} = 158.1\ \text{m/s}$$

The fundamental frequency is

$$f_{11} = \frac{\sqrt{2}c}{2a} = \frac{\sqrt{2}\times 158.1}{2\times 0.15} = 745.3\ \text{Hz}$$

The next higher tone frequency is

$$f_{12} = f_{21} = \frac{\sqrt{5}c}{2a} = \frac{\sqrt{5}\times 158.1}{2\times 0.15}\ \text{Hz} = 1.178\ \text{kHz}$$

Problem 5. *Two open pipes sounded together generate* 6 *beats* per second *between the fundamentals. If the shorter pipe is* 1 m *long, find the length of the longer pipe. Given, speed of sound in air* = 340 m/s.

Solution. For an open pipe, the fundamental frequency is $c/2l$. So, if l_1 and l_2 ($l_1 > l_2$) are the lengths of the two open pipes, the number of beats generated per second is

$$n = \frac{c}{2l_2} - \frac{c}{2l_1}$$

Given, $n = 6\,\text{Hz}$ and $l_2 = 1\,\text{m}$. So

$$6 = \frac{c}{2}\left(1 - \frac{1}{l_1}\right)$$

$$\text{or,}\quad \frac{1}{l_1} = 1 - \frac{12}{c} = 1 - \frac{12}{340} = 0.9647$$

whence $l_1 = 1.036\,\text{m}$.

Problem 6. *The air column in a closed pipe of length* 83 cm *and radius* 4 cm *resonates with a tuning fork of frequency* 100 Hz. *Calculate the speed of sound in air.*

Solution. For resonance with the fundamental frequency f_1, we have for the speed of sound (making the end correction)

$$\begin{aligned} c &= 4f_1(l + 0.6r) = 4 \times 100 \times (83 + 0.6 \times 4)\ \text{cm/s} \\ &= 341.6\ \text{m/s}. \end{aligned}$$

REVIEW QUESTIONS

1. Show that the longitudinal vibrations of a bar free at both ends contain all harmonics while those of a bar clamped at one end and free at the other contain only odd harmonics.

2. Obtain an expression for the fundamental frequency of the longitudinal vibration of a bar (i) clamped at both ends, (ii) clamped at one end, and (iii) clamped at the middle.

3. Discuss the stationary waves produced in a rod executing longitudinal vibrations.

4. Arrive at the differential equation for the transverse waves in a thin rod. Hence show that the rod behaves as a dispersive medium.

5. Discuss the simple harmonic transverse vibrations of a bar (a) free at one end and fixed at the other, and (b) free at both ends.

6. Explain the vibration of a tuning fork. Discuss the factors affecting its frequency. Why are good tuning forks made of elinvar?

7. Obtain the differential equation for the vibration of a membrane. Hence find Helmholtz equation.

8. Deduce the general equation of vibration of a rectangular membrane and obtain the frequency of vibration. Show that the emitted sound is not melodious.

9. Write the differential equation for the longitudinal vibration of an air column in a pipe. Find expressions for the displacement and the frequency of the note emitted by a pipe (i) open at both ends and (ii) open at one end and closed at the other. Why is the end correction necessary?

10. Show that the energy of a vibrating air column is distributed among the different modes.

11. Write notes on :

    (i) Aeolian tones, (ii) Jet tones, (iii) Edge tones, (iv) Reed pipe, (v) Trevelyan's rocker, (vi) Singing flame, (vii) Chladni's figures, (viii) Tuning bar.

12. Describe a flue organ pipe and explain its working. Comment on the effects of (i) overblowing or underblowing the pipe and (ii) the lateral dimension of the pipe.

13. (a) 'The vibratary motion of a Trevelyan's rocker occurs under the action of gravity, the vibrations being maintained by heat flow'. — Explain.

    (b) If a is the maximum height of a ridge of a vibrating rocker and f is the frequency, show that af^2 = constant.

PROBLEMS

1. A metal bar, free at both ends, has a fundamental frequency of 3 kHz for longitudinal vibrations. If it is clamped at one end, what will be the fundamental frequency of longitudinal vibrations?

   [**Ans.** 1.5 kHz]

2. A stretched square membrane of side 0.3 m vibrates at the fundamental frequency of 300 Hz. Find the speed of sound in the membrane.

[**Ans.** 127.3 m/s]

3. Two closed pipes of lengths 1.1 m and 1.175 m are sounded together at the fundamental modes. If the speed of sound in air is 340 m/s, determine the number of beats produced per second.

[**Ans.** 5 beats per second]

4. A long circular cylinder, open at one end and closed at the other, rests vertically on its closed face. It is being filled with water at a uniform rate. Resonance occurs at intervals of 100s with a vibrating tuning fork of frequency 300 Hz. Calculate the rate of supply of water in cm^3/s. (Radius of the cylinder = 10 cm, speed of sound in air = 330 m/s).

[**Ans.** 172.8 cm^3/s]

Chapter 8

Ultrasonics

8.1 Introduction

Acoustic energy is propagated in an elastic medium due to the vibration of a material body acting as the source. Depending on the frequency of vibration, the spectrum of acoustic energy is divided into four parts :

(i) **Infrasonics:** Sound of frequency less than about 20 Hz belongs to this region.

(ii) **Sonics:** This is the audible range of frequencies extending from about 20 Hz to 20 kHz. Though young people can hear up to 20 kHz, middleaged persons normally hear up to about 16 kHz.

(iii) **Ultrasonics:** The frequency range from about 20 kHz to 1 GHz forms this region of the spectrum.

(iv) **Hypersonics:** Frequencies above 1 GHz are said to belong to this range.

Infrasonic, ultrasonic, and hypersonic vibrations are beyond the perception of the human ear. Ultrasonic sound has assumed importance in view of its various industrial applications.

8.2 Generation of Ultrasonic Vibration

The different types of popular ultrasonic generators are as follows : (i) mechanical generator, (ii) magnetostrictive generator, and (iii) piezoelectric generator.

(i) **Mechanical generator:** The modified **Galton's whistle** (Fig. 8.1) is an ultrasonic generator of this class. An annular nozzle N is used to blow a small pipe P. The nozzle is fitted with a screw S which can alter the distance between the slit O and the edge E of the pipe. The distance OE and the air blast pressure are adjusted to set the pipe in resonant vibration. The frequency of vibration depends on the length and the diameter of the pipe. The screw-piston driven by the micrometer M changes the length of the air column in the pipe, thus altering the frequency of the emitted sound. With a careful design, this whistle can generate vibrations of constant amplitude and frequency up to about 100 kHz.

(ii) **Magnetostrictive generator:** When a ferromagnetic rod is placed in a magnetic field parallel to its length, the length of the rod increases or decreases irrespective of the direction of the field. This phenomenon is referred to as *magnetostriction* or *Joule effect*. The change of length is quite small and requires sensitive devices for its detection. Of the various ferromagnetic materials, only nickel and annealed cobalt rods behave simply and regularly. The magnetostrictive effect decreases with increasing temperature and vanishes at the Curie point.

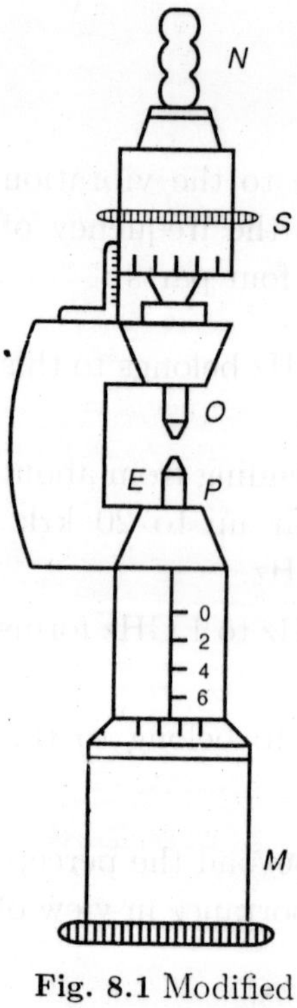

Fig. 8.1 Modified Galton's whistle

The phenomenon of magnetostriction is used to produce ultrasonic waves by means of the magnetostrictive oscillator. A nickel rod, subjected to an alternating magnetic field superimposed on a steady field, suffers a periodic change of length due to magnetostriction. If the natural frequency of longitudinal vibration of the rod equals that of the alternating current producing the alternating magnetic field, a resonance is obtained and the amplitude of oscillation becomes a maximum. Ultrasonic waves of the same frequency are emitted from the end of the rod. If the rod is clamped at the middle, the frequency of the generated waves is

$$f = \frac{1}{2l}\sqrt{\frac{Y}{\rho}},$$

where l is the length of the rod, and Y and ρ are respectively Young's modulus and the density of its material. Here the nickel rod serves as a transducer converting the electrical energy into sound energy.

Note that if the rod is subjected simply to an alternating magnetizing field parallel to its length, the rod vibrates at a frequency twice the frequency of the magnetizing field since the change in the length of the rod is independent of the direction of magnetization. To make the rod vibrate at the frequency of the alternating magnetizing field, the alternating field is superimposed on a steady field such that the resultant magnetizing field is always in the same direction.

The circuit of a magnetostrictive oscillator (which is the transistorised version of the original circuit developed by Pierce) is shown in Fig. 8.2. The ferromagnetic rod R is clamped at the middle. A coil L_1 is wound on R near one end and insulated from it. The coil L_1 and the tuning capacitor C constitute a tank circuit, the resonant frequency of which equals the fundamental frequency of the longitudinal vibration of the rod R. The tank circuit is connected to the collector of the transistor T. The other end of the rod carries the insulated coil L_2 wound on it. One end of L_2 is connected to the base of the transistor T through the coupling capacitor C_1. The turns of the coils have a diameter slightly greater than that of the rod so that the rod can vibrate freely lengthwise. The application of the collector voltage V_{cc} produces a current in L_1. The resulting magnetic field changes the length of the rod due to magnetostriction. This changes the magnetic flux through L_2, thus inducing an emf in it. If L_2 is properly connected to the base of the transistor, this induced emf will enhance the forward bias of the base-emitter junction. Therefore, the collector current increases. Thus a positive feedback required to sustain the oscillation is provided.

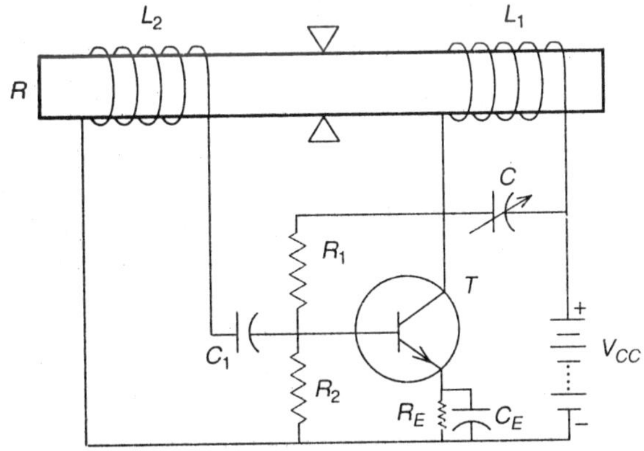

Fig. 8.2 A magnetostrictive oscillator

The resistances R_1, R_2 and R_E together with V_{cc} establish the dc operating point of the transistor. The capacitor C_E is the ac bypass capacitor. The rod is premagnetized by a permanent magnet or by the steady collector current.

The frequency range of the magnetostrictive oscillator is about 5 to 60 kHz. For higher frequencies, the length of the ferromagnetic rod becomes inconveniently short, and hysteresis and eddy current losses become prohibitively high. For such high frequencies, the piezoelectric oscillator is used.

(iii) **Piezoelectric generator:** When a plate, cut in a special manner from a single crystal of quartz, Rochelle salt, ADP etc., is compressed, the faces of the plate acquire opposite kinds of electric charge. If the plate is stretched, the signs of the charges are reversed. On the other hand, when the plate is placed in an alternating electric field, it contracts and expands periodically, thus vibrating mechanically. This phenomenon is called *piezoelectric effect*, the substances such as quartz, Rochelle salt etc. being termed piezoelectric substances. The frequency of vibration is in the ultrasonic range. Hence in the surrounding elastic medium, ultrasonic waves are set up. If the frequency of the applied electric field coincides with one of the natural frequencies of the crystal, the amplitude of vibration becomes very large due to resonance.

For maximum efficiency, the electric polarization and the mechanical strain in the crystal should be strongly coupled. This is obtained if the crystal is cut in a proper way. The unit cell of a single crystal of quartz is a hexagonal prism whose ends are bounded by two six-sided pyramids (Fig. 8.3). The long axis (the Z-axis) of the crystal is termed the *optic axis*. Suppose that a hexagonal slice is cut from the crystal by two parallel planes which are perpendicular to the optic axis. A line perpendicular to the optic axis and joining the opposite angles of the hexagon gives the direction of the *electric axis* (X-axis). Any line perpendicular to both the X-and the Z-axes is termed the *mechanical axis* (Y-axis).

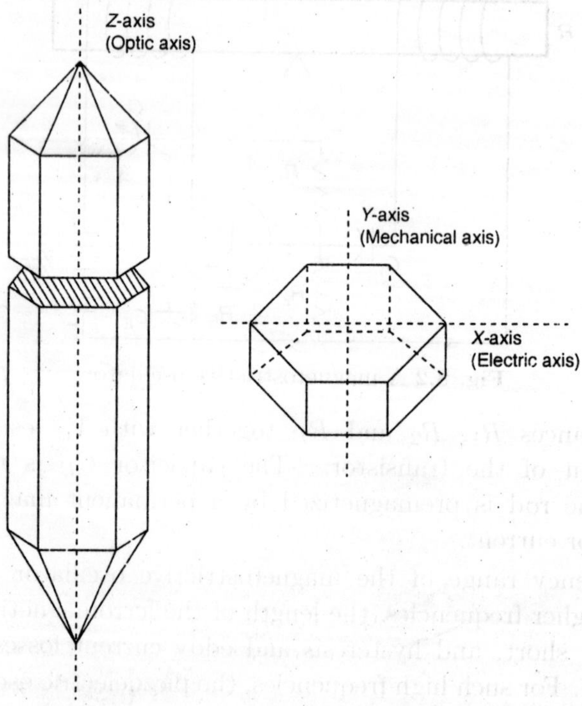

Fig. 8.3 Optic, electric, and mechanical axes of a quartz crystal

If the two largest faces of a quartz plate are perpendicular to the X-axis, the plate is referred to as an *X-cut crystal.* If the faces are perpendicular to the Y-axis, the plate is a *Y-cut crystal.* In ultrasonic work, the X-cut crystal is frequently employed. If an X-cut quartz crystal is placed in an electric field parallel to its X-axis, the plate expands along the Y-axis and contracts along the X-axis, or exhibits contraction along the Y-axis and expansion along the X-axis, depending on the direction of the electric field. If the electric field is alternating, the plate executes mechanical vibrations along both X-and Y-axes. The vibrational mode along the Y-axis (which is along the length) is termed the *length vibration*, whereas that along the X-axis (i.e., the thickness direction) is known as the *thickness vibration.* When the frequency of the electric field is equal to any of the natural frequencies for either mode of vibration of the plate, the amplitude of vibration of the plate becomes very large.

The dimension l of the plate in the direction of vibration controls the natural frequency of vibration. The frequencies of the length vibration are smaller than those of thickness vibration. The wavelength for the fundamental mode of vibration in the crystal is $\lambda = 2l$. If Y is Young's modulus along the vibrations and ρ is the density of the crystal, the compressional waves in the crystal have the velocity $c = \sqrt{\frac{Y}{\rho}}$.

For length vibration, the frequency is $272/l$ kHz, whereas for thickness vibration, it is $287/l$ kHz, l being in cm. Since the radiating surface is much greater in thickness vibration, the electro-acoustic efficiency is larger for such vibrations. Therefore, the thickness vibration is usually used in generating ultrasonic waves. Another advantage of the thickness mode is that the generated ultrasonic beam has a very small divergence.

For the excitation of the natural modes of vibration of a quartz crystal, the requisite alternating voltage is applied to it by electronic circuits. For the connection of electrodes, a thin metal film is deposited on each face of the crystal.

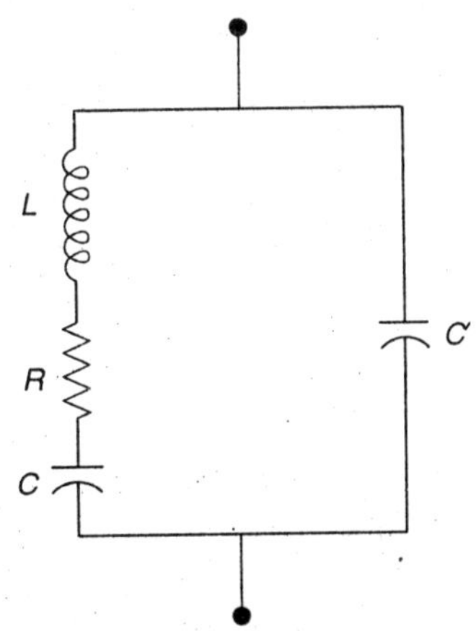

Fig. 8.4 Equivalent electrical circuit of a vibrating quartz crystal

The electrical equivalent circuit of a vibrating quartz crystal consists of a resonant circuit (Fig. 8.4). The capacitance C' is the electrostatic capacitance between the crystal electrodes when the crystal does not vibrate. The capacitance C represents the elasticity of the crystal; it is the electrical equivalent of the effective mechanical compliance. The inductance L is the electrical equivalent of the mass of the crystal. The resistance R is the electrical analogue of the resistance to the mechanical vibration caused by friction. The values of the circuit parameters are determined by the dimensions of the crystal. Since the inductance L is much larger than the resistance R, the quality factor Q of the resonant circuit is very high.

The circuit of Fig. 8.4 has two resonant frequencies f_s and f_p. f_s is the series resonant frequency of the series LCR branch :

$$f_s = \frac{1}{2\pi\sqrt{LC}}.$$

At this frequency, a low impedance is offered to the external terminals. At a frequency greater than f_s the series LCR branch offers an inductive reactance. If L' is the corresponding equivalent inductance of this branch, the equivalent circuit reduces to that of Fig. 8.5. At a frequency f_p ($> f_s$), we have

$$\omega_p L' = \frac{1}{\omega_p C'} \text{ or, } \omega_p L - \frac{1}{\omega_p C} = \frac{1}{\omega_p C'}$$

$$\text{or, } \quad f_p = \frac{\omega_p}{2\pi} = \frac{1}{2\pi}\sqrt{\frac{C+C'}{LCC'}}.$$

Thus f_p is the parallel resonant frequency of the circuit of Fig. 8.5. At this frequency, a high impedance is offered to the external terminals. Since C'/C is very large, the frequencies f_p and f_s do not differ greatly. Typically, $f_p - f_s \simeq 0.01 f_p$.

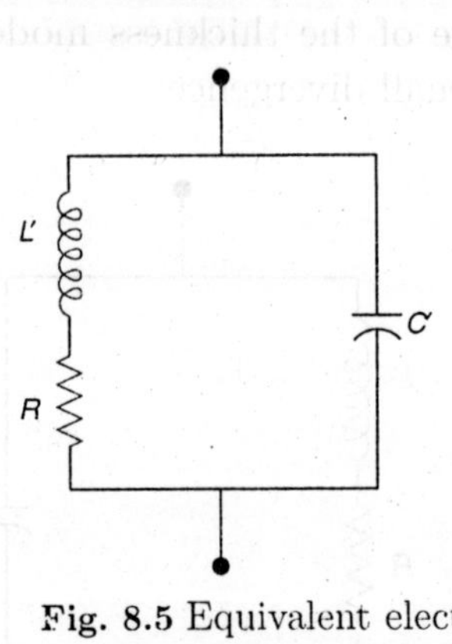

Fig. 8.5 Equivalent electric circuit for frequencies greater

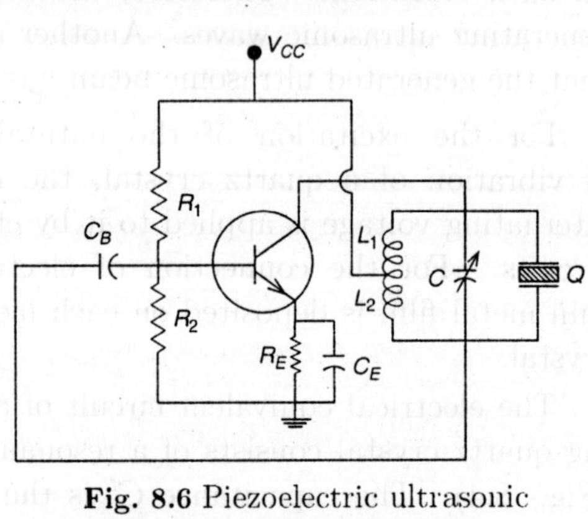

Fig. 8.6 Piezoelectric ultrasonic generator

The circuit of a piezoelectric generator, shown in Fig. 8.6, is a modified Hartley oscillator. The electrical oscillations maintain the mechanical vibration of the quartz crystal Q which is connected across the variable capacitor C of a tuned circuit. The capacitor C is adjusted such that the frequency of the resonant circuit coincides with the parallel resonant frequency f_p of the crystal. The feedback voltage for maintaining the oscillations is obtained by connecting the positive terminal of collector supply V_{cc} to a tap in the inductor of the tank circuit. Ultrasonic vibrations from a few hundred kHz to about 10 MHz can be generated by this circuit.

8.3 Detection of Ultrasonics

As the magnetostrictive and the piezoelectric effects are reversible, they can be used also to detect ultrasonic waves.

(i) **Magnetostrictive detector :** The magnetostrictive detector commonly employed in under-water applications, is shown in Fig. 8.7. The device contains several nickel tubes (N) rigidly connected to a plate (P). A coil of wire (C) is wound over each nickel tube. The plate is set into vibrations by the ultrasonic waves incident on it. As a result, stresses and strains are produced in the nickel tubes, and the magnetization in them due to the permanent magnets (M) fluctuates. Consequently, alternating voltages are induced in the surrounding coils (C) which are connected in series. The resultant voltage is amplified by high-gain amplifiers and is detected by an oscilloscope or a meter.

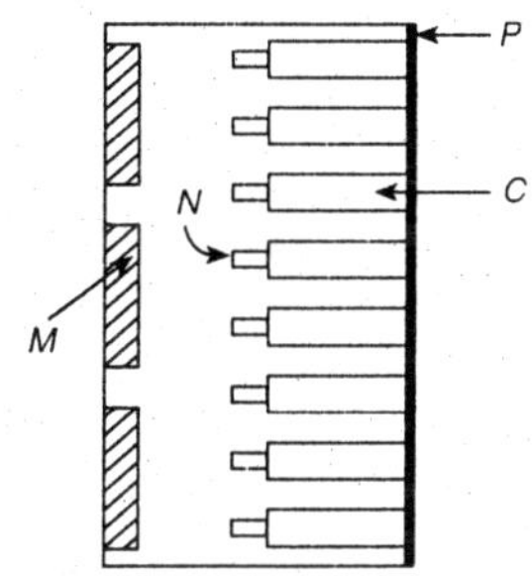

Fig. 8.7 Magnetostrictive ultrasonic detector

(ii) **Piezoelectric detector :** A quartz crystal can also serve as the detector of ultrasonic waves. The incident waves produce alternating mechanical stresses in the crystal. Due to piezoelectricity, alternating voltages are developed across the faces of the crystal plate. The voltage is amplified and fed to a meter to detect the oncoming waves.

The voltage across the crystal faces will be a maximum when the frequency of the incident wave matches the natural frequency of the crystal, whereby resonance is produced. Using high-gain amplifiers, piezoelectric receivers can be also operated at frequencies away from their resonant frequencies. Although the voltage produced across the crystal by the incident wave is small in such cases, the high-gain amplifiers enhance the voltage to the desired level for detection.

8.4 Uses of Ultrasonics

(i) **Measurement of velocity of sound:** Since ultrasonic waves are short waves, the effects of the walls of the experimental vessels in the determination of its velocity are negligible unlike the case of sonic waves for which the wavelengths are much longer. Ultrasonics have been employed to measure the velocity of sound in different media. The *optical diffraction method* used for the determination of sound velocity in liquids works on the following principle. When an ultrasonic wave propagates in a liquid, the density of the liquid acquires a periodic fluctuation in the direction of the propagation of the wave. The alternate high and low density regions cause the liquid to behave like an optical diffraction grating. Hence light waves passing through the liquid at right angles to the direction of propagation of the ultrasonic wave are diffracted. If θ is the deviation of the nth order spectrum of light of wavelength λ and $(a+b)$ is the grating constant, then from physical optics we can write $(a+b)\sin\theta = n\lambda$. For the acoustic grating, the grating constant is the wavelength of sound λ_s consisting of a compression and a rarefaction. Thus $\lambda_s \sin\theta = n\lambda$. By finding the angle θ with the standard optical technique, λ_s can be determined. If f is the frequency of the ultrasonic wave, its velocity is $c = f\lambda_s$. Since λ_s is now known, c can be readily found.

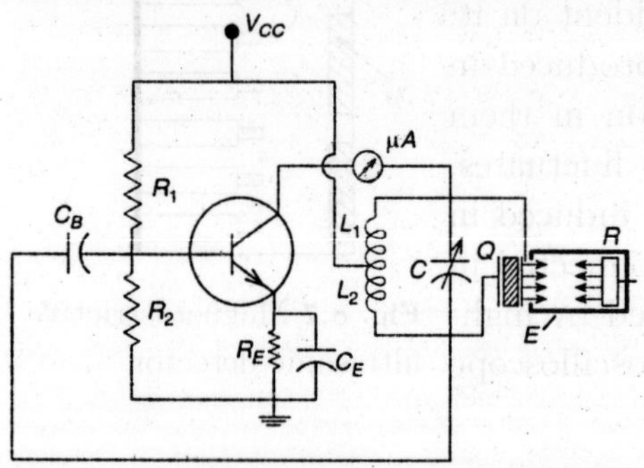

Fig. 8.8 Experimental set up for the measurement of sonic velocity in a gas

The velocity of sound in a gas or liquid using ultrasonic waves can also be found by the *acoustic interferometer method*. The experimental arrangement for a gas or air is shown in Fig. 8.8. An ultrasonic generator using a quartz crystal Q sends out ultrasonic waves in the gas under test, kept in the enclosure E. The quartz crystal is mounted in front of a smooth reflector R which can be moved parallel to itself by a micrometer screw. The ultrasonic waves are reflected by R and form standing waves. Depending on the phase of the reflected waves incident on Q, the microammeter (μA) records current maxima and minima (Fig. 8.9). The phase of the incident wave is changed

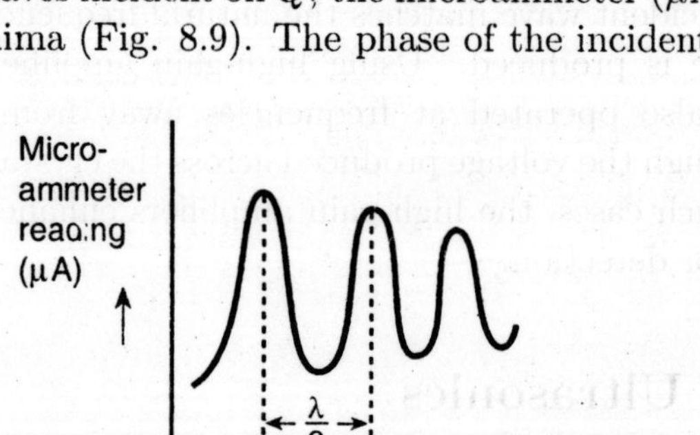

Fig. 8.9 Variation of the microammeter reading with the position of the reflector

by moving the reflector towards the crystal. If there is absorption of the ultrasonic waves in the medium, the change of the current between maxima and minima decreases with increasing distance of the reflector. The distance between two successive current peaks gives the distance $\lambda/2$ between two nodes (or antinodes), λ being the wavelength of the ultrasonic wave. Since the frequency of vibration of the crystal, f, is known, the sonic velocity $c(= f\lambda)$ can be found.

The same method with some modifications can be employed to determine the velocity of ultrasonic waves in a liquid, the arrangement being shown in Fig. 8.10. Q is the crystal vibrator, and G represents the associated electronic circuit. Owing to the large impedance mismatch, the

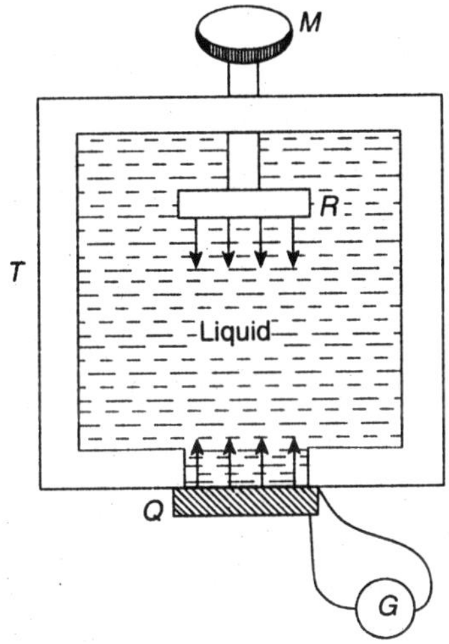

Fig. 8.10 Experimental set up for the determination of sonic velocity in a liquid

face of the crystal in contact with air emits very little energy. Most of the energy is transmitted into the liquid in the form of longitudinal waves. These waves move up the liquid column in the tank T and are reflected from the reflector R. The superposition of the incident and the reflected waves produces standing waves in the liquid. The position of the reflector can be changed by means of the micrometer screw M. As the distance x of the reflector from the face of the vibrating crystal is varied, the collector current I in the electronic circuit shows maxima and minima (Fig. 8.11) depending on the phase of the oncoming waves incident on the crystal. The distance between two successive maxima (or minima) gives half the

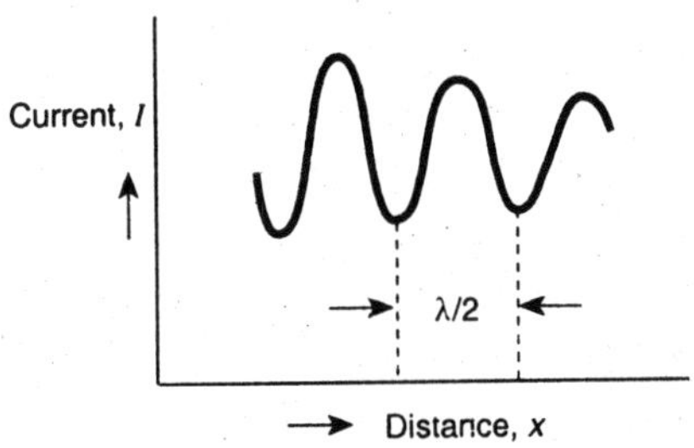

Fig. 8.11 Variation of current with the distance of the reflector

wavelength, $\lambda/2$. The frequency f being known, the velocity of the ultrasonic waves in the liquid, c, is at once determined from the relationship $c = f\lambda$.

The amplitude of vibration of the crystal should not be too large to avoid any heating of the medium. Also, the crystal and the reflector surfaces must be parallel. The crystal diameter must be considerably greater than the wavelength to ensure plane waves.

If the medium is *absorptive*, the amplitude of the current oscillations drops off with increasing distance x as $\exp(-\beta x)$ where β is the *attenuation constant*. By reading off the successive current peaks, β can be found. The viscosity and structural properties of the medium, heat conduction, and other mechanisms are responsible for the absorption of ultrasonic waves. The attenuation constant resulting from absorption depends on the frequency. At frequencies of strong absorption, the velocity of the wave changes, producing *dispersion*.

In the *pulse amplitude method*, a short-duration pulse of ultrasonic waves is sent out by the crystal towards a reflector. The reflected pulse received by the crystal is observed on a synchroscope. In the time interval τ between the emission of the pulse and its reception by the crystal, the ultrasonic sound travels twice the distance d between the crystal and the reflector. Hence the sonic velocity is found from the relationship $c = 2d/\tau$.

(ii) **Under-water applications:** Radio communication between submerged submarines in a sea is not possible due to the high attenuation of the electromagnetic waves in sea water. However, sound can travel long distances in water because the attenuation is low. The communication with under-water submarines is therefore achieved by means of ultrasonic waves.

The wavelength being short, the ultrasonic waves are highly directional and are conveniently used in *echo-depth sounding*. In this method, an ultrasonic sound pulse is sent into the sea. It is reflected from the bottom or from an intervening object like a school of fish, and is received at the surface. The time interval between the emission and the reception of the reflected pulse is noted. The velocity of sound in sea being known, the depth of the object can be precisely found.

Sound **na**vigation and **r**anging (SONAR) is a system of undersea signalling mainly used for the detection of submarines. The principle is similar to that of echo-depth sounding. An ultrasonic pulse is transmitted and the time that elapses between the transmission and the reception of the reflected pulse is recorded. This enables the distance of the target to be determined. The angular position of the transmitter can be changed. The beam being highly directional, the knowledge of the angular position of the transmitter for each received echo, and that of the distance of the target help in the determination of the speed and the direction of motion of the target.

By means of SONAR, a tracking submarine can detect a ship passing by. However, the ship can outwit the submarine by releasing a large number of tiny air bubbles in its trail. The bubbles scatter the ultrasonic SONAR beam so that the reflected beam reaching the submarine is too feeble to be detected.

An interesting *natural example* of the directive property of the ultrasonic waves is found in the *flight of bats*. Ultrasonic waves are emitted by the bats and the reflected echos from obstacles in front are received by them. Thus they can sense the obstacles and change their course accordingly.

(iii) **Medical applications:** The heating effect of concentrated ultrasonic waves can destroy diseased tissues in the body of an animal. Ultrasonic waves have been used in sterilizing equipment. Milk is freed from bacteria by such equipment.

(iv) **Chemical applications:** When exposed to intense ultrasonic beams, the particles of colloidal material in a liquid are shattered and produce very fine emulsions. Ultrasonic waves can disintegrate polymerized molecules and break the chain of starch molecules into a few fragments. Ultrasonic waves are used to coagulate smoke; the heavier particles so formed cannot remain suspended in air. Thus smoke, mist, and dust particles can be removed from air and other gases.

(v) **Flaw detection:** The *pulse-echo method* is commonly applied to test the internal flaws in a metal structure. A short duration pulse of ultrasonic waves is launched into the specimen. The pulse is reflected from an internal flaw, if any. The time interval between the transmission of the pulse and the reception of its echo gives the depth of the flaw in the specimen. A strong echo indicates a large flaw whereas a weak echo indicates a small flaw.

Generally, in the flaw detection and in the investigation of the physical properties of materials low-power ultrasonic waves are used. High power waves are applied in emulsification, drilling, cleaning etc.

8.5 Properties of Ultrasonic Waves

(i) Ultrasonic waves have frequencies lying in the range 20 kHz to 1 GHz. **(ii)** They propagate in an elastic medium, and are highly directional as the wavelength is short. **(iii)** The waves can be reflected and scattered. **(iv)** The absorption of ultrasonic waves in a medium depends on the frequency. In the region of strong absorption, the velocity of the wave changes so as to produce dispersion. **(v)** The attenuation of ultrasonic waves in sea water is much less than that of radio waves. **(vi)** When focussed, ultrasonic waves can heat up a target considerably. **(vii)** Ultrasonic waves can shatter the particles of collidal material in a liquid, disintegrate polymerized molecules, coagulate smoke, etc.

8.6 Solved Problems

Problem 1. *What is the length of the nickel rod, clamped at the middle, in a magnetostrictive oscillator generating ultrasonic waves of frequency* 25 kHz*? Young's modulus* = 20.6×10^{10} Pa *and mass density* = 8900 kg/m^3 *for nickel.*

Solution. The frequency is given by $f = \dfrac{1}{2l}\sqrt{\dfrac{Y}{\rho}}$

So, the desired length of the rod is $l = \dfrac{1}{2f}\sqrt{\dfrac{Y}{\rho}}$

where $f = 25\times10^3$ Hz, Y = Young's modulus = 20.6×10^{10} Pa, and ρ = density = 8900 kg/m^3. Thus

$$l = \frac{1}{2\times25\times10^3}\sqrt{\frac{20.6\times10^{10}}{8900}}\ \text{m} = 9.6\ \text{cm}$$

Problem 2. *A piezoelectric quartz oscillator exhibiting thickness vibration generates ultrasonic waves of frequency* 1.5 MHz. *Calculate the dimension of the quartz plate in the direction of vibration.*

Solution. If l is the required dimension in cm, the the frequency is

$$f = \frac{287}{l}\ \text{kHz}$$

Here $f = 1.5\times10^3$ kHz. Hence $l = \dfrac{287}{1.5\times10^3}$ cm = 1.91 mm

Problem 3. *A properly cut quartz crystal of dimension* 2.8 mm *in the direction of vibration produces ultrasonic waves. For quartz, Young's modulus* = 8×10^{10} Pa *and density* = 2650 kg/m^3. *Calculate the velocity of the waves in the crystal and the frequency.*

Solution. The required velocity is

$$c = \sqrt{\frac{Y}{\rho}} = \sqrt{\frac{8\times10^{10}}{2650}}\ \text{m/s} = 5.49\times10^3\ \text{m/s}$$

The wavelength in the fundamental mode of vibration is $\lambda = 2l$ where l is the given dimension of the crystal. The frequency is

$$f = \frac{c}{\lambda} = \frac{c}{2l} = \frac{5.49\times10^3}{2\times2.8\times10^{-3}}\ \text{Hz} = 981\ \text{kHz}.$$

REVIEW QUESTIONS

1. Give the acoustic spectrum and show the range of ultrasonic frequencies. What are the different types of popular ultrasonic generators?
2. Discuss how ultrasonic waves are produced by a modified Galton's whistle.

3. What is magnetostriction? Explain the operation of a magnetostrictive ultrasonic oscillator. What is the frequency range over which it can be used? Why can it not be used at higher frequencies? Why is a steady magnetizing field required?

4. (a) Explain how ultrasonic waves can be produced by a quartz plate. What determines the frequency of these waves? **[cf. C.U. 2007]**

   (b) How can ultrasonic waves be detected? **[C.U. 2006]**

5. Describe a simple method for the generation of ultrasonic waves. **[Burd. U. 1998]**

6. (a) Discuss the different applications of ultrasonics.

   (b) What is a piezoelectric crystal? Indicate briefly how is it used to generate ultrasonic waves. **[C.U. 2010]**

7. How can the flying bats locate obstacles on their path and change their trajectory accordingly?

8. Give the principle of the optical diffraction method for the determination of the speed of the ultrasonic waves in a liquid.

9. How can the acoustic interferometer method be employed to determine the wavelength and the velocity of ultrasonic waves in (i) a gas, (ii) a liquid?

10. Mention the under-water applications of ultrasonics.

11. Explain the principle of echo-depth sounding and SONAR using ultrasonic waves.

12. How can ultrasonic waves be employed to detect internal flaws in a solid specimen?

13. What are ultrasonic waves? How can they be generated and detected in the laboratory? Discuss some practical applications of ultrasonics. **[N. Beng. U. 2001]**

14. What is the range of frequencies for ultrasonic waves? Outline the principle of any method for the generation of ultrasonic waves. **[C.U. 2005]**

15. Discuss the properties of ultrasonic waves. **[C.U. 2008]**

PROBLEMS

1. A nickel rod of length 6 cm and clampled at its middle is used in a magnetostrictive oscillator to generate ultrasonic waves. Find the

frequency of the generated waves if, for nickel, $Y = 20.6 \times 10^{10}$ Pa and density = 8900 kg/m^3.

[**Ans.** 40.1 kHz]

2. If the dimension of a quartz plate in the direction of vibration is 3.8 mm, what is the frequency of length vibration?

[**Ans.** 715.8 kHz]

Chapter 9

Musical Sound, Speech, and Noise

Human ear is a natural receiver of sound and hearing is the natural method of detection of sound waves. The sensation of hearing comes from the brain. The characteristics of hearing is thus subjective in nature. The subjective aspects of sound which we consider in this chapter, belong to *physiological acoustics*.

9.1 Musical Sound

Musical sound is a pure tone or a complex note having a regular periodicity. Pitch, loudness, and quality are the subjective sensations related to a musical sound. They are not physical quantities since sensations cannot be precisely measured. Pitch, loudness, and quality are broadly linked with frequency, intensity, and waveform, respectively.

A *musical scale* is a scale of relative frequencies where the frequency of a note is expressed in terms of the frequency of a standard reference note, termed the *key note* or *tonic*. In physics, the key note is taken as a note of frequency $256 (= 2^8)$ Hz. However, the musicians' concert pitch employs a frequency of 264 Hz.

The human ear recognizes the ratio in which the frequencies of the musical notes alter, but not their numerical difference. In this subjective perception, the absolute frequencies of the notes in a musical scale are unimportant. The ratio of the frequencies of any two notes is referred to as the *musical interval* between them. Two or more notes are said to be *in unison* when they have the same frequency. Whatever be the actual frequencies, if the ratio of the frequencies of two notes is $2:1$, the interval between the two notes is termed an *octave*. Therefore, a note of frequency 400 Hz is an octave higher, and that of frequency 100 Hz is an octave lower, than a note of frequency 200 Hz. If two notes have frequencies in the ratio 2 : 3, the interval between them is known as a *fifth*. For a frequency ratio 3 : 4, the interval is a *fourth*. In Table 9.1, the frequency ratios and the corresponding names of the intervals are shown.

TABLE 9.1
Frequency ratios and musical intervals

Frequency ratio	1 : 2	2 : 3	3 : 4	4 : 5	5 : 6
Interval	octave	fifth	fourth	major third	minor third

Let three notes have the frequencies f_1, f_2, and f_3 respectively, where $f_1 > f_2 > f_3$. The interval between two successive notes will be f_1/f_2 and f_2/f_3. The interval between f_1 and f_3 is

$$\frac{f_1}{f_3} = \frac{f_1}{f_2} \times \frac{f_2}{f_3}. \tag{9.1}$$

Thus the interval between two notes is the product of the component intervals. This is also true for more than two component intervals. From Eq. (9.1), we obtain

$$\log\frac{f_1}{f_3} = \log\frac{f_1}{f_2} + \log\frac{f_2}{f_3}, \tag{9.2}$$

i.e., the logarithm of the interval between two notes is the sum of the logarithms of the component intervals. This is true for any number of component intervals.

9.2 Consonance and Dissonance

When two or more tones arriving at the ear simultaneously give a pleasant sensation in the ear, the sounds are said to be *consonant* and produce *consonance*. On the other hand, if the tones give an unpleasant sensation in the ear, the sounds are called *dissonant*, producing *dissonance*. Helmholtz found that the combination of two musical tones is generally dissonant due to the resulting beats and the upper partials. However, for certain ratios between the frequencies, no beats are produced or the beats are so feeble in intensity that they do not give any unpleasant effect in the ear. Such cases produce consonance.

The number of beats giving dissonance depends on the frequency of the component tones. For frequencies between 250 and 500 Hz, the maximum dissonance appears at a beat frequency of 33 Hz. Just as a flickering light is disturbing to the eye, so also the beats in the united sound are unpleasant to the ear.

When the interval between the component tones in a musical sound is an octave, the beat frequencies are integral multiples of the lower frequency. No dissonance is then produced. Thus *an octave gives a perfect consonance.*

An interval less than an octave does not generate a perfect consonance. Suppose that the musical interval between two frequencies f_1 and f_2 is a fifth, i.e., $f_2/f_1 = 3/2$. If $f_1 = 250$ Hz, then $4f_1 = 1000$ Hz and $5f_1 = 1250$ Hz, while $3f_2 = 1125$ Hz. Thus, $3f_2$ lies between $4f_1$ and $5f_1$, so that some dissonance is created. However, the unpleasantness in this case is tolerable. Hence the interval appears to be nearly consonant. For $f_2/f_1 = 4/3$ or $5/4$, the dissonance will increase but may not be annoying if the beating partials are weak.

When the number of beats increases beyond that giving the maximum discomfort, the sensation of harshness falls and finally disappears. For example, when the frequency of the lower tone is 256 Hz, the maximum dissonance appears at a beat frequency of 18.8 Hz. If the beat frequency increases above 47 Hz, the dissonance disappears.

In general, when the component frequencies are in the ratio of small integers, the united sound is consonant. The smaller the integers, the greater the concord. The irritating effect of the beats between the fundamental frequencies of the notes or between the different overtones that may be present is responsible for dissonance. Beat frequencies between about 30 and 130 Hz cause irritation in the middle of the musical scale.

9.3 Harmony and Melody

Three notes of frequencies in the ratio 4 : 5 : 6, when sounded together give a pleasant or concordant combination. These three notes are said to form a musical *triad.* If a triad and an octave to the lowest note of the triad are sounded together, the combination is said to produce a *chord.* Musical concords consist of chords. The series of notes forming concords, when sounded together, produce a pleasing effect which is referred to as *harmony.* The pleasing effect due to two or more concordant notes sounded in succession is termed *melody.* European music is based on harmony whereas Indian music is based on melody. Clearly, a musical scale consisting of notes that can produce both harmony and melody is highly desirable. Unfortunately, a scale suitable for one may not suit the other because the requirements for the two are not the same.

9.4 Diatonic Scale

The musical scale formed by dividing the octave into seven intervals by introducing six notes between a given note and its octave, is known as the *diatonic scale.* The intervals are in the ratio of small integers, such that the notes are consonant. The eight notes in the diatonic scale are denoted by the letters C, D, E, F, G, A, B and C', respectively. These notes are collectively known as an octave. A few such octaves on either side of the key note are needed in music. The last member of an octave forms the first member of the following octave. In Table 9.2, we show the vocal notations, symbols and frequency relations of the notes between the key note and its octave in the diatonic scale. In the table, the key note frequency is taken as 264 Hz.

TABLE 9.2

Musical symbols and frequency relations in the diatonic scale

Symbol or scale note	C	D	E	F	G	A	B	C′
Vocal notation (European)	DO	RE	MI	FA	SOL	LA	SI	do
Vocal notation (Indian)	SA	RE	GA	MA	PA	DHA	NI	sa
Actual frequency with C = 264	264	297	330	352	396	440	495	528
Relative frequency	1	9/8	5/4	4/3	3/2	5/3	15/8	2

The frequency range from about 32 to 4000 Hz can be covered roughly by 7 octaves. Usually 3 octaves are sufficient in a musical scale. The diatonic scale gives the best harmonious combinations. No other scale could be designed to replace it. The notable defect is that it is not flexible. Modern music changes the key note frequently to avoid monotony. But this change of the key note is not permitted in the diatonic scale because the frequency ratios are not equal. This defect is pointed out by saying that the diatonic scale lacks the power of modulation.

9.5 Tempered Scale

A convenient scale must satisfy the following requirements : (i) flexibility or power of modulation, i.e., the scale must be so designed that the key note can be changed, (ii) practical convenience, i.e., noninclusion of too many notes, and (iii) consonance, i.e., agreeability while playing two or more notes simultaneously. A compromise between these conflicting requirements is termed *temperament*, and is achieved in the *tempered scale* at the cost of the simple frequency ratios of the diatonic scale.

In the tempered scale of equal temperament, there are twelve notes to the octave, the intervals between successive notes being equal. This equal interval is known as a *semitone*. Since the interval I between two successive notes is the ratio of their frequencies, we have here $I^{12} = 2$, so that

$$I = 2^{\frac{1}{12}} = 1.059463. \tag{9.3}$$

Hence in the tempered scale of equal temperament, the interval between any two successive notes is $2^{1/12}$ or 1.059463. The relative frequencies of the different musical notes in the diatonic and the tempered scales are compared in Table 9.3. The table shows that the intervals in the two scales agree well within 1%. So, a tempered scale can give the concords of a diatonic scale. The five additional notes in the tempered scale with relative frequencies $2^{1/12}$, $2^{3/12}$, $2^{6/12}$, $2^{8/12}$ and $2^{10/12}$ which are not included in Table 9.3, have the symbols or scale notes $C\#, D\#, F\#, G\#$, and $A\#$, respectively. They are called C sharp, D sharp, etc. In the tempered scale, the advantage is that any of the notes can be considered to be the key note and still all the invervals required for concord are obtained. This cannot be achieved in the diatonic scale. Harmonium, piano, etc. use the tempered scale with fixed intervals. The white keys in a piano or a harmonium give the scale notes $CDEFGABC'$, whereas the black keys generate C sharp, D sharp, etc.

TABLE 9.3

Comparison of the relative frequencies of the musical notes in diatonic and tempered scales

Symbol or scale note	C	D	E	F	G	A	B	C′
Vocal notation (European)	DO	RE	MI	FA	SOL	LA	SI	do
Relative frequency (Diatonic)	1	9/8 =1.125	5/4 =1.250	4/3 =1.333	3/2 =1.500	5/3 =1.667	15/8 =1.875	2
Relative frequency (Tempered)	1	$2^{2/12}$ = 1.122	$2^{4/12}$ = 1.260	$2^{5/12}$ = 1.335	$2^{7/12}$ =1.498	$2^{9/12}$ =1.682	$2^{11/12}$ =1.888	2

9.6 Speech Sound

When air is forced from the lungs of an animal through the vocal slit and the different oral and nasal cavities, the speech sound is generated. The velocity and so the pressure of the air stream carrying the acoustic energy is modulated to give the audible sound. When this modulation is achieved with the vocal cords we get *voiced sounds*. If the modulation is done without the vocal cords, *unvoiced* or *breath sounds* are produced.

In the voiced sound, the vibrations of the vocal cords interrupt the airflow at a frequency varying from about 75 to 500 Hz. This frequency range is determined by voice quality, sex, and intonation. A person can change the fundamental frequency of his voice over a range of about 2 octaves. The velocity and pressure variations of the outgoing airstream caused by the vocal cords contain a number of harmonics. The pressure wave is further modulated by the resonating oral and nasal cavities. By controlling the shape and size of these cavities, various kinds of voiced sounds can be produced.

In whispered speech or unvoiced sound, as the breath passes through the different irregular vocal tracts, a turbulence is set up in the air, producing a *hissing sound.* This sound can be modulated by the lips, tongue, and teeth. Unvoiced sounds are distributed over a quasicontinuous frequency band lying in the upper region of the audible range.

All the vowels are voiced sounds, whereas the consonants may be voiced or unvoiced. Both voiced and unvoiced sounds are employed in an actual speech which contains vowels as well as consonants. While sounding the vowels, the mean fundamental frequency in the voice of an adult male is about 125 Hz and that for an adult female voice is about 245 Hz. Any given vowel sound consists of two or three definite groups of partials. In a speech, the energy distribution between different frequencies can change considerably for short periods. All the vowels can be recognized considering the frequency range 400 to 2000 Hz, whereas many consonants can be recognized in the frequency range 700 to 3000 Hz. To recognize all the sounds distinctly, the range should be extended to about 8000 Hz. For the transmission of short duration sounds, a wider frequency band is required.

9.7 Noise

Any unwanted sound is defined to be *noise.* When somebody is taking a rest or listening to a speech, even music, however sweet it may be, turns out to be noise. Noise has both subjective and physical characters. Sometimes a pure single tone appears to be a noise if it has disagreeable pitch or loudness. The combined sound output of many sound sources of different frequencies often produces a noise. Thus noise is unpitched sound.

Tremendous noise has irritating effects and can cause health hazards including hearing impairment. Hence care has to be taken to avoid 'noise pollution'. Noise can interfere with and mask the desired sound, and so it has to be minimized in conference rooms and in auditoria. For comfort and safety, the noise power level should be below 30 dB. Long exposure to noise above 45 dB can cause sleep disorders, headache, hypertension, high blood sugar, and cardiac problems.

The noise arising in a room due to a running fan or motor has a direct component coming from the source and a reflected component from the walls. Near the source, the sound absorption by the walls is not effective in reducing the noise. At some distance from the source, however, the absorption by the walls decreases the sound. A 3 dB reduction in the noise is possible by doubling the wall absorption. Also, the conversation of a large number of people in a room is difficult unless the wall absorption is large.

Noise can enter into a room from outside through the walls, doors, and windows. The transmission loss through the walls increases, thus reducing the incoming noise, if the effective area of the wall is increased. The noise transmitted through the doors can be minimized if a double door is used and if the edges of the doors are fitted with felt or rubber strips. In this way, the air gaps at the border of the doors allowing the entry of noise are plugged; the absorption of sound is increased; and the vibrations of the body of the doors are reduced. The entry of noise through the windows can be minimized using double windows mounted in separate frames with a gap of 10 cm. The transmission of noise through the floor can be decreased by constructing a wooden floor on fibre glass quilts laid over joists.

9.8 Some Features of Hearing

The *audible frequency range* extends from about 20 Hz to 20 kHz. The frequencies below 20 Hz can however be perceived as sound by some people, if it is sufficiently intense. For young people, the upper limit of audibility is 20 kHz, but for middle-aged persons, it is about 16 kHz.

Within the audible range of frequencies, a sound is not heard unless its intensity exceeds a certain value, known as the *threshold audibility.* For a pure tone of 1 kHz, the rms sound pressure at the threshold audibility is $2 \times 10^{-5}\,\text{N/m}^2$. This value is taken as a standard, though the threshold audibility depends on the frequency, on the listener, and on several other factors. The threshold audibility increases with age, the rise being greater for men than for women.

The normal human ear is most *sensitive* at about 3 kHz. This is partly due to resonance in the auditory canal at this frequency.

When the intensity of sound of a given frequency exceeds an upper limit, called the *threshold of feeling* or *tolerance*, an irritation is felt in the ear. This threshold is also a function of frequency. At 1 kHz, the rms sound pressure at the threshold of feeling is about $20\,\text{N/m}^2$.

The *loudness* of a sound is a judgment of the brain and is related to the intensity. The unit of loudness level is *phon* (see Sec. 5.15).

The subjective loudness, measured in *sone*, shows how much louder one sound is than another when their loudness levels in phon are known. The relationship between loudness and loudness level is shown graphically in Fig. 9.1. The unit *sone* for loudness gives the loudness of a pure 1 kHz tone 40 dB above the threshold audibility, i.e., of a sound whose loudness level is 40 phon. It is noted that equal increase of loudness level in phon does not give equal increase of loudness in sone.

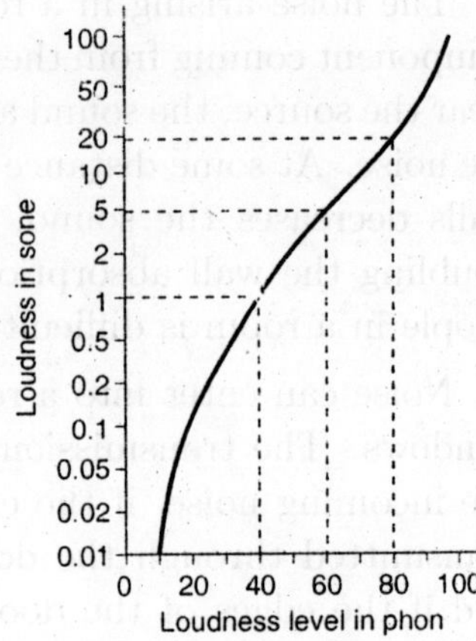

Fig. 9.1 Loudness versus loudness level

The *pitch* of a sound is a judgment of the brain and is mainly determined by the frequency, but it is not the same as frequency. The subjective unit of pitch is *mel*, derived from the word *melody*. The pitch of a pure 1 kHz tone at a sound pressure level of 60 dB is referred to as 1000 mel. The variation of pitch in mel with frequency in Hz is depicted in Fig. 9.2. When the frequency is doubled from 1 to 2 kHz, the pitch is not doubled but increases from 1000 to 1500 mel.

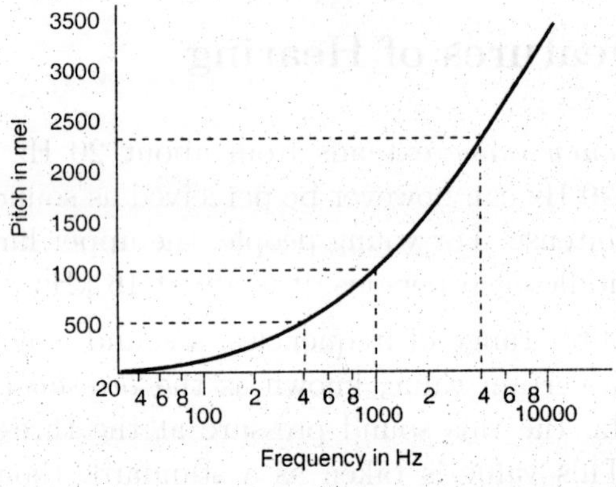

Fig. 9.2 Pitch versus frequency curve

The *quality* or *timbre* of a complex musical sound is the subjective characteristic that enables us to distinguish it from another note of the same loudness and fundamental frequency. The quality is primarily determined by the waveform of the note. A set of subjective units of timbre has not been established, however.

Observations

(1) To a normal listener, the pitch of a pure tone of a given frequency appears to depend on loudness. At frequencies below 1000 Hz, the pitch falls with increasing loudness, the effect being most noticeable

between 70 and 300 Hz. Above 1000 Hz, the drop of pitch with increasing loudness is not significant.

(2) If the fundamental does not appear in a mixed note consisting of several harmonics of the fundamental, the pitch of the note corresponds to the fundamental for psychophysiological reasons. Thus, when a note contains the frequencies 400, 600, and 800 Hz, the pitch appears to be that of a 200 Hz tone. If the frequencies 500 and 700 Hz are added to the note in question, the pitch will drop to that of a 100 Hz tone.

The *missing fundamental* supplied in the ear plays a useful part in small loudspeakers as employed in domestic radio receivers. Such loudspeakers being inefficient radiators of low frequencies, their sound output is expected to be richer in higher frequencies. But the harmonic distortion at lower frequencies gives harmonics of such frequencies, so that the missing fundamental is supplied in the ears of the listeners.

9.9 Solved Problems

Problem 1. *A note of frequency* 300 Hz *is an octave lower than a note of frequency* f_1. *The interval between the note of frequency* f_1 *and another note of frequency* f_2 $(< f_1)$ *is a fifth. Find the frequencies* f_1 *and* f_2.

Solution. We have $f_1/300 = 2$; hence $f_1 = 2 \times 300 = 600$ Hz.

Also $f_2/f_1 = 2/3$. So, $f_2 = 2 \times 600/3 = 400$ Hz.

Problem 2. *Calculate the frequencies in a tempered scale for the key note frequency of* 264 Hz

Solution. The interval between two successive notes is $2^{1/12} = 1.059463$.

The key note frequency is $f_0 = 264$ Hz (given). The other frequencies are

$f_1 = 1.059463 \times f_0 = 279.7$ Hz; $f_2 = 1.059463 \times f_1 = 296.3$ Hz;

$f_3 = 1.059463 \times f_2 = 314$ Hz; $f_4 = 1.059463 \times f_3 = 332.6$ Hz;

$f_5 = 1.059463 \times f_4 = 352.4$ Hz; $f_6 = 1.059463 \times f_5 = 373.4$ Hz;

$f_7 = 1.059463 \times f_6 = 395.6$ Hz; $f_8 = 1.059463 \times f_7 = 419.1$ Hz;

$f_9 = 1.059463 \times f_8 = 444$ Hz; $f_{10} = 1.059463 \times f_9 = 470.4$ Hz

$f_{11} = 1.059463 \times f_{10} = 498.4$ Hz; and $f_{12} = 2 \times 264 = 528$ Hz

REVIEW QUESTIONS

1. What do you mean by a musical scale and a musical interval? Name some important intervals. How is the interval between two notes related to the component intervals?

2. What are consonance and dissonance? How do they originate? Discuss the consonance and dissonance of an octave and a major fifth having an interval 3/2. How is dissonance related to the number of beats?

3. Explain the following : Musical triad, chord, harmony, melody.

4. What is a diatonic scale? Write the musical symbols and the frequency relations in this scale. Mention the advantages and the disadvantages of the diatonic scale.

5. What are the requirements of a convenient musical scale? Define temperament and the tempered scale. Describe the tempered scale of equal temperament and compare it with the diatonic scale. Which instruments do use the tempered scale?

6. What is speech sound? Name the different types of speech sound and discuss their features.

7. What is noise? Mention the safe noise power limit. How effectively does the wall absorption reduce the noise in a room? In what ways can noise enter into a room from outside? What steps can be taken to reduce the entry in each case?

8. (a) At what frequency is the human ear most sensitive? (b) Explain the following terms: (i) threshold audibility, (ii) threshold of feeling, (iii) sone, (iv) mel, and (v) timbre.

PROBLEMS

1. The numerical difference between the frequencies of two notes is 150 Hz and the interval between them is a major third. Calculate the frequencies.

   [**Ans.** 750 Hz, 600 Hz]

2. A note of frequency f_1 is an octave higher than a note of frequency f_2. The interval between the note of frequency f_2 and another note of frequency f_3 ($< f_2$) is a fourth. If $f_2 \sim f_1 = 400$ Hz, calculate f_1, f_2, and f_3.

   [**Ans.** 800 Hz, 400 Hz, 300 Hz]

Chapter 10

Microphones, Loudspeakers, and Hearing Aids

A *transducer* is a device which transforms a signal of some form of energy into another. Microphones and loudspeakers are electroacoustic transducers. Microphones are used to convert an acoustic signal into an electric signal. Loudspeakers do the reverse, i.e., transform an electric signal into an acoustic signal. The salient features of microphones and loudspeakers will be discussed in this chapter. We shall also deal with several types of hearing aids later on. The topics incorporated in this chapter can be grouped under *technical* or *industrial acoustics.*

10.1 Microphone

A *microphone* gives a varying or alternating voltage when sound waves are incident on it. It forms the first element in telephone, broadcasting, and sound recording. A microphone operates in air. If operated in water, it is referred to as a *hydrophone.*

A microphone has the following important characteristics.

(i) **Sensitivity:** The *sensitivity* or the *response* of a microphone is the open-circuit voltage produced by a unit sound pressure at a given frequency.

(ii) **Fidelity:** If the response of a microphone is the same at all frequencies and varies linearly with pressure, the microphone is said to possess a *high fidelity.*

(iii) **Dynamic range:** The range of intensities over which the output voltage varies linearly with the acoustic pressure, is known as the *dynamic range* of the microphone.

(iv) **Self noise:** The random motion of the electrons in the electric resistance of the microphone and that of the molecules in contact with the diaphragm of the microphone produce an output voltage even in the absence of an input acoustic signal. This output voltage is referred to as the inherent noise or the *self noise* of the microphone.

(v) **Directivity:** The *directivity* of the microphone refers to the change of response with a change in the angle between the source direction and the microphone axis.

(vi) **Harmonic distortion:** If the vibration amplitude is large, the restoring force is not strictly proportional to the displacement. The resulting nonlinearity generates harmonics of the input frequency, thus giving *harmonic distortion* of the signal at the output. The maximum input sound pressure is thus limited by the tolerable harmonic distortion.

The commonly used microphones can be broadly divided into two groups : (i) pressure microphones and (ii) velocity microphones. The *pressure microphones* respond to the changes of the sound pressure whereas the *velocity microphones* are actuated by the particle velocity arising from the propagation of the sound wave through the medium. Changing resistance or carbon microphone, electrostatic or capacitor microphone, moving coil or electrodynamic microphone, and piezoelectric or crystal microphone are different types of pressure microphone. Ribbon microphone belongs to the category of velocity microphone.

The *requirements of a good microphone* are as follows :

(i) high sensitivity,

(ii) low noise,

(iii) no amplitude and frequency distortions,

(iv) a directivity appropriate for its use, and

(v) characteristics independent of external disturbances like temperature and humidity changes.

No single microphone can meet all these requirements in practice.

10.2 Carbon Microphone

The carbon microphone is a changing-resistance pressure microphone in which the electric current is modulated by the change of resistance of loosely packed carbon granules acted on by the pressure variation of the incoming sound waves. Carbon microphones find applications in telephony.

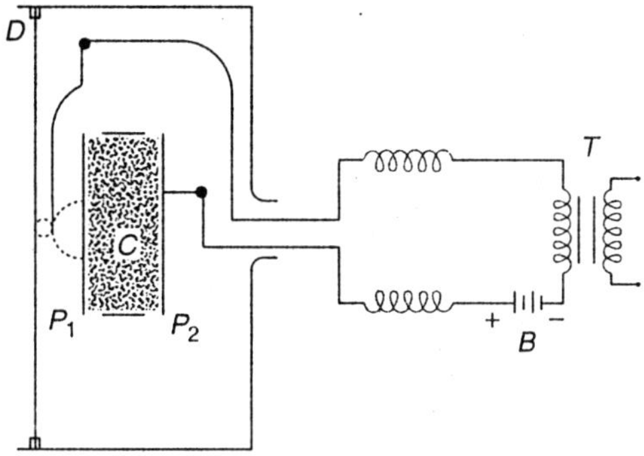

Fig. 10.1 Carbon microphone

Figure 10.1 shows schematically a carbon microphone. Carbon granules (C) are contained in a box, the front and the back of which are polished carbon plates (P_1, P_2). The front plate P_1 is provided with a boss attached to the centre of the diaphragm (D). As sound waves fall on D, it vibrates together with P_1. As a result, the pressure on the carbon granules fluctuates. Accordingly, the electric resistance of the carbon granules changes, because this resistance is determined by the tightness of packing. An increase of pressure decreases the resistance, and *vice versa*. The diaphragm is suitably damped with the help of flannel or cotton-wool washers to avoid resonance effects.

The battery B drives an electric current through the carbon granules. As the resistance of the carbon granules changes due to the impinging sound waves, the electric current in the circuit changes correspondingly. This fluctuating current passes through the primary of the step-up transformer T. The voltage induced across the secondary terminals is the electrical replica of the incident acoustic wave. The *source* of power in the circuit is the battery B, the sound vibrations merely acting as the *modulating agent.*

Mathematical analysis

Let R_0 be the resistance of the granular path in the absence of any acoustic wave. Assuming that the change of resistance of the granular path is linearly related to the displacement of the diaphragm, the electric resistance of the path in the presence of an incident acoustic wave is given by

$$R = R_0 + kx, \tag{10.1}$$

where x is the instantaneous displacement of the diaphragm due to the impinging wave, and k is a constant giving the resistance change per unit displacement of the diaphragm.

Away from resonance, the amplitude of oscillation of the diaphragm is small and the oscillations are stiffness controlled. If s is the stiffness of the diaphragm, then the restoring force is sx, and it can be equated to the force on the diaphragm exerted by the sound waves. This force is $Ap_0 \cos\omega t$, where A is the area of the diaphragm, p_0 is the pressure amplitude, and ω is the angular frequency of the incident wave. Thus

$$sx = Ap_0 \cos\omega t$$

$$\text{or,} \qquad x = x_0 \cos\omega t, \tag{10.2}$$

where

$$x_0 = \frac{Ap_0}{s}. \tag{10.3}$$

Substituting for x from Eq. (10.2) into Eq. (10.1), we get

$$R = R_0 + kx_0 \cos\omega t. \tag{10.4}$$

If V_0 is the battery voltage, the instantaneous current in the microphone circuit is

$$\begin{aligned} i &= \frac{V_0}{R} = \frac{V_0}{R_0 + kx_0 \cos\omega t} \\ &= \frac{V_0}{R_0}\left(1 + \frac{kx_0}{R_0}\cos\omega t\right)^{-1} \\ &= \frac{V_0}{R_0}\left(1 - \frac{kx_0}{R_0}\cos\omega t + \frac{k^2x^2}{R_0^2}\cos^2\omega t + \cdots\right). \end{aligned} \tag{10.5}$$

Since $kx_0 \ll R_0$, the terms higher than the linear in kx_0/R_0 in Eq. (10.5) are negligible. These higher order terms give the harmonics of the frequency ω. For example, the $\cos^2\omega t$ term generates the second harmonic since $\cos^2\omega t = \frac{1}{2}(1 + \cos 2\omega t)$. Neglecting these insignificant terms we obtain

$$i \simeq \frac{V_0}{R_0}\left(1 - \frac{kx_0}{R_0}\cos\omega t\right). \tag{10.6}$$

Thus the current has an alternating component at the frequency of the incident wave superimposed on the steady component V_0/R_0. The alternating component of the current is

$$i_{\text{ac}} = -\frac{V_0kx_0}{R_0^2}\cos\omega t. \tag{10.7}$$

The alternating current i_{ac} can be considered to arise from an alternating emf v_{ac} generated internally in the microphone, where

$$v_{\text{ac}} = i_{\text{ac}}R_0 = -\frac{V_0kx_0}{R_0}\cos\omega t. \tag{10.8}$$

The peak value of v_{ac} is

$$V_p = \frac{V_0 k x_0}{R_0} = \frac{V_0 k A p_0}{R_0 s}.$$

The *sensitivity* or the *response* of the microphone is expressed by

$$S = \frac{V_p}{p_0} = \frac{V_0 k A}{R_0 s}.$$

Thus the sensitivity increases if the battery voltage V_0 and the diaphragm area A are increased, or if the granule resistance R_0 and the stiffness of the diaphragm s are reduced. But increasing V_0 produces excessive heating and self noise, and increasing k/R_0 increases the amplitude of the second harmonic. Thus there are practical limitations to the enhancement of the sensitivity.

The peak output voltage across the transformer secondary is $\omega M V_0 k x_0 / R_0^2$, where M is the mutual inductance between the primary and the secondary of the transformer.

The response is usually expressed in dB with reference to a standard response S_0. Thus, the response in dB is given by $20 \log_{10}(S/S_0)$. Normally, S_0 is taken as 1 volt per microbar, i.e., $S_0 = 1$ volt per dyne/cm^2 = 1 volt per 0.1 N/m^2 = 10 V/Pa.

Discussions

The sensitivity of the carbon microphone is higher than all other microphones. As it is employed almost exclusively in telephony, the frequency range is confined to 100 to 5,000 Hz. The frequency response of the carbon microphone is displayed in Fig. 10.2. The peak response takes place near 2,000 Hz and is related to the fundamental resonant frequency of the diaphragm. The uneven response at higher frequencies is accounted for by the break up of the diaphragm into different overtone modes of vibration.

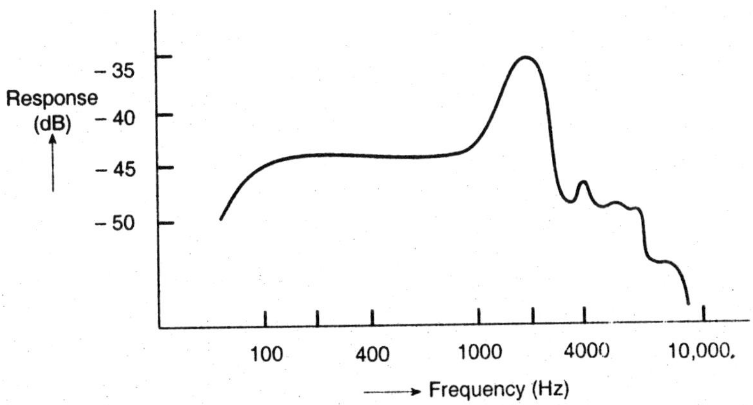

Fig. 10.2 Frequency response of a carbon microphone

The local heating at the points of contact of the carbon granules gives a large self noise in carbon microphones. Furthermore, the sensitivity is reduced by the packing tendency of the carbon granules. The resistance variation proportional to the diaphragm displacement, as assumed in the theory, is true only for small sound pressures. High acoustic pressures introduce harmonic distortions. In some special designs of carbon microphone, the harmonics are automatically reduced, and a flat frequency response comparable to the best microphones of other varieties can be achieved.

10.3 Capacitor Microphone

The capacitor microphone or the electrostatic microphone utilizes the change of capacitance of a parallel plate capacitor caused by the vibration of a diaphragm under the action of the sound waves. A schematic diagram of the capacitor microphone is given in Fig. 10.3. A steel diaphragm D, about 0.005 cm thick, stretched almost to its elastic limit, forms one plate of a parallel plate capacitor. The other plate of the capacitor is the insulated steel disc P which is separated from the diaphragm by an air gap of about 0.002 cm. The stretching of the diaphragm enhances its natural frequency to about 10 kHz, so that any mechanical resonance is avoided. The back plate P is grooved to control the natural frequency and to provide the requisite damping to the diaphragm. The capacitor is connected in series with a battery of voltage V_0 and a high resistance R (a few meg ohm).

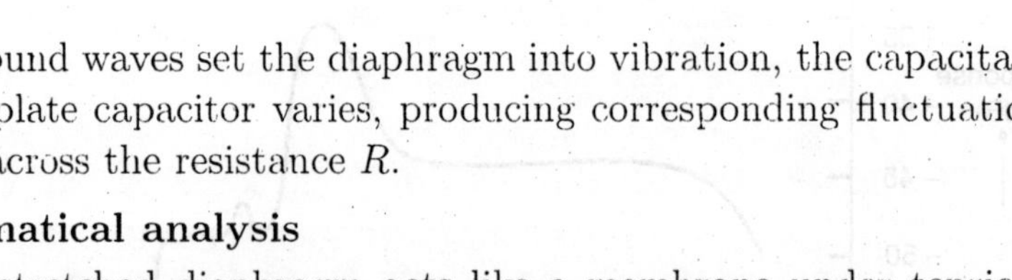

Fig. 10.3 Capacitor microphone

As sound waves set the diaphragm into vibration, the capacitance of the parallel plate capacitor varies, producing corresponding fluctuations in the voltage across the resistance R.

Mathematical analysis

The stretched diaphragm acts like a membrane under tension. When an excess pressure p is exerted on the diaphragm, its displacement is approximately expressed as

$$x = \frac{pA}{8\pi T} \tag{10.9}$$

where A is the area of the diaphragm and T is the tension per unit length of the boundary of the diaphragm.

In the absence of the incident sound wave, let d_0 be the distance between the plates of the parallel plate capacitor. Then the capacitance is (in SI units)

$$C_0 = \epsilon_0 \frac{A}{d_0}, \tag{10.10}$$

where ϵ_0 is the permittivity of free space. In the presence of the sound wave of angular frequency ω, the instantaneous distance between the plates is

$$d = d_0 - x = d_0 - \frac{p_0 A}{8\pi T} \cos \omega t, \tag{10.11}$$

where p_0 is the excess pressure amplitude. The instantaneous capacitance is

$$C = \frac{\epsilon_0 A}{d_0 - x} = \frac{\epsilon_0 A}{d_0 \left(1 - \frac{x}{d_0}\right)} = \frac{C_0}{1 - x/d_0}. \tag{10.12}$$

The initial charge stored in the capacitor is $C_0 V_0$. If V is the instantaneous voltage across the capacitor when the capacitance is C, the charge stored is CV. *Under open circuit conditions*, the capacitor charge is conserved, i.e.,

$$CV = C_0 V_0$$

$$\text{or,} \quad V = \frac{C_0}{C} V_0 = \left(1 - \frac{x}{d_0}\right) V_0, \tag{10.13}$$

where Eq. (10.12) has been used. The change in the voltage is

$$\Delta V = V_0 - V = \frac{V_0}{d_0} x = \frac{V_0 p_0 A}{8\pi T d_0} \cos \omega t. \tag{10.14}$$

The peak voltage change is

$$\Delta V_p = \frac{V_0 p_0 A}{8\pi T d_0}. \tag{10.15}$$

The *sensitivity* or the *response* is given by

$$S = \frac{\Delta V_p}{p_0} = \frac{V_0 A}{8\pi T d_0}. \tag{10.16}$$

The fluctuating voltage across the capacitor is small and so it must be amplified. Special moisture-resistant coatings are applied to the insulators to prevent charge leakage due to humidity. The microphone has little background noise and its response is uniform over the frequency range from about 100 to 4000 Hz (Fig. 10.4). There is a small rise of the response near the fundamental resonant frequency of 5000 Hz. The sensitivity of the microphone is not high. This type of microphone is mainly used to measure acoustic pressures and as a studio microphone for high-fidelity (Hi-Fi) pickup of music.

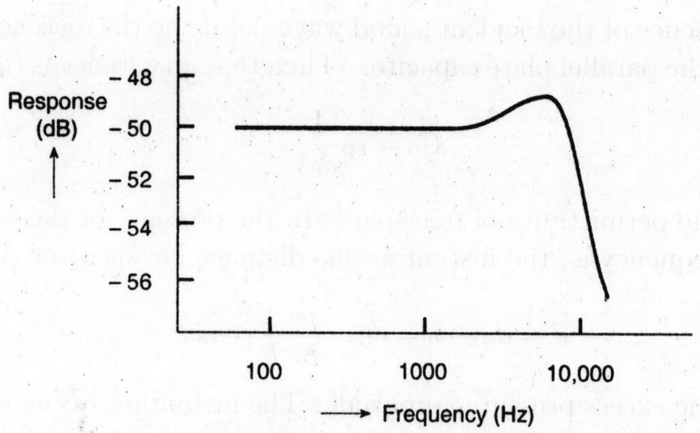

Fig. 10.4 Frequency response of a capacitor microphone

Expression for the output voltage across R

The instantaneous capacitance C is given by Eq. (10.12). If $x = x_0 \cos \omega t$, we have

$$C \simeq C_0 \left(1 + \frac{x}{d_0}\right) = C_0 + C_1 = C_0 + C_{10} \cos \omega t, \tag{10.17}$$

since $x \ll d_0$. Here $C_{10} = C_0 x_0 / d_0$. The Kirchhoff's voltage low equation for the circuit of Fig. 10.3 is

$$R\frac{dq}{dt} + \frac{q}{C_0 + C_1} = V_0, \tag{10.18}$$

where q is the instantaneous charge stored on the capacitor. We write $q = q_0 + q_1$, where $q_0 (= C_0 V_0)$ is the steady charge and q_1 is the fluctuating charge. Hence $\frac{dq}{dt} = \frac{dq_1}{dt}$ and

$$\frac{q}{C_0 + C_1} = \frac{q_0 + q_1}{C_0 + C_1} = \frac{q_0(1 + q_1/q_0)}{C_0(1 + C_1/C_0)} \simeq V_0 \left(1 + \frac{q_1}{q_0}\right)\left(1 - \frac{C_1}{C_0}\right)$$

$$\simeq V_0 \left(1 + \frac{q_1}{q_0} - \frac{C_1}{C_0}\right).$$

Hence Eq. (10.18) simplifies to

$$R\frac{dq_1}{dt} + V_0 \left(1 + \frac{q_1}{q_0} - \frac{C_1}{C_0}\right) = V_0$$

$$\text{or,} \quad C_0 R\frac{dq_1}{dt} + q_1 = C_1 V_0 = C_{10} V_0 e^{j\omega t}. \tag{10.19}$$

Taking $q_1 \sim e^{j\omega t}$, Eq. (10.19) yields $(1 + j\omega C_0 R) q_1 = C_{10} V_0 e^{j\omega t}$ whence

$$q_1 = \frac{V_0 C_{10} e^{j\omega t}}{1 + j\omega C_0 R}.$$

So, the ac output voltage across R is

$$v_{ac} = R\frac{dq_1}{dt} = \frac{j\omega C_{10}RV_0e^{j\omega t}}{1+j\omega C_0R} = \frac{\omega C_{10}RV_0}{\sqrt{1+\omega^2C_0^2R^2}}e^{j(\omega t+90^\circ-\theta)}$$

where $\tan\theta = \omega C_0R$. Taking the real part, we have for the physical solution

$$v_{ac} = \frac{\omega C_{10}RV_0}{\sqrt{1+\omega^2C_0^2R^2}}\cos(\omega t+\phi), \qquad (10.20)$$

where $\phi = 90^\circ - \theta$.

Comparison of capacitor microphones with carbon microphones

(i) The sensitivity of the carbon microphone is higher than that of the capacitor microphone. However, the frequency response of the capacitor microphone is flatter over a wider frequency range.

(ii) With increasing battery voltage, the sensitivity of both the microphones increases. For the carbon microphone, the enhanced battery voltage causes excess heating at the contact points. For the capacitor microphone, this effect is absent.

(iii) In capacitor microphones, long cables are not used since that would enhance the stray capacitance. This restriction does not apply to carbon microphones.

(iv) Charge leakage may occur in capacitor microphones. This effect is of no concern in carbon microphones.

(v) Carbon microphones are widely used in telephony, whereas capacitor microphones are used to measure acoustic pressures and in studios for high-fidelity pickup of music.

(vi) The carbon microphone has considerable background noise arising from the heating at the contacts of the carbon granules. The capacitor microphone has little background noise.

Observation

Recently, a high-quality ultra-small *silicon microphone* of the capacitor type has been fabricated by replacing the steel diaphragm and the steel disc of the conventional capacitor microphone by single-crystal silicon. The fabrication method uses two silicon wafers bonded together, the desired portions of which are chemically etched. Silicon microphone has several advantages :

(i) Single-crystal silicon has a tensile strength 15 times greater than that of steel. This results in a wide dynamic range, and excellent fidelity and sensitivity.

(ii) It has a long life and can withstand high-temperature and high-humidity conditions.

(iii) Sensitivity and directionality can be increased by placing multiple silicon microphones in an array on a silicon wafer.

Silicon microphones are used in many applications including programme production and cellular phones.

10.4 Moving-coil Microphone

This is an electrodynamic microphone acting on the principle that when the magnetic flux linked with a conductor changes with time, an emf is induced in it. The device consists of a permanent 'pot' magnet having a central north pole (N) and a south pole (S) on a surrounding ring (Fig. 10.5). The magnetic field lines in the air gap between the poles are radial. In the annular space between the poles, there is a circular coil (C, C) attached to the lower face of a diaphragm D.

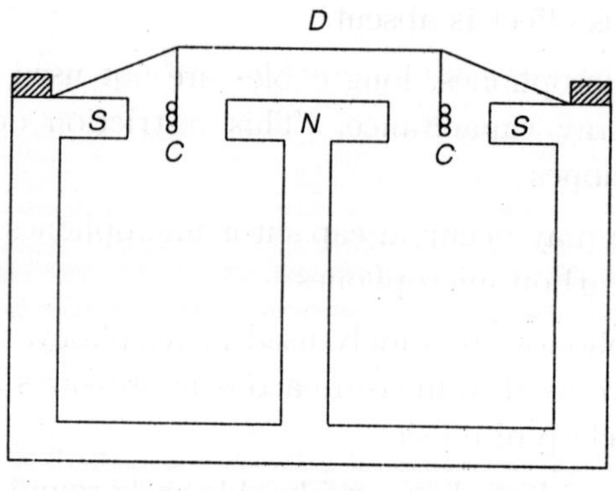

Fig. 10.5 Moving-coil microphone

As sound waves impinge on the diaphragm D, the latter vibrates and the coil (C, C) cuts across the magnetic field lines. Hence an electromotive force is induced in the coil. This emf is stepped up by a transformer and fed to an amplifier.

Mathematical analysis

If B is the magnetic induction between the poles, l is the length of the coil, and v is the velocity of the coil due to the incident acoustic pressure, the induced emf in the coil is $\mathcal{E} = Blv$. The equation of motion of the moving system acted on by the sound wave is

$$m\frac{d^2x}{dt^2} + R\frac{dx}{dt} + sx = P_0\alpha \cos \omega t, \tag{10.21}$$

where m is the mass of the vibrating system, R is the mechanical resistance, s is the stiffness constant, x is the displacement at time t, α is the area of the diaphragm, and $P_0 \cos \omega t$ is the incident sound pressure. Putting $\cos \omega t = \text{Re}(e^{j\omega t})$ and $x = \text{Re}(x_0 e^{j\omega t})$, where 'Re' stands for the 'real part', we have from Eq. (10.21)

$$(-m\omega^2 + j\omega R + s)x_0 = P_0\alpha,$$

whence

$$x = \frac{P_0 \alpha e^{j\omega t}}{j\omega Z_m}, \tag{10.22}$$

where Z_m is the mechanical impedance of the moving system :

$$Z_m = R + j\left(\omega M - \frac{s}{\omega}\right). \tag{10.23}$$

The velocity of the coil is

$$v = \dot{x} = \frac{P_0 \alpha e^{j\omega t}}{Z_m}.$$

The physical solution is

$$v = \frac{P_0 \alpha}{|Z_m|} \cos(\omega t - \theta) = v_0 \cos(\omega t - \theta), \tag{10.24}$$

where $|Z_m| = \sqrt{R^2 + (\omega M - \frac{s}{\omega})^2}$ and $\tan\theta = \frac{\omega M - s/\omega}{R}$.

The peak value of the induced emf is $\mathcal{E}_0 = Blv_0 = BlP_0\alpha/|Z_m|$. Hence, the *sensitivity* or the *response* is given by

$$S = \frac{\mathcal{E}_0}{P_0} = \frac{Bl\alpha}{|Z_m|}. \tag{10.25}$$

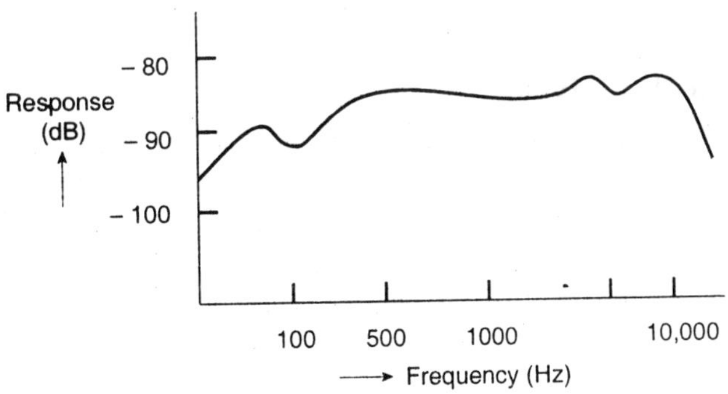

Fig. 10.6 Frequency response of a moving-coil microphone

The frequency response of the electrodynamic microphone is depicted in Fig. 10.6. The response is fairly flat between 50 and 8000 Hz. The sensitivity is reasonably good and the background noise is quite small. The uniform response over a wide frequency range makes this microphone suitable for use in broadcasting.

10.5 Crystal Microphone

The crystal microphone is of the piezoelectric variety, and is based on the principle that when a mechanical stress is exerted in a certain direction of a suitably cut slab of a piezoelectric crystal, such as Rochelle salt, a potential difference appears across the electrodes in contact with the two faces of the slab. If a plate with 45° X-cut of Rochelle salt crystal expands or contracts under a longitudinal stress, electric charges appear on its flat faces.

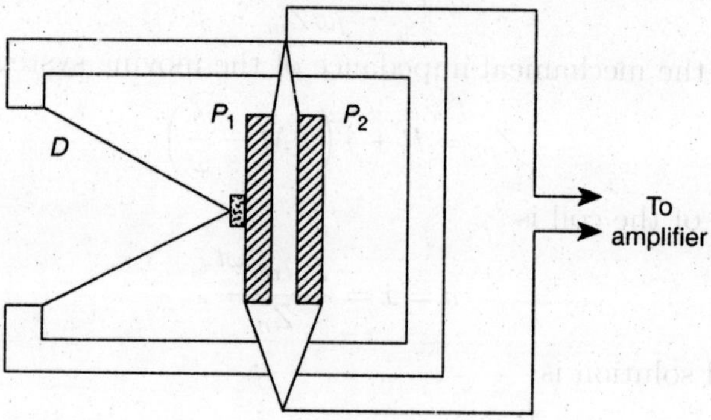

Fig. 10.7 Crystal microphone

The crystal microphone is schematically shown in Fig. 10.7. Two plates of Rochelle salt crystal (P_1, P_2) are employed. As the diaphragm D is set into vibration under the action of the incident sound waves, one of the Rochelle salt plates is stretched while the other is compressed. The plates are so mounted and the metallic foil leads are so connected to them that the voltage developed across the plates are added to give the potential difference between the leads. This voltage is fed to an amplifier.

Mathematical analysis

When a properly cut piezoelectric plate is subjected to a longitudinal strain $d\xi/dx$, a surface density of charge σ appears on its two faces, where $\sigma = k\frac{d\xi}{dx}$, k being the piezoelectric coupling constant. If A is the area of a face of the plate, the potential difference across the two faces will be $\mathcal{E} = \sigma A/C$, where C is the capacitance between the faces. Thus

$$\mathcal{E} = \frac{kA}{C}\frac{d\xi}{dx}. \tag{10.26}$$

The acoustic pressure for the incident sound wave is $p = P_0 \cos \omega t$, so that the strain is $\frac{d\xi}{dx} = -p/Y$, where Y is the appropriate Young's modulus. The negative sign is needed since a positive acoustic pressure compresses the crystal. So,

$$\mathcal{E} = -\frac{kAP_0}{CY} \cos \omega t. \tag{10.27}$$

The peak voltage is $\mathcal{E}_m = \frac{kAP_0}{CY}$, so that the *sensitivity* is

$$S = \frac{\mathcal{E}_m}{P_0} = \frac{kA}{CY}. \tag{10.28}$$

The performance of a Rochelle salt crystal changes with temperature, the maximum safe temperature being about 45°C. Low humidity dehydrates the crystal while a high humidity dissolves it. So, protective coatings must be applied to the crystal.

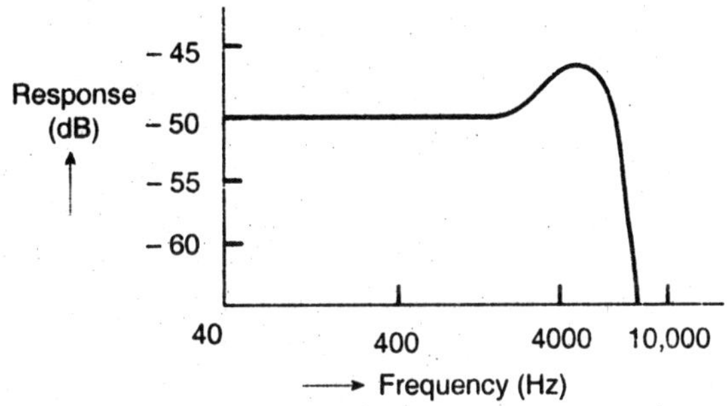

Fig. 10.8 Frequency response of a crystal microphone

Crystal microphones have practically nondirectional response. The response is quite flat : it does not differ by more than 6 dB over the frequency range 40 to 8000 Hz (Fig. 10.8). The rise of the response between 2000 and 5000 Hz originates from the higher strains developed by a given pressure as the applied frequency approaches the fundamental frequency of resonance of the system. The frequency response is similar to that of a capacitor microphone. A great advantage of the crystal microphone is that it is simple, inexpensive, and is available in small sizes. It is used in hearing aids, sound-level meters, and in low-priced public address systems. However, such microphones are not suitable for precisional work.

10.6 Ribbon Microphone

The ribbon microphone is operated by the *pressure gradient*, the driving force on the diaphragm being proportional to the difference of the acoustic pressures on the two sides of the diaphragm. Since this pressure gradient is proportional to the time derivative of the particle velocity in the medium, the ribbon microphone is referred to as a *velocity microphone*.

The ribbon microphone consists of a thin corrogated strip of aluminium foil or ribbon (R) suspended in a slot in a baffle (Fig. 10.9). The ribbon system lies between the poles N, S of a strong permanent magnet. The instantaneous pressure difference between the front and the back of the baffle, produced by the impinging sound waves, causes the ribbon to vibrate. The ribbon serves as its own diaphragm, and its vibration in the magnetic field induces a voltage across its ends. This voltage is fed to the transformer T. The natural frequency of the ribbon, kept under tension, is less than the lowest frequency of the incident sound wave. The sound goes round the baffle to reach the back of the ribbon.

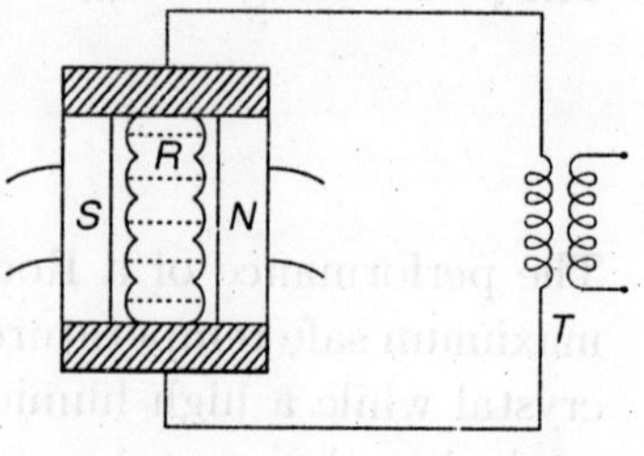

Fig. 10.9 Ribbon microphone

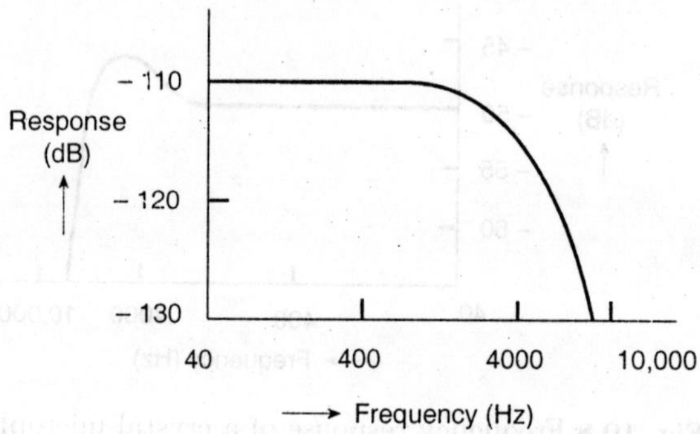

Fig. 10.10 Frequency response of a ribbon microphone

The ribbon microphone has a remarkable directivity. It is most sensitive to sound from the front and does not respond to sound propagating in the plane of the diaphragm. If the effective distance l between the front and the back of the diaphragm equals the wavelength λ of the sound, the driving force becomes zero. Hence the response of the microphone is poor for frequencies in the neighbourhood of c/l, where c is the velocity of sound. For best results, l is chosen to be less than $\lambda/2$ corresponding to the highest frequency of interest. A typical frequency-response characteristic of a ribbon microphone is shown in Fig. 10.10. The response remains quite uniform up to a frequency of about 2000 Hz but drops off rapidly beyond this frequency. The velocity microphone is used in broadcasting.

10.7 Loudspeaker

The function of a loudspeaker is opposite to that of a microphone. A microphone transforms an acoustic signal into an electric signal, whereas a loudspeaker converts an electric signal into an acoustic signal.

The requisite *properties of a good loudspeaker* are as follows :

(i) response independent of frequency over the whole audible range,

(ii) uniform output over a wide angle,

(iii) no harmonic or other distortions in the output,

(iv) high efficiency,

(v) high power handling capability, long life, small size, and low cost.

A practical loudspeaker cannot meet all these requirements. The two main types of loudspeakers are the *direct radiator loudspeaker* and the *horn loudspeaker*. In the direct radiator speaker, the diaphragm or the vibrating surface launches the sound waves directly into the air, while in the horn speaker, a properly shaped horn is used for a more effective matching of the loading impedance of the air to that of the massive vibrating diaphragm. Thus the horn type loudspeakers have higher efficiencies. The direct radiator speakers are commonly used in domestic radio and TV receivers, magnetic tape reproducers, and intercommunication systems, where the acoustic power requirements are not high. The horn type finds applications in public address systems in a large gathering, where a large acoustic power output is required. Both direct radiator and horn speakers are commonly of the *moving coil* or *electrodynamic* variety.

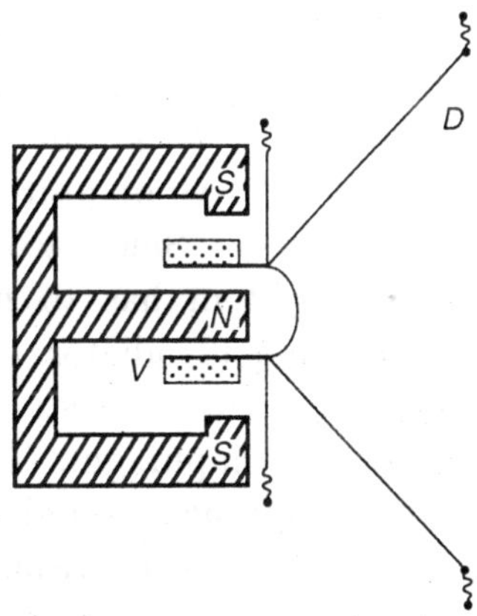

Fig. 10.11 A direct radiator moving coil loudspeaker

(i) **Direct radiator moving coil loudspeaker:** The device consists of a pot type *permanent magnet* of alnico having a central north pole (N) surrounded by a ring like south pole (S) (Fig. 10.11). A strong magnetic field exists in the narrow annular gap between the N and the S poles. A small and light multilayer coil of fine wire wound on a small paper cylinder can move freely along the axis in the gap between the poles. The coil is referred to as the *voice coil* (V). The paper cylinder on which the voice coil is mounted, is connected to the narrow end of a light paper cone, called the *diaphragm* (D). Flexible annular strips of leather or rubber support the diaphragm near the voice coil and at the outer periphery.

Alternating currents corresponding to the sound waves to be reproduced pass through the voice coil. As the voice coil is in a magnetic field, it experiences a magnetic force as electric currents flow through it. As a result, the voice coil and the diaphragm vibrate in accordance with the current fluctuations. The vibrations of the diaphragm generate the desired sound wave in the surrounding air.

The flexibility of the diaphragm surface somewhat broadens the directional pattern. The effect arises from the fact that the motion of the outer portions of the cone lag behind that of the central part and the voice coil. At high frequencies, the cone does not vibrate as a unit, rather, it vibrates in distinct zones separated by nodal circles. The vibration amplitudes of the outer zones are quite small, so that only the central portion can be regarded as the effective source of sound. This tendency of the cone at higher frequencies is augmented by providing the cone with a number of circular corrugations.

The ratio of the generated acoustic power and the supplied electric power is known as the *efficiency* of the loudspeaker. Direct radiator loudspeakers have low efficiencies, below 10%. Hence these speakers are not suitable in situations where a large acoustic power is required.

The vibrating diaphragm acts like an *acoustic doublet*, producing sound waves on either its front side or its back side. At the surface of the diaphragm, these two wave systems have opposite phases, and so produce a feeble output upon superposition. Since acoustic doublets are poor generators of sound, one wave system has to be suppressed. This is accomplished by mounting the direct radiator loudspeaker in a *cabinet enclosure* or in a *baffle* which serves as an obstacle to isolate acoustically the front and the back sides of the diaphragm. The distance between the front and the back sides of the diaphragm round the baffle must exceed a quarter wavelength of the sound wave for a satisfactory result.

The most common type of cabinet enclosure of a loudspeaker is the *open-back cabinet* [Fig. 10.12(a)]. Here the loudspeaker is mounted on one wall of a box, the opposite wall being absent. The box introduces a path length between the front and the back sides of the diaphragm. A troublesome feature appears at the fundamental resonant frequency of the cabinet. The increased sound output resulting from the enhanced motion of the speaker cone at this frequency introduces a 'boomy' quality which interferes with the naturalness of sound. Furthermore, at frequencies less than the fundamental resonant frequency, the path length provided by the box is inadequate. The sound from the back of the cone and that from the front then combine to give the doublet radiation.

The simplest type of cabinet is the *closed back* type where the loudspeaker is mounted on one wall of a closed box [Fig. 10.12(b)]. Here the energy is transferred to the external medium from the front of the speaker cone. At high frequencies, the sound from the back of the cone tends to produce standing wave patterns inside the box. The effect gives resonances, to avoid which the box is lined with a sound-absorbing material.

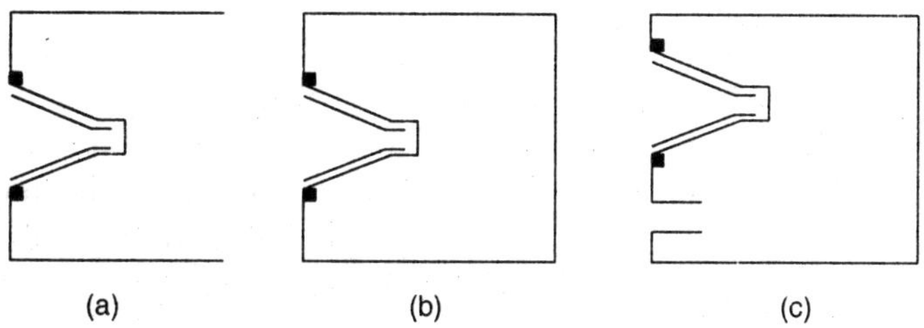

Fig. 10.12 Loudspeaker cabinets: (a) open-back cabinet, (b) closed-back cabinet, (c) phase-inverter cabinet

The response of the direct-radiator loudspeakers at low frequencies is limited by the inefficient coupling between the cone and the air. One way to increase this coupling is to use a larger cone. Another way is to use a *phase inverter cabinet* [Fig. 10.12(c)]. Here the sound from the back of the cone superimposes on that from the front in phase, thus increasing the acoustic output. The simplest phase inverter cabinet is the vented box having a hole on the same wall on which the speaker is mounted. The phase-inverter cabinet gives a better low-frequency response than the open-back cabinet. Also its response is smoother, and free from the boomy quality of the open-back cabinet. The phase-inverter cabinet is therefore a very popular baffle for high-fidelity (Hi-Fi) reproduction of sound.

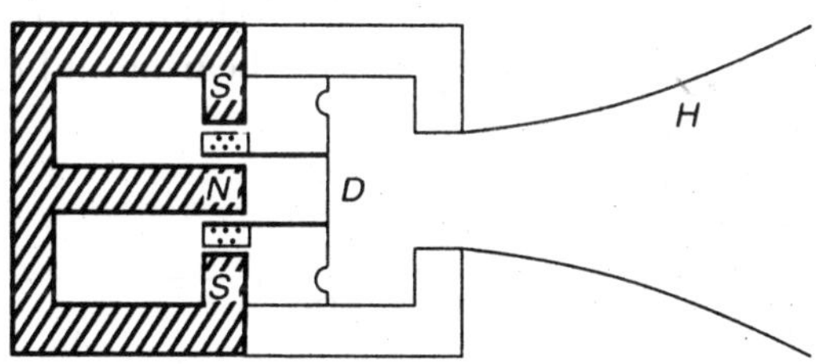

Fig. 10.13 Horn loudspeaker

(ii) **Horn loudspeaker** : A horn loudspeaker contains an electrically driven diaphragm (D) coupled to a horn (H) (Fig. 10.13). The horn acts as an acoustic transformer to match the loading impedance of the light air effectively to that of the massive diaphragm. The efficiency of the horn loudspeakers lies within 20 to 50%; it is much larger than that of the direct radiator speakers. The horn also provides the emitted sound with directional properties. With suitable designs, the horn speakers give a satisfactory response over a wide audio frequency range.

Horns can have different shapes like conical, hyperbolical and exponential. The most common type is the *exponential horn* whose cross-sectional area increases exponentially with distance from the narrow end, i.e., the throat. Thus the cross-sectional area S_x at a distance x from the throat is $S_x = S_0 e^{mx}$, where S_0 is the area of the throat, and m is a constant known as the *flare constant*. The cut-off frequency, i.e., the lowest frequency at which the exponential horn gives a significant acoustic power output, decreases with decreasing flare constant. The length of the horn must be sufficient to avoid appreciable reflection of the incident energy at the mouth. The response of an exponential horn is improved by using multiple flare sections (Fig. 10.14).

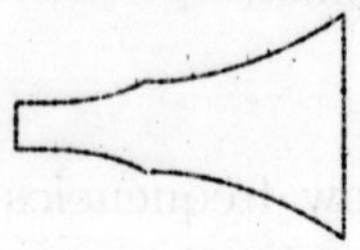

Fig. 10.14 A multiple-flare exponential horn

The limited frequency response of a single speaker is overcome in a high-fidelity (Hi-Fi) system by using *multiple-speaker* assemblies. A two-way speaker consists of a large, low-frequency speaker (the *woofer*) and a small, high-frequency speaker (the *tweeter*). The tweeter supplements the frequency range of the woofer, so that the entire audio range is covered in a smooth way.

10.8 Hearing Aids

A hearing aid is a miniature, portable reproducing system that increases the sound pressure over that usually received by the ear. A person with impaired hearing may be benefited by using a hearing aid. Conventional simplest electroacoustic hearing aids are head-on devices consisting of a carbon microphone (C), a battery (B), an attenuator (A), and a telephone receiver (R), as shown in Fig. 10.15(a). This hearing aid can be used if the hearing loss is about 20 dB.

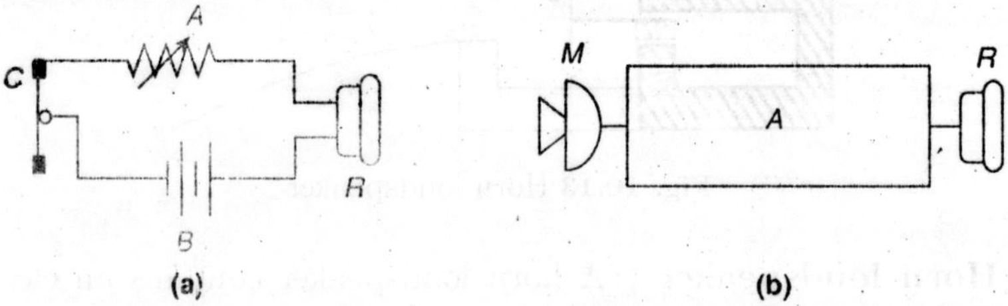

Fig. 10.15 Hearing aids : (a) simple carbon microphone hearing aid; (b) schematic diagram of a hearing aid using an electronic amplifier

For greater hearing losses and for a better performance, hearing aids incorporating electronic audio amplifiers with volume and tone controls, are employed. In the arrangement, compensation circuits can be incorporated to complement the characteristics of the ear. Figure 10.15(b) shows schematically the components of an electronic hearing aid. The electronic amplifier with volume and tone controls is denoted by the symbol A. M is a diaphragm crystal type microphone; R is an air conduction type receiver or a bone conduction type receiver. The system includes a battery to supply power.

The cross-sectional view of the air conduction insert type telephone receiver is depicted in Fig. 10.16(a). The components of the receiver are the ear piece E, the diaphragm D, the coil C, and the magnet M. A moulded plug fits the ear cavity to hold the receiver in place. The leak of sound at the ear is then very small. The low-frequency response as shown in Fig. 10.16(b), is quite good.

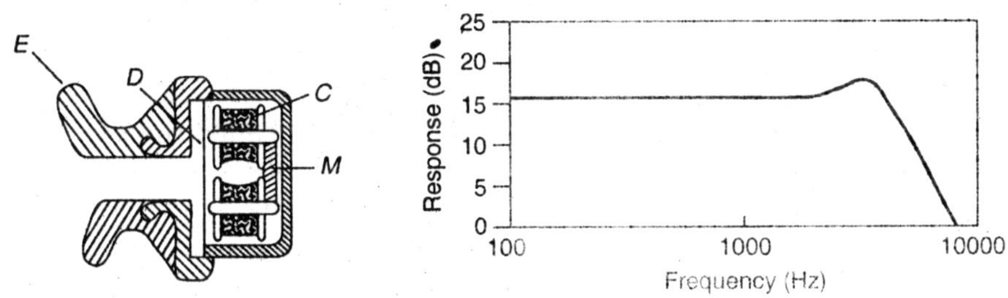

(a) (b)

Fig. 10.16 (a) Cross-sectional view of an insert type telephone receiver; (b) its frequency response characteristic

In some class of deafness, the middle ear containing a set of bones carrying sound to the inner ear, is damaged, but the inner ear consisting of nerves, is normal. In this case, sound can be carried to the inner ear through the bones of the head using a bone conduction receiver. The face of the bone conduction receiver is normally placed against the mastoid bone behind the ear. The cross-sectional view of a bone coduction receiver is shown in Fig. 10.17(a); its frequency response characteristic is given in Fig. 10.17(b).

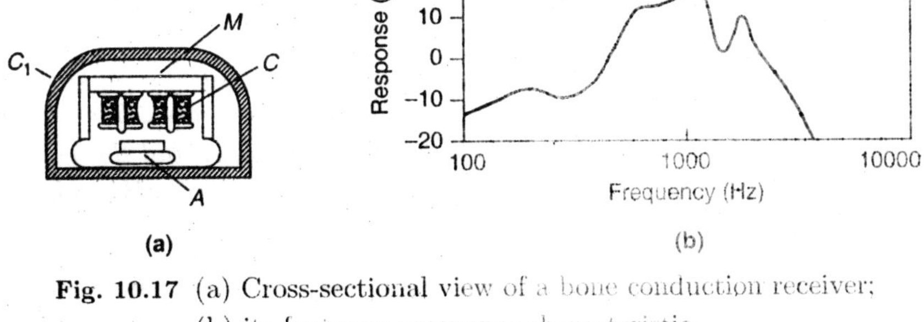

(a) (b)

Fig. 10.17 (a) Cross-sectional view of a bone conduction receiver; (b) its frequency response characteristic

The coil C, the magnet M, and the armature A of the receiver are housed inside the case C_1. The frequency response characteristic is inferior to that of the insert type receiver.

The poor speech recognition feature of conventional hearing aids can be improved by assisting devices which enhance the signal-to-noise ratio. Holding the detached microphone closer to the sound source also improves the signal-to-noise ratio. The microphone output signal can be fed to the amplifier by wire, or by radio frequency or infrared signals.

In recent developments, cochlear implants transform acoustic signals into electric signals, and after processing, deliver them to the nerve fibres of the inner ear by means of electrodes which are introduced surgicallly into the inner ear.

Earphone and Headphone

An *earphone* (Fig. 10.18) contains a coil of many turns of wire wound on a small permanent magnet (NS). A thin sheet of iron, called the *diaphragm*, is held in front of the magnet inside a plastic case.

When an alternating current flows through the coil, the magnetic flux created by it is alternately added to and subtracted from the flux of the permanent magnet. Hence the diaphragm vibrates in accordance with the signal current in the coil. These vibrations, in turn, alternately compress and rarefy the air in contact with the diaphragm, and generate sound waves. Two pieces of earphones can be held by a support cn a human head and pressed against the ears. The arrangement is called the *headphone*. The sound produced by the vibrations of the diaphragm is then readily perceived by the ears.

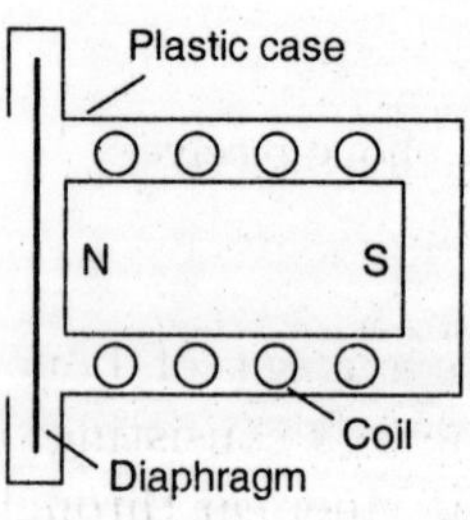

Fig. 10.18 Earphone

The headphone is commonly used as a *null detector* in audio-frequency bridges. Some impedances in the bridge arms are adjusted for a minimum or a vanishing sound in the headphone. This corresponds to a minimum current ın the earphones, indicating the null point or that the bridge is balanced.

10.9 Solved Problems

Problem 1. *A capacitor microphone has an aluminium diaphragm of thickness* 0.004 cm *and radius* 2 cm, *stretched to a tension* 2.5×10^4 N/m. *The gap between the diaphragm and the plate is* 0.004 cm *and the polarizing voltage is* 300 V. *Calculate the open-circuit response in* dB, *taking the reference level as* 1 volt per dyne/cm^2.

Solution. The open-circuit response is $S = \frac{V_0 A}{8\pi T d_0} = \frac{V_0 a^2}{8 T d_0}$, since $A = \pi a^2$, where a is the radius of the diaphragm. Here $V_0 = 300$ V, $a = 2$ cm,

$T = 2.5 \times 10^4\,\mathrm{N/m} = 2.5 \times 10^7$ dyne/cm and $d_0 = 0.004$ cm. Hence

$$S = \frac{300 \times 2^2}{8 \times 2.5 \times 10^7 \times 0.004} = 1.5 \times 10^{-3} \frac{\text{volt}}{\text{dyne/cm}^2}.$$

As the reference is $S_0 = 1$ volt per dyne/cm^2, the response in dB is

$$n = 20 \log_{10}(S/S_0) = 20 \log_{10} 0.0015 == -56.5 \text{ dB}$$

Problem 2. *The diaphragm of a moving-coil microphone has an area of* 75 cm^2. *The coil of wire length* 40 cm *moves in a magnetic field of* 1.6 T. *At a frequency of* 1 kHz, *the mechanical impedance of the moving system is* 20 ohm. *Determine the sensitivity of the microphone in* dB *at* 1 kHz *with respect to the reference of* 1 volt per microbar (*i.e.*, 10 V/Pa).

Solution. The sensitivity is

$$S = \frac{Bl\alpha}{|Z_m|},$$

where B = magnetic field = 1.6 T, l = wire length of the coil = 0.4 m, α = diaphragm area = 75×10^{-4} m^2, and $|Z_m|$ = magnitude of the mechanical impedance = 20 ohm.

So,

$$S = \frac{1.6 \times 0.4 \times 75 \times 10^{-4}}{20} = 2.4 \times 10^{-4} \text{ V/Pa}$$

The sensitivity in dB is

$$n = 20 \log_{10} \frac{2.4 \times 10^{-4}}{10} = -92.4 \text{ dB}$$

REVIEW QUESTIONS

1. What is a transducer? What type of transducer is a microphone? Name the important characteristics of a microphone.

2. What are the commonly used varieties of microphone? Describe the construction and the working principle of a carbon microphone.

3. Derive an expression for the sensitivity of a carbon microphone. What are the practical limitations to the increase of the sensitivity? Draw the frequency-response characteristics of the carbon microphone.

4. Describe with a diagram the principle of operation of a capacitor microphone. Derive an expression for its open-circuit response and also for the ac output voltage across the resistance.

5. (a) Compare between a capacitor microphone and a carbon microphone.

(b) Which type of microphone would you use in the following cases : (i) high-fidelity pickup of music, (ii) telephony, (iii) hearing aid.

(c) What are the requirements of a good microphone?

6. Explain with a diagram the operation of an electrodynamic microphone. Find an expression for its response.

7. What is a crystal microphone? Explain its principle of operation, and name the factors affecting its response. Draw its frequency-response characteristic.

8. Why is a ribbon microphone called a velocity microphone? Explain the working of such a microphone and draw its frequency response curve. Where is this type of microphone used?

9. What are the two main types of commonly used loudspeakers? Show with a diagram the essential parts of a direct radiator moving-coil loudspeaker, and explain its working. Can you use this loudspeaker to address a large gathering in a field?

10. (a) What are the requirements of a good loudspeaker?

(b) What is a 'baffle' and why is it used for a loudspeaker? Name the different types of loudspeaker cabinets and compare between them.

11. What is a horn loudspeaker? Explain the function of the horn. What is the usual type of horn employed in speakers? Define flare constant.

12. Write notes on (i) hearing aids, and (ii) Earphones and headphones.

PROBLEMS

1. The diaphragm of a capacitor microphone of radius 2.2 cm is stretched by a tension of 2.2×10^4 N/m. The distance between the diaphragm and the plate is 0.02 mm and the polarizing voltage is 350 V. Find the open circuit response in dB, taking the reference as 10 V/Pa.
[**Ans.** −46.4 dB]

2. The radius of the diaphragm of a carbon microphone is 1.2 cm and its stiffness is 5×10^5 N/m. It is operated by a battery of voltage 12 V. The granule resistance is 100 ohm, and the resistance change for 1 mm displacement of the diaphragm is 4×10^5 ohm. Taking the reference as 1 volt per microbar, determine the sensitivity of the microphone in dB.
[**Ans.** −47.2 dB]

Chapter 11

Recording and Reproduction of Sound

Sound can be reproduced instantaneously or subsequently. An instantaneous reproduction is required at a short distance from the source in a public address system, whereas a long distance reproduction is necessary in telephony. For a subsequent reproduction, i.e., reproduction at a later time, as required in a tape, film or gramophone, the sound must be first recorded, i.e., its waveform must be stored in a device.

In the instantaneous reproduction, the sound from the source is transformed into corresponding electrical signals by a microphone. The microphone output is amplified and fed into a loudspeaker. The diaphragm of the loudspeaker vibrates following the characteristics of the electrical signal. Consequently, vibrations are set up in the surrounding air following those produced by the original sound, which is thereby reproduced.

For recording, the sound is also converted into electrical signals by a microphone. The amplified microphone output then actuates a recording device, in which some features of the recording material is altered in accordance with the waveform of the microphone output. Thus, a permanent record of the sound waveform is obtained. In reproduction, mechanical or electrical signals of the same waveform are produced from the record, and subsequently transformed into sound by feeding into a loudspeaker.

In modern times, digital audio recording and reproduction as used in the compact disc (CD) system, has become very popular.

Recording and reproduction of sound is a subject-matter of *technical* or *industrial acoustics*.

11.1 Disc Recording and Reproduction

Here the waveform of the sound is stored for subsequent reproduction by impressing a fine groove on a cylinder or disc of wax. The variation in the depth of the groove or its lateral amplitude represents the waveform. The second method forms the basis of *gramophone recording.* Here the record is a flat disc containing a spiral of groove of fixed depth, the waveform being represented by a lateral cut indicating the transverse undulations.

In the recording process, the disc made of soft wax is placed on a heavy turntable rotated uniformly. A needle called the *record head* is placed on the disc near its edge. As the disc rotates with the turntable, the needle is slowly driven radially inwards. Thus the needle cuts a uniform spiral groove in the disc.

The sound to be recorded is transformed into electrical signals by a microphone. These signals are amplified and passed through filters before being applied to the record head. The amplified electrical signals force the needle to execute corresponding lateral vibrations. Consequently, instead of a uniform spiral, a wavy spiral is marked on the disc. The lateral amplitudes or the transverse undulations of the wavy spiral groove follow the electrical signal output of the microphone, and thus give a replica of the sound waveform. In this way, the sound is recorded on the disc.

The disc must be *processed* subsequently by first dusting the surface of the recorded wax with fine bronze powder and then polished. A strong deposit is then formed by electroplating the surface with copper. The 'negative' so obtained is called the *master*. It is peeled off, and the face previously in contact with wax is freshly deposited with copper. This is again peeled off to give a 'positive' known as the *mother*. The process is repeated to yield another negative, termed the *working matrix*. This is employed to produce impressions on the discs which give the final *record*. If the working matrix is damaged, it is replaced by another one made from the mother. If the mother is damaged, another one is developed from the master.

The disc records are made of vinyl plastic, and impressions are made on them when they are hot. Subsequent cooling gives a hard smooth disc having a groove of the same form as the original cut wax.

While reproducing, the disc must be rotated at the same speed as during recording. Otherwise, the reproduction is unclear and highly distorted. Sound is reproduced from the disc mechanically by the gramophone, and electrically by the radiogram.

(a) Gramophone: The gramophone consists of a cylindrical box (B) called the *sound box*, whose aperture is closed by a thin, circular metal diaphragm (D) (Fig. 11.1). The diaphragm is corrugated to improve the reproduction of low frequency sound. The outer edge of the diaphragm is held against rubber washers. The other end of the cylindrical box is connected to a *horn*. The passage to the horn is known as the *tone-arm*.

The record is mounted on a *turntable* rotating at a constant speed. The *needle* (N) is placed in the groove of the record. As the record rotates, the needle vibrates laterally along the wavy track. These vibrations are magnified by the metal lever (L) which carries the needle at one end, F being the fulcrum of the lever. The other end of the lever is connected to the middle of the diaphragm. The magnified vibrations of the needle are thus carried to the diaphragm. As the diaphragm vibrates, similar vibrations are set up in the air inside the tone-arm, which are transmitted through the horn to the outside air.

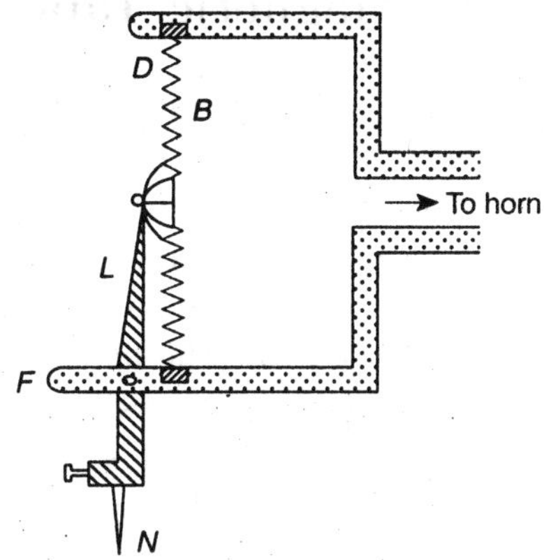

Fig. 11.1 The gramophone

The horn increases the intensity of sound by disturbing a larger mass of air. The shape of the horn also determines to some extent the quality of the sound. The diameter of the mouth of the horn and its flare control the lowest frequency of the emitted sound. The sound generated by a short, thick needle is louder than that produced by a long, thin needle.

(b) Radiogram: In a radiogram, the movement of the needle in the wavy track of the record actuates a microphone serving as the *electrical pickup*. In the *magnetic pickup*, a fixed coil (C_1, C_2) is placed in a horse-shoe magnet provided with four pole pieces N_1, N_2, S_1, S_2 (Fig. 11.2). The needle N is attached to a pivoted iron armature A which is suitably damped by rubber pads. The magnetic flux linked with the coil changes as the needle moves sidewise in the record groove. Consequently, an emf is induced in the coil. This induced voltage is amplified and then fed to the amplifier for reproduction.

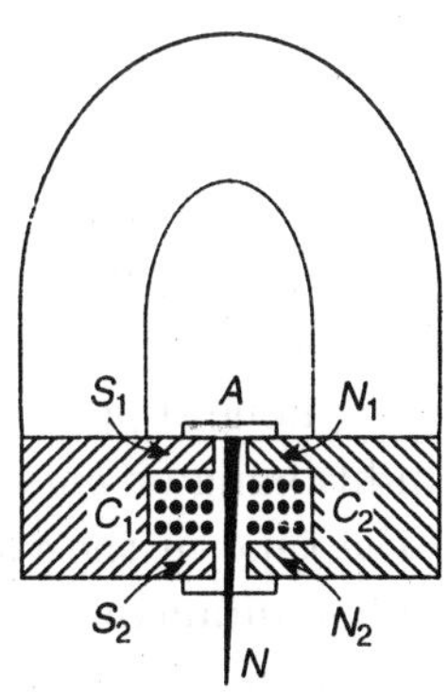

Fig. 11.2 Magnetic pickup

In the *piezoelectric* or *crystal pickup*, the vibration of the needle produces by torsion a strain in a piezoelectric crystal like Rochelle salt. The resulting voltage generated between the opposite faces of the crystal is fed to the loudspeaker after proper amplification.

In the *electrodynamic pickup*, the movement of the needle in the groove moves the coil of an electrodynamic microphone. The emf induced in the coil is amplified and fed to the loudspeaker.

11.2 Magnetic Tape Recorder

In this device, the tape is made of thin plastic coated with some magnetic material. The tape is unwound from one reel and wound over another by means of an electric motor running at a uniform speed (Fig. 11.3). These two reels are called the *supply reel* and the *take-up reel*, respectively. The tape passes over three ring-shaped magnetic heads termed the *erase head*, the *record head*, and the *playback head*, respectively. While passing the erase head, the tape encounters a strong, high-frequency ($\sim$ 100 kHz) oscillating magnetic field. This field is produced by sending a corresponding alternating current through a coil of wire wound over the erase head which serves as an electromagnet. The magnetic field at the small air gap of the erase head links with the portion of the tape in its vicinity and wipes out any magnetism the portion of the tape may have. Therefore, this magnetic field is known as the *erase field.*

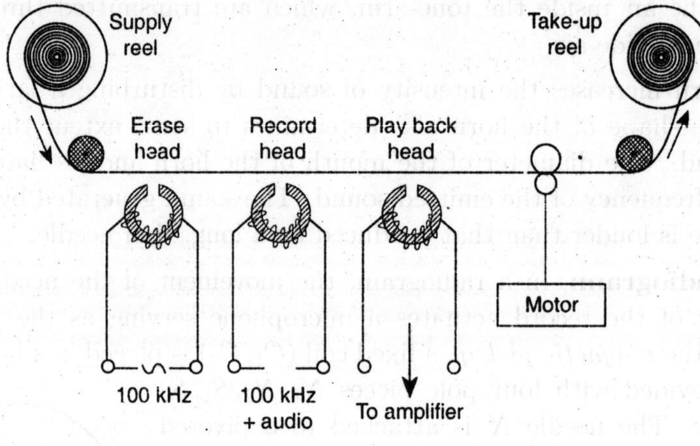

Fig. 11.3 Schematic representation of a tape recorder

The tape then passes over the small air gap of the record head which also serves as an electromagnet. It is excited by the amplified current from the microphone superimposed on a high-frequency current required for magnetic conditioning or *biasing.* The biasing makes the magnetization of the portion of the tape in the vicinity of the air gap of the record head proportional to the instantaneous audio signal.

Finally, the tape is drawn over the small air gap of the playback head which is also an electromagnet with a coil called the *pickup coil,* wound over it. The different portions of the tape have a *longitudinal variation* of magnetization which follows the audio signal. These variations of the magnetic flux induce a time varying voltage in the pickup coil of the playback head. This voltage has the waveform of the original sound. It is amplified and transmitted to the loudspeaker for the reproduction of the sound.

While recording, both the erase and the record heads are simultaneously operated. The playback head is then operated for reproduction after rewinding the tape. If the recorded programme is to be wiped out, only the erase head is operated.

The tape recorder has several advantages. Long programmes can be recorded faithfully, and can be reproduced immediately afterwords. The same tape can be used for other recordings after wiping out the undesired previous record. The recording can be interrupted at suitable moments, thus leaving out the irrevalent parts of the programme. The machine is compact, portable, and very convenient for outdoor recording.

Erasing technique: The tape is demagnetized by the erase field before recording is done by the record head. The demagnetization is performed by subjecting the tape to a number of magnetizing cycles with gradually diminishing field intensity. The air gap of the erase head is wide enough to force about 8 cycles of magnetization on the tape. This magnetization is sufficiently strong to saturate the tape. As the tape moves out of the air gap, the magnetic field intensity drops off in the successive cycles and finally falls to zero. In this way the tape is demagnetized.

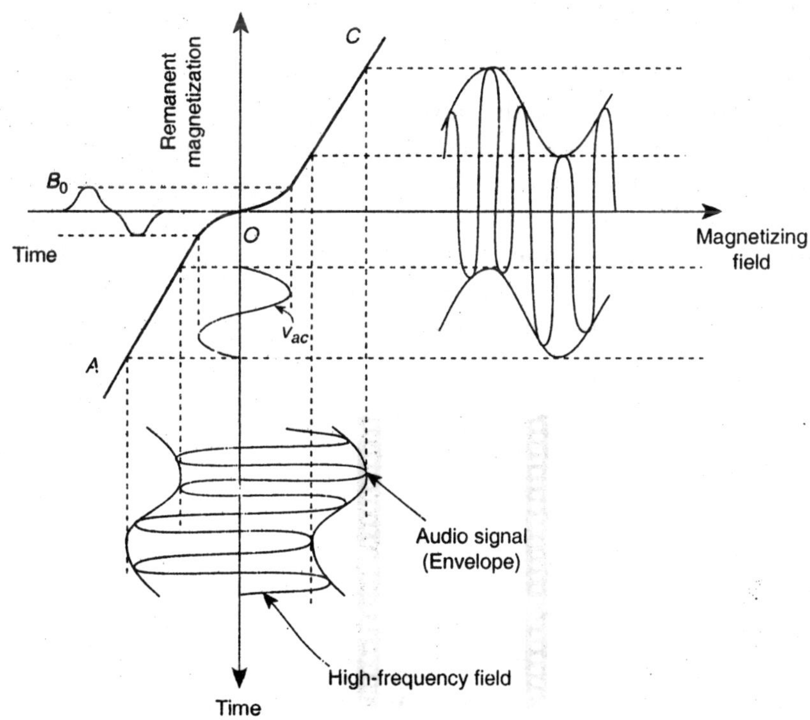

Fig. 11.4 Action of the biasing field

Observation

The action of the biasing field can be understood by referring to Fig. 11.4. The curve AOC shows the variation of the remanant magnetization of the tape with the magnetizing field. If the audio signal v_{ac} operates alone, the output magnetization B_0 is small and highly distorted due to the nonlinearity of the curve AOC near the origin O. However, if the audio signal is superimposed on a large-amplitude high-frequency field, the envelope of the resulting field shifts to the linear portion of the curve AOC. Consequently, for a sinusoidal sound, the average remanant magnetization will be sinusoidal, following the sound faithfully. As the tape is drawn away from the record head, its residual magnetization follows the envelope function. For high frequencies, the wavelength is small so that the demagnetization is very effective for the high-frequency magnetization, which is therefore, not permanent. Also, such high frequencies are not audible.

11.3 Recording on Films

In film recording employed in cinema, the sound waveform is stored as a variation of the transparency of the film. In the *variable density method*, the image of a narrow slit illuminated by a light source is cast on the space of the film alloted for the sound track. The intensity of light in the image of the slit can be varied by a light valve actuated by the amplified microphone current. As the film runs at a uniform speed, the slit images form a series of transverse parallel lines (Fig. 11.5). The transparencies of these lines follow the microphone current, i.e., the sound to be recorded. On 'developing', the regions of the least transparency of the film will correspond to the maximum intensity of light, i.e., to the largest microphone current. When a 'positive' is made for reproduction, the case is just the opposite.

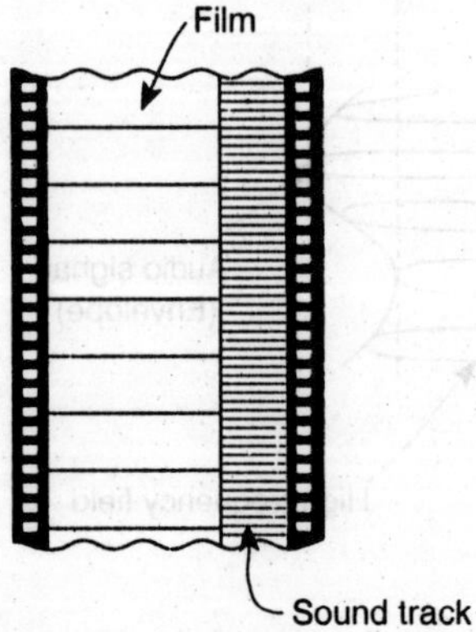

Fig. 11.5 Series of parallel lines on the sound track of a film in the variable density method

In another method, called the *variable area method*, light from a source L passes through a triangular aperture A, referred to as the *mask*. A mirror M mounted on a moving iron armature-type galvanometer, casts the image of the mask on a horizontal slit S (Fig. 11.6). The amplified microphone current passing through the galvanometer turns the mirror M about a central axis, so that the triangular image falling on the slit moves perpendicular to the slit. The image of the slit is formed on the sound track of the moving film F. Since the illuminated portion of the slit is controlled by the microphone current, the image of the slit blackens regions of varying width on the film (Fig. 11.7). The width of the blackened portion varies as the deflection of the mirror and hence as the microphone current. In the 'positive' film, the black areas become proportionately transparent.

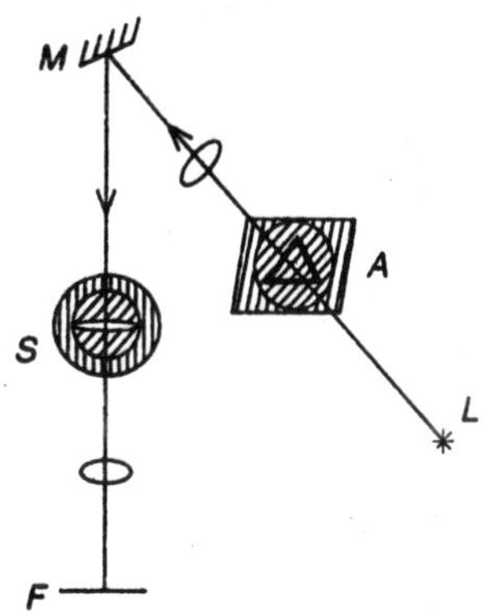

Fig. 11.6 Variable area method

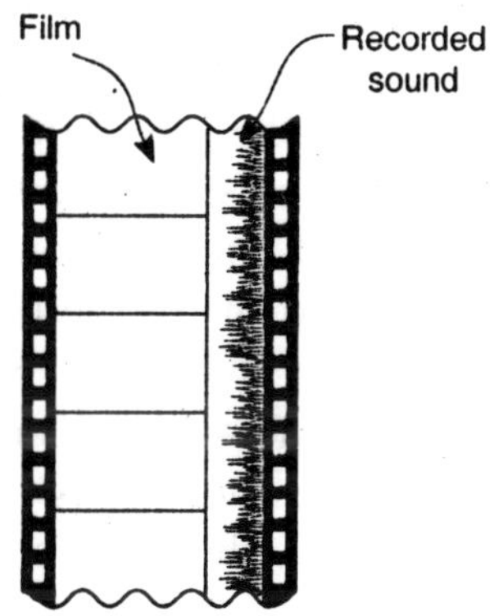

Fig. 11.7 Sound track record in the variable area method

Reproduction: For both the variable density and the variable area records, the reproduction technique is essentially the same. Light from the source L passes through a narrow slit S (Fig. 11.8). The Lens system L_1 produces an image of the slit in the plane of the sound track of the film F which moves at the same speed as that during recording. The light transmitted through the sound track is picked up by the lens system L_2 and focussed on the light sensitive surface of the photoelectric cell C. The

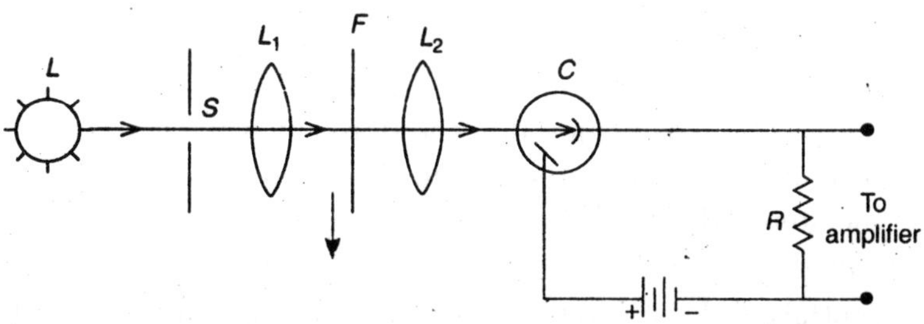

Fig. 11.8 Reproduction of sound from film records

number of photoelectrons emitted in the photo cell is determined by the intensity of the incident light. These photoelectrons produce a time-varying current in the resistance R. The corresponding voltage across R varies as the intensity of the light falling on the photo cell. Therefore, the time variation of the voltage has the waveform of the recorded sound. This voltage is amplified and transmitted to a loudspeaker for reproduction.

11.4 High-Fidelity Reproduction of Sound

The high-fidelity (abbreviated Hi-Fi) reproduction attempts to recreate by electronic and electromechanical means the full range of the audible sound from about 20 Hz to 20 kHz without distortion. The Hi-Fi reproduction thus closely approaches the original sound. Exact high-fidelity cannot be achieved due to uncontrollable factors not present in the original, such as the acoustics of the room, the reproduction level etc. In a practical Hi-Fi system, a compromise is sought among various conflicting factors to achieve a remarkably good reproduction. Improvements in the design of high-quality amplifiers and in the electromechanical devices like loudspeakers, records, and tape recorders have made significant advances in the Hi-Fi reproduction.

A simple Hi-Fi system consists of at least one input source such as a record player, an audio amplifier and a loudspeaker with a bafle. In a more elaborate system, there are many input sources, different amplifiers, and a multiple speaker.

11.5 Stereophonic Reproduction

The stereophonic ('Stereo' for short) reproduction of sound has significantly advanced the field of Hi-Fi. Stereophonic systems need recording and reproduction of sound over two completely different channels. This simulates the features of hearing with two ears. Hence the stereophonic system is sometimes referred to as the *binaural system*. The two ears enable us to sense the direction and the depth of sound. The *monaural system* that reproduces sound over a single channel fails to maintain the spatial relationships or the depth of sound. Such a reproduction does not give a perception as to whether the sound is coming from the front or the back, the left or the right, the upper or the bottom portion of the source. These spatial characteristics of depth and direction are achieved in the stereo by using two separate sound channels for recording and reproduction. The stereo may not be Hi-Fi, but of course the stereo Hi-Fi offers a superb technique of sound recording and reproduction.

The essentials of the stereo technique are schematically given in Fig. 11.9, where a 'live' concert is considered. When a listener sits at the centre of the room facing the orchestra, the sound arriving at his two ears would be different. Since the sound path lengths from the instruments at the right of the listener to his right ear are shorter than those to his left ear, these instruments are heard by the right ear a little sooner than by the left ear. The opposite is the case for the instruments to the left of the listener. These instruments are heard sooner by the left ear. Only the sound from the instrument played right in front of the listener, arrives at the two ears simultaneously. The difference in the time taken by the sound to reach the two ears produces the perception of the direction of the source of sound. Furthermore, the reverberated sounds produced by the wall reflections are sensed by the two ears and blended with the direct sounds. This enables the listener to make an impression about the size of the concert hall and about the depth of the sound.

In the stereo system, the effects of binaural hearing are simulated by placing two microphones M_1 and M_2 in front of the orchestra, M_1 taking the place of the left ear and M_2 that of the right ear. This is shown by the different path lengths of sounds arriving at M_1 and M_2 in Fig. 11.9. Of course, several microphones can be used for each ear. The sound received by each microphone is processed in its own recording and reproducing channel which may be a tape recorder. It is then reproduced by the particular loudspeaker connected to the channel. The listener O positioned at the

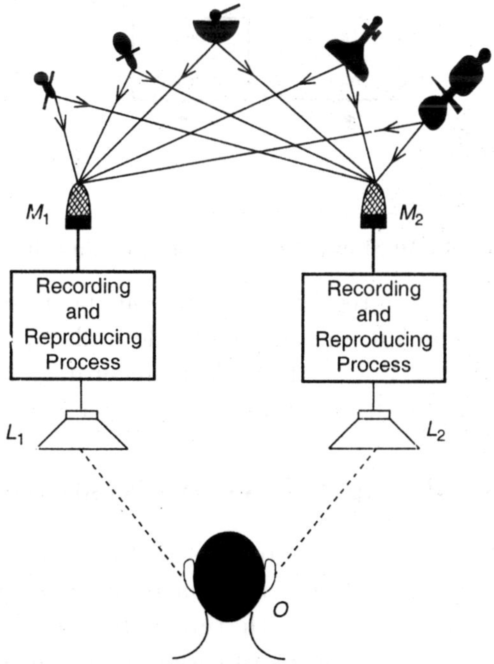

Fig. 11.9 Schematic representation of stereo technique

centre between the loudspeakers L_1 and L_2 hears the differences between the two sounds, just as if he were listening the original sound in the concert hall.

The stereo sound recording is schematically shown in Fig. 11.10. The sound is picked up by two or more microphones positioned suitably in front of the sound source. The sound received by each of the microphones M_1 and M_2 is recorded on two different magnetic tracks of a single tape T. Both the tracks begin and end together on the same reel of tape. These parallel sound tracks are replayed at the same time through two different electronic amplifier systems and separate loudspeakers. Thus the stereo is reproduced.

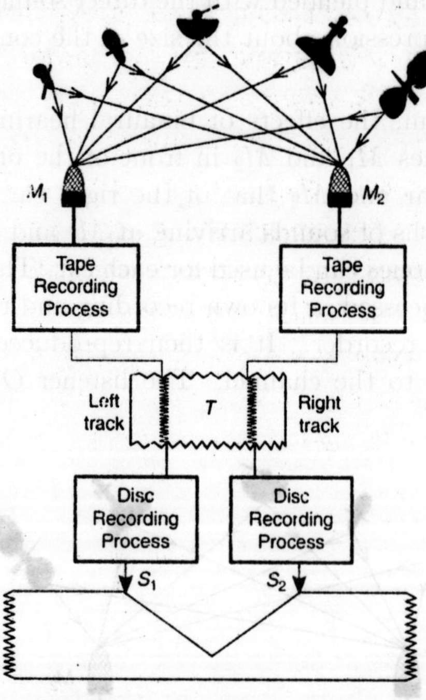

Fig. 11.10 Stereo recording on tape and disc

The two tracks of a stereo master tape can be rerecorded by a disc recording process on the left and the right side walls S_1 and S_2 of a single V-shaped record groove (Fig. 11.10). After recording, the master disc is processed to produce quantity recordings for commercial distribution.

11.6 Outline of Digital Audio Systems

The techniques described in the previous sections are analog in nature. For instance, the voltage output of a microphone is an analog of the air pressure or velocity varying with time. In the analog system, the information is carried by the continuous variation of some parameter such as the microphone output voltage. The analog gramophone recording, invented in the second half of the nineteenth century, had a lifetime of nearly 100 years. It was then supplemented by the *compact disc* (CD) which appeared in 1982. The CD is a commercially successful digital audio system.

Despite their good quality and wide popularity, analog audio recordings suffer from several limitations. In the analog system, the signal is degraded by noise introduced at each stage through which the signal passes. Such degradation cannot be separated from the original signal. Consequently, the signal-to-noise ratio is relatively low (about 50 dB). Also, the dynamic range is limited to about 70 dB. By contrast, the dynamic range of orchestral music is about 100 to 120 dB. So, in the analog recording of the music, at low music level, noise will be audible. In order to prevent the noise, if recording is done at high music level, saturation will occur.

Digital audio recording and playback improves the fidelity of the recorded music by enhancing the signal-to-noise ratio and the dynamic range. Also, digital recordings usually have a long life and they do not deteriorate with playing time, in contrast to analog recording. Table 11.1 compares some of the important features of analog and CD systems. The table clearly reflects the advantages of the latter.

TABLE 11.1

Comparison of analog and CD systems

Feature	Analog system	CD system
Frequency response	30 Hz to 20 kHz ±3 dB	20 Hz to 20 kHz +0.5 dB to −1 dB
Dynamic range	70 dB 1 kHz	> 90 dB
Harmonic distortion	1 to 2 per cent	0.005 per cent
Signal-to-noise ratio	60 dB	> 90 dB
Stylus life	500 to 600 hours	5000 hours
Performance	HF response degrades with playing	Degradation of response with playing is not perceived

In a digital audio system, the audio signal is represented digitally. An almost universal way of this representation is the *pulse code modulation* (PCM). Here the time axis is represented in discrete steps. The signal voltage is measured at regular intervals. This process is referred to as *sampling*. The frequency at which the samples are collected is known as the *sampling frequency* or *the sampling rate*. Usually the sampling rate is fixed and is the same for all signal frequencies. The samples ought to be taken at fixed time intervals. If there is a time-base error, the instants of the arrival of the samples are changed and the effect can be detected. To get rid of the effect of an irregular time-base, the samples are stored temporarily in a memory and read out using a stable, locally generated clock. This process is termed *time-base correction.* Clearly, a digital recorder does not have any flutter.

In a system of infinite bandwidth, the sampling misses the portion of the waveform between the samples. But a practical analog audio signal does not have infinite bandwidth. There is a frequency response limit of human ears and of all analog signal sources from microphones, pickup cartridges etc. For a signal of finite bandwidth, its rate of change is limited, and the manner of its change is predictable. Since the waveform can change in a known way between the samples, the original waveform can be recovered from the samples.

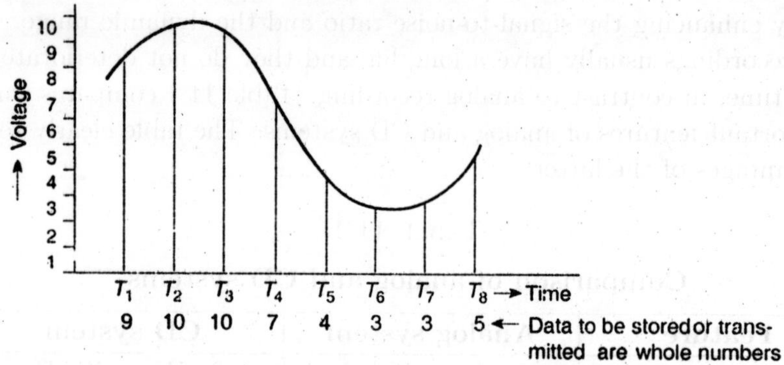

Fig. 11.11 Samples in the PCM of an analog waveform

Figure 11.11 shows the PCM of an analog voltage waveform. The voltage of each sample is represented by a whole number. This is an approximation and is called *quantizing.* The magnitude of the error is controllable till it is negligible consistent with the allowable resolution for the audio. The advantage of using whole numbers is that the original information can be described in a manner that can strongly resist unwanted changes.

Clearly, the digital audio expresses the original waveform numerically. The sample number is an analog of time, and the sample height, i.e., the sample magnitude is an analog of the pressure at the microphone. Thus both the axes of the digitally expressed waveform are discrete, and the waveform can be restored from these numbers. Obviously, for greater accuracy, more numbers will be required. The quality of the digital audio is determined by the sampling rate.

The whole number denoting the sample length in a digital audio system is expressed in the binary form in which two digits 0 and 1 are used. These digits are called *bits.* A word or number of eight bits is called a *byte* which is coined from 'by eight'. To avoid large numbers, units of kilobyte, megabyte, and gigabyte are used. With the binary expression, the signals have two states and they change at fixed times given by a stable clock. If there is some noise in the trasmitted signal, it will be rejected by the receiver, since all voltages above a level are labelled 1, and those below are labelled 0. This process is called *slicing*. The advantage of the digital audio is that the quality of reproduction is very good, and the conversion of the audio to the digital domain offers many opportunities which are not possible for analog signals. Digital technology is not expensive, and digital recordings require less space than analog recording. Tape costs and storage costs are remarkably reduced. Manufacturing cost of digital circuitry is also less. Binaries are handled by switching circuits which can be integrated more densely than analog circuits in an IC chip.

When audio is in the digital domain, it becomes data indistinguishable from any other form of data. Low-cost disc drives and memories used in computers can be employed in digital audio systems for storing data. Communication networks can be used to carry the digital audio over long distances without any quality loss. These techniques are used in *digital audio broadcasting* (DAB) to eliminate the problems of interference, fading, and multipath reception encountered in analog broadcasting.

The essence of a digital audio system is that here the analog sound signal is converted into samples of binary numbers by an analog-to-digital converter (ADC). The data are transmitted to the other end of the line where the noise introduced into the signal is rejected by a slicer. The whole sample is read out and transferred to the digital-to-analog converter (DAC) for reconversion of the sample to an analog voltage.

Figure 11.12 shows a general block diagram of the elements of a CD digital audio system. In recording [Fig. 11.12 (a)], the two signals from the left and the right microphones (M_1 and M_2) in a recording studio or a concert hall pass through an ADC where they are sampled and digitized. The sampling rate, by Nyquist's criterion, should be about 40 kHz which is two times the maximum frequency of audible sound. Allowing some safety guard band of frequencies, the sampling rate in a CD system is chosen at 44.1 kHz.

The output of the *digital recorder* contains a sequence of binary digits which are passed through the *channel encoder*. The channel encoder introduces some controlled redundancy in the binary information sequence, which can be used at the playback system to eliminate the effects of noise encountered in the transmission of the signal. Thus the redundant bits introduced in the information sequence helps the playback system in decoding the required information. From the channel encoder, the signal passes through the *modulator*, where digital control and display information are added together with a table of contents of the disc. This information is required for the programmability of the CD player. The digital signal from the modulator is *optically recorded* using a *laser* on the surface of a *glass disc* coated with photoresist. A master disc is thus produced. A number of CDs are obtained from the master disc by a series of processes that ultimately transform the information into tiny pits on the plastic disc. The disc is coated with a reflective aluminium coating and then with a protective lacquer.

In the CD playback system [Fig. 11.12 (b)], a laser optically scans a track on the CD at a constant speed, and hence reads the digitally recorded signal. The two signals from M_1 and M_2 are demodulated in the *demodulator*. They then pass through the *channel decoder* which reconstructs the original information sequence following the code used by the channel encoder and the redundancy contained in the received data.

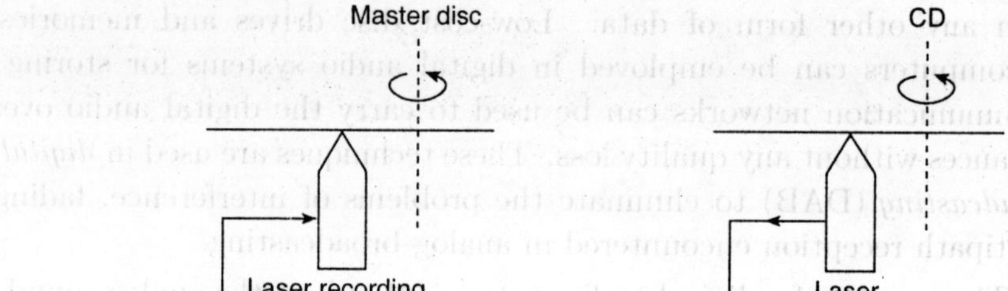

Fig. 11.12 Blocks of a CD digital audio system

Before the digital audio signals are fed to the DAC for reconversion into analog audio signals to excite the loudspeakers L_1 and L_2, they are passed through the *up-sampling and interpolation* block. The reason is that the two audio signals from M_1 and M_2 are quantized by using uniform PCM with 16 bits/sample. The digital-to-analog conversion of the two 16-bit signals at the 44.1 kHz sampling rate is practically very expensive. However, a cheaper

DAC with 12-bit (or less) precision is easily available. Thus the problem is to find a method for D/A conversion which uses low precision and hence employ a cheap DAC, while maintaining the 16-bit precision of the digital signal. A practical solution to this problem is to widen the bandwidth of the digital audio signal by up-sampling through interpolation and digital filtering before analog conversion.

REVIEW QUESTIONS

1. Explain how sound is recorded on a disc and reproduced subsequently.
2. Describe with a diagram the construction of a gramophone and explain its operation. What is a radiogram?
3. Show with a diagram the essential parts of a magnetic tape recorder. How is sound recorded on a magnetic tape and reproduced? Mention the advantages of the tape machine.
4. Explain the principle of recording of sound on films and its reproduction.
5. What are Hi-Fi and stereophonic sound? Explain the operation of a stereo system.
6. Mention the important features of a digital audio system. What are the advantages of this system?
7. Show the basic blocks of a CD digital audio system.

Chapter 12

Acoustics of Buildings

The subject of *acoustics of buildings* or *architectural acoustics* deals with the design of conference rooms, concert halls, etc. in buildings from an acoustical standpoint. The design is aimed at obtaining optimum listening conditions and reducing noise. Scientific developments in this field began with the experimental studies of the acoustics of a room by Sabine in 1895. He found that the acoustical properties of a room are markedly affected by its reverberation characteristics.

12.1 Reverberation

Suppose that a source is producing a sound continuously in a closed hall. Some of the sound energy is absorbed by the walls and the rest is reflected. Hence the energy density in the enclosure reaches a steady state where the time rate of energy generated by the source equals the time rate of energy lost through absorption. If the source suddenly stops sounding, the sound does not disappear instantaneously, but gradually dies out through absorption. The gradual decay of sound in an enclosure after the removal of the source due to absorption at the wall reflections is called *reverberation*. Following Sabine, the time for the sound intensity to drop to the threshold audibility from an initial intensity 10^6 times as high, is termed the *reverberation time*.

For the sound to be heard *distinctly* in an auditorium, each separate syllable must have sufficient intensity and it should not interfere with the syllables following it. The reflection from the walls increases the intensity and hence the loudness, but too much reflection affects the clarity of the syllables. The design of an auditorium seeks a *compromise* between *loudness* and *clarity*. If the absorption is enhanced, the reverberation time decreases, making the syllables distinct but reducing the loudness. On the contrary, if absorption is reduced, a longer reverberation time results and so loudness increases at the cost of clarity. An *optimum* reverberation time is thus attempted in the design of the auditorium. An empirical formula for the optimum reverberation time T_0 (in seconds) is $T_0 = (0.0036V^{1/3} + 0.107)n$, where V is the volume of the room in cubic feet. The factor n is 4, 5, and 6 for speech, music, and choral works, respectively. Typically, T_0 is about 1 s for an auditorium of volume 64000 ft^3 used for speech. For music, T_0 is about 1.3 s for the same volume.

The reverberation time is affected by the number of people in the hall. To remedy the unwanted effects in a less packed hall, the seats are made of materials which are good sound absorbers. There should be a good reflector behind the sound source and a good absorber behind the audiance to obtain satisfactory results. Care must be taken that the enclosure does not resonate at frequencies of interest.

Thus the design of an auditorium is concerned with the control of reverberation, interference, and resonance. Curved walls and domed ceilings are avoided since they tend to focus sound. Resonance effects of sound boards, walls etc. are eliminated by proper damping.

Summing up, the *requirements of a good auditorium* can be stated as follows :

(i) The sound arriving at the ears of a listener should be sufficiently loud.

(ii) The successive syllables must be clearly heard.

(iii) The sound must not be distorted.

12.2 Sabine's Formula

We consider a sound source of constant output in an enclosure. On incidence on the walls of the enclosure, the sound is partly reflected and partly absorbed. After a number of reflections, the sound is diffused and the energy density in the enclosure becomes uniform. The time rate of growth of the acoustic energy density and its ultimate value are governed by the absorptive power of the walls. To find the relevant relationships, the following assumptions are made :

(i) The source emits sound uniformly in all directions at a constant rate whatever be the energy density in the enclosure.

(ii) There is no interference of sound waves.

(iii) Sound energy is uniformly distributed in the enclosure.

(iv) Absorption coefficients are independent of the intensity of the incident sound.

(v) Energy losses can be treated as continuous.

Let E be the energy density of the sound field in the enclosure at any instant of time t. We consider a surface element dS on the wall of the enclosure and a volume element dV in the enclosure, where $dV = rd\theta \cdot dr \cdot r\sin\theta d\phi$ in spherical coordinates. r is the distance of the volume element from dS, the angle between r and the normal to dS being θ (Fig. 12.1). The energy contained in dV is EdV. If $d\Omega$ is the solid angle subtended by dS at the centre of dV, the energy reaching dS from dV is

$$\begin{aligned} dW &= EdV\frac{d\Omega}{4\pi} \\ &= \frac{Erd\theta \cdot dr \cdot r\sin\theta d\phi dS\cos\theta}{4\pi r^2} \\ &= \frac{EdS\sin\theta\cos\theta d\theta d\phi dr}{4\pi}. \end{aligned} \tag{12.1}$$

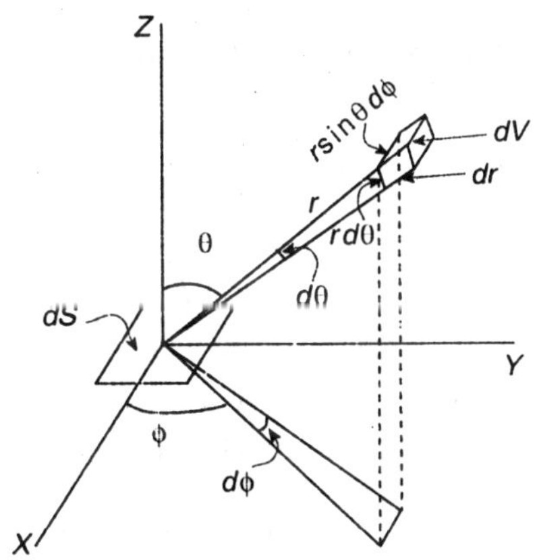

Fig. 12.1 Volume element dV and a surface element dS in the enclosure

The energy incident on dS per second from the enclosure is that contained in a hemisphere of radius c centred on dS, where c is the speed of sound. This energy W is obtained from Eq. (12.1) by integrating over θ from 0 to $\pi/2$, over ϕ from 0 to 2π, and over r from 0 to c. Hence

$$W = \frac{EdS}{4\pi}\int_{\theta=0}^{\pi/2}\sin\theta\cos\theta d\theta\int_{\phi=0}^{2\pi}d\phi\int_{r=0}^{c}dr = \frac{1}{4}EcdS \tag{12.2}$$

The sound energy incident on a unit area from all directions per second is $Ec/4$. This is referred to as the *intensity I of the diffuse sound.* Thus $I = Ec/4$.

Let a be the absorption coefficient of the material of area dS. Then the time rate of absorption of energy by the walls is

$$\frac{1}{4}Ec\sum adS = \frac{1}{4}EcA \tag{12.3}$$

where $A(=\sum adS)$ is the total absorption.

If P be the power of the source of energy, then the energy balance requires that the rate of production of energy by the source is equal to the sum of the rate of absorption of energy by the walls and the rate of increase of energy in the enclosure. That is,

$$P = \frac{1}{4}EcA + \frac{d}{dt}(VE), \tag{12.4}$$

where V is the volume of the enclosure. Equation (12.4) gives for a constant volume V

$$V\frac{dE}{dt} = P - \frac{1}{4}EcA$$

$$\text{or,} \quad dt = \frac{VdE}{P - \frac{EcA}{4}}. \tag{12.5}$$

Putting $x = P - \frac{EcA}{4}$, we get $dE = -\frac{4dx}{cA}$, so that

$$dt = -\frac{4V}{cA}\frac{dx}{x}.$$

Integration gives

$$t = -\frac{4V}{cA}\ln\left(P - \frac{EcA}{4}\right) + K,$$

where K is the integration constant. Assuming that the sound source starts at time $t = 0$, we get $E = 0$ at $t = 0$. Hence $K = \frac{4V}{cA}\ln P$, so that

$$t = -\frac{4V}{cA}\ln\left(\frac{P - \frac{EcA}{4}}{P}\right)$$

$$\text{or,} \quad \ln\left(1 - \frac{EcA}{4P}\right) = -\frac{cAt}{4V}$$

$$\text{or,} \quad E = \frac{4P}{cA}\left(1 - e^{-\frac{cA}{4V}t}\right). \tag{12.6}$$

Since the intensity I is given by $I = Ec/4$, Eq. (12.6) can be written as

$$I = \frac{P}{A}\left(1 - e^{-\frac{cAt}{4V}}\right). \tag{12.7}$$

Clearly the intensity I attains its maximum value

$$I_m = \frac{P}{A} \tag{12.8}$$

as $t \to \infty$. The *time constant* of the growth of the intensity is $\tau = \frac{4V}{Ac}$. If the absorption A is small, the time constant is large, and a fairly long time is needed for the intensity to reach a certain percentage of its maximum value.

When the energy density or the intensity attains its maximum value in practice, let the source be removed. Then the wall absorption will cause the energy density or the intensity to decrease. This *decay*-process can be studied from Eq. (12.4) by putting $P = 0$. The differential equation then reduces to

$$V\frac{dE}{dt} = -\frac{1}{4}EcA$$

$$\text{or:}\quad \frac{dE}{E} = -\frac{cA}{4V}dt. \tag{12.9}$$

Integrating, we obtain

$$\ln\frac{E}{E_m} = -\frac{cAt}{4V},$$

where E_m is the integration constant. Thus

$$E = E_m e^{-\frac{cAt}{4V}} = E_m e^{-\frac{t}{\tau}}, \tag{12.10}$$

where $\tau \left(= \frac{4V}{cA}\right)$ is the time constant of the decay. In terms of the intensity $I(= Ec/4)$, Eq. 12.10) becomes

$$I = I_m e^{-\frac{t}{\tau}}. \tag{12.11}$$

At $t = 0$, $I = I_m$, the maximum value of I. The growth and the decay of the intensity with time are displayed in Fig. 12.2.

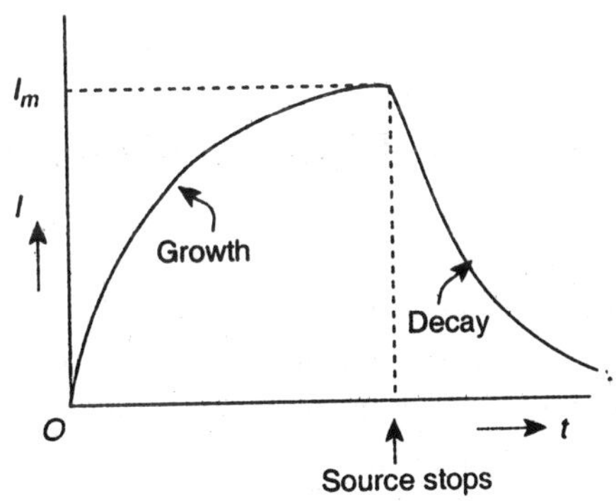

Fig. 12.2 Growth and decay of diffuse sound intensity in the enclosure

Reverberation time: Equation (12.11) shows that the intensity decays exponentially with time after the removal of the source. As already stated, according to Sabine. the *reverberation time* is the time required by the intensity to fall to the minimum audibility from an initial value 10^6 times as large. Thus, if T is the time of reverberation, and I is the intensity at the threshold audibility, we have

$$\frac{I_m}{I} = 10^6 = e^{\frac{T}{\tau}} = e^{\frac{cAT}{4V}}. \tag{12.12}$$

Clearly, $10\log_{10}(I_m/I) = 60$, so that T is the time for the intensity to drop by 60 dB. Equation (12.12) gives

$$T = \frac{4V}{Ac}\ln 10^6 = \frac{4V}{Ac} \times 2.303 \times 6 = \frac{55.3V}{Ac}. \quad (12.13)$$

If V is in m^3, the areas involved in A are in m^2, then putting $c = 343$ m/s,

$$T = \frac{0.161V}{A}. \quad (12.14)$$

Often the absorption A is measured in *sabin* which is the unit of absorption corresponding to 1 sq.ft of a perfectly absorbing surface like an open window. The absorption A can be found in sabin if the areas are expressed in sq.ft. Measuring V in cubic feet and using these units Eq. (12.13) reduces to

$$T = \frac{0.049V}{A}. \quad (12.15)$$

Equation (12.15) is referred to as *Sabine's formula* for the reverberation time.

Experimentally it is found that Sabine's formula is applicable for small absorption coefficients not exceeding 0.4. The absorption being low, the rate of growth or decay of the intensity of the diffuse sound is slow. Such rooms are known as *live rooms*. Thus Sabine's formula holds in a live room where T is large.

12.3 Reverberation Time in a Dead Room

If the absorption coefficient of the walls is unity, the reverberation time should be zero. But Sabine's formula [Eq. (12.15)] yields in this case $T = 0.049\,V/S$, where S is the area of the inner surface of the room. Thus, Sabine's formula is incorrect for large absorption coefficients. In fact, it is not valid for absorption coefficients greater than 0.4. Such rooms where the absorption coefficients exceed 0.4 are called *dead rooms*. In a dead room, the growth and decay of sound are faster and the reverberation time is small.

In deriving Sabine's formula, the absorption of sound was taken to be continuous and a uniform energy density in the enclosure was assumed. In a dead room, these simplifications are not possible. In Eyring's analysis, the absorption is taken to be discontinuous, occurring at each reflection on the walls. Let $\bar{a}$ be the average absorption coefficient of the materials of the inner walls of the room. Each time the sound is incident on the wall, a fraction $\bar{a}$ of its intensity is absorbed, the fraction $(1 - \bar{a})$ being reflected. After successive reflections, the intensity I_0 becomes $I_0(1 - \bar{a})$, $I_0(1 - \bar{a})^2$,

$I_0(1-\bar{a})^3$, etc. Hence after the sound source is shut off, the intensity drops to $I_0(1-\bar{a})^n$ after the nth reflection. If the threshold audibility is attained after the nth reflection, we have

$$(1-\bar{a})^n = 10^{-6}$$

or, $$n\ln(1-\bar{a}) = -2.303\times 6 = -13.82$$

or, $$n = -\frac{13.82}{\ln(1-\bar{a})}. \qquad (12.16)$$

If T be the reverberation time and c is the speed of sound, then the distance traversed by the sound in n reflections is cT. The average distance traversed between successive wall reflections in cT/n. This distance is $4V/S$, where V is the volume of the room and S is the total area of the walls. Hence,

$$n = \frac{cTS}{4V} = -\frac{13.82}{\ln(1-\bar{a})}$$

or, $$T = \frac{55.3V}{cS[-\ln(1-\bar{a})]} = \frac{55.3V}{cS\ln\left(\frac{1}{1-\bar{a}}\right)}. \qquad (12.17)$$

Using SI units with $c = 343$ m/s, Eq. (12.17) gives for the reverberation time (in s)

$$T = \frac{0.161V}{S\ln\left(\frac{1}{1-\bar{a}}\right)}. \qquad (12.18)$$

This is referred to as *Eyring's formula*. If $\bar{a}$ is small, we have $\ln(1-\bar{a}) \simeq -\bar{a}$, so that Eq. (12.18) gives

$$T = \frac{0.161V}{S\bar{a}} = \frac{0.161V}{A} \qquad (12.19)$$

which is *Sabine's formula*. Eyring's formula is more general and includes Sabine's formula. In dead rooms, the former gives much better agreement with the measured reverberation times. Also, Eq. (12.18) yields $T = 0$ when $\bar{a} = 1$, i.e., when the absorption is complete. A major weakness of Eyring's theory is that it is strictly valid when the walls have the same absorption coefficient.

12.4 Measurement of Absorption Coefficient

Let P_1 and P_2 be the powers of two sound sources. One of the sources (say, P_1) is operated in the enclosure, the average absorption coefficient of the material of the boundaries of which is to be determined. After a sufficiently long time, the energy density in the enclosure reaches its maximum value of

$4P_1/(cA)$. The source is then shut off, and the time T_1 in which the energy density decays to the threshold audibility is determined either by the ear or by the oscillograph. Similarly, the corresponding time T_2 is found for the source P_2. Hence, if V is the volume of the enclosure, Eq. (12.10) yields

$$\frac{4P_1}{cA}e^{-\frac{cAT_1}{4V}} = \frac{4P_2}{cA}e^{-\frac{cAT_2}{4V}}$$

each side being equal to the energy density at the threshold audibility. Thus

$$\frac{P_1}{P_2} = e^{\frac{cA}{4V}(T_1 - T_2)}$$

$$\text{or,} \quad \ln\left(\frac{P_1}{P_2}\right) = \frac{cA}{4V}(T_1 - T_2) = \frac{c\bar{a}S}{4V}(T_1 - T_2),$$

where $\bar{a}$ is the average absorption coefficient and S is the total wall area of the enclosure. Thus

$$\bar{a} = \frac{4V}{cS(T_1 - T_2)}\ln\left(\frac{P_1}{P_2}\right). \tag{12.20}$$

Knowing P_1/P_2 and measuring T_1 and T_2, the average absorption coefficient $\bar{a}$ can be found. This technique is known as the *reverberation method* of measuring the absorption coefficient.

Table 12.1 lists the absorption coefficients of some commonly used wall materials for a sound of frequency 512 Hz. For most of the materials, the absorption coefficient generally increases with increasing frequency.

TABLE 12.1

Absorption coefficients

Material	Absorption coefficient
Acoustic plaster	0.5
Unpainted brick wall	0.03
Felt	0.56
Plaster	0.05
Glass	0.05
Wood paneling	0.1
Glazed tile	0.01
Concrete floor	0.02
Carpeted floor	0.37

12.5 Solved Problems

Problem 1. *A auditorium of volume* 70,000 cubic feet *has a reverberation time* 0.9 s *when empty. Find the reverberation time when* 250 *persons are* *present in the auditorium. Given, absorption coefficient of a person* = 0.4, *average absorbing area of a person* = 6 sq ft.

Solution. Let A be the total absorption of the empty auditorium of volume $V = 70{,}000$ cubic feet. Using Sabine's formula $T = 0.049V/A$, we have $A = 0.049V/T = 0.049 \times 70{,}000/0.9 = 3811$ sabin. The extra absorption for 250 persons is $A_1 = 250 \times 0.4 \times 6 = 600$ sabin. So the total absorption is $A_2 = A + A_1 = 3811 + 600 = 4411$ sabin, and the new reverberation time is $T' = 0.049 \times 70{,}000/4411 = 0.78\,\text{s}$.

Problem 2. *A rectangular room has dimensions* 30 ft × 15 ft × 10 ft. *Its inner walls have an average absorption coefficient* 0.1. (i) *What is the reverberation time of the room?* (ii) *If there is a sound source of output* 15μW *in the room, calculate the maximum intensity in the room.*

Solution. The volume of the room is $V = 30 \times 15 \times 10 = 4{,}500\,\text{ft}^3$. The area of the inner walls is $S = 2(30 \times 15 + 15 \times 10 + 30 \times 10) = 1800\,\text{ft}^2$. The total wall absorption is $A = 1800 \times 0.1 = 180$ sabin. By Sabine's formula, the reverberation time is

$$T = \frac{0.049 \times V}{A} = \frac{0.049 \times 4{,}500}{180} = 1.23\,\text{s}.$$

The maximum intensity for a source of power $P(= 15\mu\text{W})$ is

$$I_m = \frac{P}{A} = \frac{15}{180} = 0.083\mu\text{W/ft}^2.$$

Problem 3. *A hall of volume* 2,500 m^3 *has a total absorption equivalent to* 300 m^2 *of open window. There are* 10 *persons in the room, each equivalent to* 4 sabin *absorption.* (i) *What is the reverberation time of the room?* (ii) *If, on the average, two persons are talking, each with* 5μW *output, find the maximum intensity in the room.*

Solution. (i) The reverberation time, by Sabine's formula, is

$$T = \frac{0.161V}{A}.$$

Here $V = 2{,}500\,\text{m}^3$ and $A = 300\,\text{m}^2 + 10 \times 4$ sabin $= 300 + 10 \times 4 \times (0.3048)^2\,\text{m}^2 = 303.7\,\text{m}^2$. Hence

$$T = \frac{0.161 \times 2{,}500}{303.7} = 1.32\,\text{s}.$$

(ii) The power output of two talking persons, $P = 2 \times 5\mu\text{W} = 10\mu\text{W}$. The maximum intensity is

$$I_m = \frac{P}{A} = \frac{10}{303.7} = 0.033\mu\text{W/m}^2.$$

Problem 4. *The interior walls of an auditorium* 250 × 60 × 30 ft *have an average absorption coefficient* 0.2. *The carpeted floor has an absorption coefficient* 0.3 *and the ceiling has a coefficient* 0.4. *What is the reverberation time? Find the power output of a source required to produce an intensity level of* 70 dB. (*Given, reference intensity level* = 10^{-16} W/cm^2).

Solution. The volume of the auditorium is $V = 250 \times 60 \times 30 = 4{,}50{,}000\,\text{ft}^3$. The area of the ceiling (or floor) = $250 \times 60 = 15{,}000\,\text{ft}^2$. The area of the four walls = $2(250 \times 30 + 60 \times 30) = 18{,}600\,\text{ft}^2$. The absorption of the boundaries, $A = 0.4 \times 15{,}000 + 0.3 \times 15{,}000 + 0.2 \times 18{,}600 = 14{,}220$ sabin. The reverberation time is

$$T = \frac{0.049 \times V}{A} = \frac{0.049 \times 4{,}50{,}000}{14{,}220} = 1.55\,\text{s}.$$

If I is the desired intensity, $10 \log_{10}(I/I_0) = 70$, whence $I = I_0 \times 10^7 = 10^{-16} \times 10^7 = 10^{-9}\,\text{W/cm}^2 = 10^{-9} \times (30.48)^2\,\text{W/ft}^2 = 0.929\mu\text{W/ft}^2$. If P is the power output of the source, we have

$$I = \frac{P}{A},$$

whence

$$P = IA = 0.929 \times 14{,}220 = 13210\mu\text{W} = 13.2\,\text{mW}.$$

Problem 5. (a) *The absorption coefficient of the wall material of a hall is* $a = 0.55$. (i) *Find the number of reflections suffered at the walls by the sound of initial intensity* 10^6 *times the threshold intensity before it drops to the threshold intensity.* (ii) *The volume of the hall is* 2000 m^3 *and the area of the walls is* 360 m^2. *Calculate the reverberation time using both Eyring's formula and Sabine's formula. Which value is more reliable and why?* (iii) *Is the hall 'live' or 'dead'?*

(b) *Repeat* (a) *for* $a = 0.05$.

Solution. (a) (i) The number of reflections is

$$n = -\frac{13.82}{\ln(1-a)} = -\frac{13.82}{\ln 0.45} = 17.3,$$

which is rounded off to 18 i.e., the next higher integer.

(ii) By Eyring's formula, the reverberation time is

$$T = \frac{0.161\,\text{V}}{S \ln\left(\frac{1}{1-a}\right)} = \frac{0.161 \times 2000}{360 \ln\left(\frac{1}{0.45}\right)} = 1.12\text{s}$$

By Sabine's formula,

$$T = \frac{0.161\,\text{V}}{Sa} = \frac{0.161 \times 2000}{360 \times 0.55} = 1.63\text{s}$$

Eyring's value is more reliable because Sabine's formula does not hold for $a > 0.4$.

(iii) The hall is 'dead' since the absorption coefficient is greater than 0.4.

(b) (i) For $a = 0.05$, $n = -\dfrac{13.82}{\ln 0.95} = 269.4$, which is rounded off to 270.

(ii) By Eyring's formula,

$$T = \frac{0.161 \times 2000}{360 \ln\left(\frac{1}{0.95}\right)} = 17.44\text{s}$$

By Sabine's formula, $$T = \frac{0.161 \times 2000}{360 \times 0.05} = 17.89\text{s}.$$

Eyring's and Sabine's values are very close here since a is very small, and either value is acceptable.

(iii) The hall is 'live' since $a \ll 0.4$.

Problem 6. *The volume of a closed room is* $2200\,\text{m}^3$ *and the surface area of the inner walls is* $350\,\text{m}^2$. *A sound source is operated in the room for a long time and then removed. The energy density in the room drops to the minimum audibility in* 1.4s. *When the source power is halved, the corresponding time is* 1.1s. *If the speed of sound in air is* $343\,\text{m/s}$, *calculate the average absorption coefficient of the inner walls of the room.*

Solution. Let the power of the first source be P_1. The power of the second source is $P_2 = \frac{1}{2}P_1$. The required absorption coefficient is

$$\bar{a} = \frac{4\,\text{V}}{cS(T_1 - T_2)} \ln\left(\frac{P_1}{P_2}\right) = \frac{4 \times 2200}{343 \times 350 \times (1.4 - 1.1)} \ln 2$$
$$= 0.2443 \ln 2 = 0.17.$$

Problem 7. *Show that after a sound source operating in a live room is shut off, the intensity level decays linearly with time at the rate* $372\frac{A}{V}$ dB/s *where* A *is the total absorption in* m^2 *and* V *is the volume of the room in* m^3.

Solution. If I_m is the intensity at time $t = 0$ when the source is shut off, the intensity at a subsequent time t is

$$I = I_m e^{-\frac{cAt}{4V}}$$

where c is the speed of sound. If I_0 is the reference intensity at threshold audibility, the intensity level in dB is

$$IL = 10 \log_{10}\left(\frac{I}{I_0}\right) = 10 \log_{10}\left(\frac{I_m}{I_0}\right) + 10 \log_{10} e^{-\frac{cAt}{4V}}$$
$$= 10 \log_{10}\left(\frac{I_m}{I_0}\right) + \frac{10}{2.303} \ln e^{-\frac{cAt}{4V}}$$
$$= 10 \log_{10}\left(\frac{I_m}{I_0}\right) - \frac{10}{2.303} \times \frac{cAt}{4V}$$

This shows that IL decays linearly with time t. The time rate of decay is

$$-\frac{d(IL)}{dt} = \frac{10}{2.303} \times \frac{cA}{4V} = \frac{10 \times 343\ A}{2.303 \times 4\ V} \quad \text{(taking } c = 343 \text{ m/s)}$$

$$= 372\frac{A}{V} \text{ dB/s.}$$

REVIEW QUESTIONS

1. (a) What is reverberation? Discuss the importance of reverberation in the acoustics of an auditorium.

   (b) On which factors does the reverberation time depend? What do you mean by 'optimum reverberation time'?

2. What are the requirements of a good auditorium? Which aspects should be emphasized in its design?

3. What do you understand by the acoustical properties of an auditorium? Deduce an expression for the time of reverberation of a hall in terms of the volume and surface of the hall and the average sound absorption coefficient. **[Burd. U. 1996, '99]**

4. Deduce Sabine's formula for the reverberation time in an auditorium. What do you mean by a 'live room' and a 'dead room'? What is sabin?

5. Derive expressions for the growth and the decay of the acoustic energy density with time in an enclosure. Give Sabine's definition of reverberation time.

6. Derive Eyring's formula for the reverberation time and obtain Sabine's formula from it.

7. Discuss the principle of measuring the absorption coefficient of a material on the basis of reverberation.

PROBLEMS

1. A hall has dimensions 70 × 45 × 20 ft. Its reverberation time is 1.3 s when empty. Find the reverberation time when an audience of 300 persons is present. Assume that the sound absorption by each person is 4.5 sabin.

   [**Ans.** 0.83 s]

2. An office room of rectangular shape has a dimensions $30 \times 20 \times 15$ ft. The average absorption coefficient of the floor is 0.2, that of the ceiling is 0.4, and that of the four walls is 0.03. Five persons, each having an absorption of 4 sabin, are sitting in the room. Three type writers, each generating $0.1 \mu W$ of sound energy, are operated simultaneously in the room. Determine the reverberation time and the intensity developed in the room.

[**Ans.** 1.04 s, $0.706\,nW/ft^2$]

3. A sound source of $10\,\mu W$ operates in a live room of volume $1800\,m^3$. If the reverberation time is 1.06s, determine the maximum intensity in the room.

[**Ans.** $0.0366\,\mu W/m^2$]

Chapter 13

Acoustic Measurements

13.1 Measurement of Frequency

The frequency of a tuning fork which serves as a source of sound in the laboratory can be measured in a simple way by the *sonometer method.* Here, resonance is achieved between the vibrating fork and a stretched sonometer wire normally vibrating in its fundamental mode. If the tension, the vibrating length, and the mass per unit length of the wire are known, the frequency of the fork can be determined. Since the vibrations of the fork are not maintained over a long time, the method does not give very accurate results. If the vibrations of the fork are electrically maintained, the stroboscopic or the phonic wheel method can be used to determine its frequency accurately.

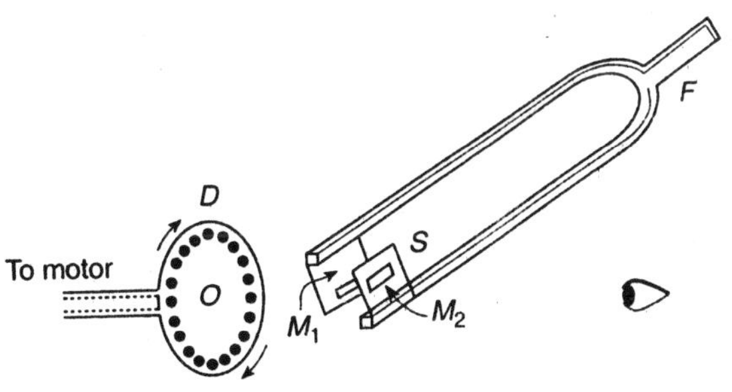

Fig. 13.1 Stroboscopic method

(i) Stroboscopic method: A stroboscope is a device which causes a vibrating or a rotating body to appear stationary. In the stroboscopic method, two light, thin metal plates M_1 and M_2 are attached to the prongs of the tuning fork F under test (Fig. 13.1). Each metal plate has a slit S in it, such that when the prongs are at rest the slits exactly face each other.

Thus a window is formed by the slits in this position, and light can pass through the window. When the fork is set into vibration, the window opens two times in a complete cycle. A vertical, circular white disc D is placed behind S. The disc is mounted on the spindle of an electric motor, and has a number of equidistant black dots around its circumference. These dots can be viewed through the window formed by the slits S.

The vibrations of the fork are electrically maintained, and the angular speed of the disc is gradually increased. At some minimum speed of the disc, the dots appear stationary. This happens when a succeeding dot occupies the position of a preceding dot in the time interval between the two successive openings of the window. If n is the number of dots on the disc and N is the number of revolutions of the disc per second, then the time required for one dot to occupy the position of its previous dot is $1/(nN)$ sec. If T is the time period of vibration of the fork, then $T/2 = 1/(nN)$, since two glimpses of the disc occur in one full cycle of vibrations of the fork. Thus the frequency of the fork (in Hz) is

$$f = \frac{1}{T} = \frac{1}{2}nN. \tag{13.1}$$

The error in the method is not more than 1 in 10^3. If should be noted that the disc also appears stationary when its speed of rotation is an integral multiple of the minimum speed required to give stationary dots.

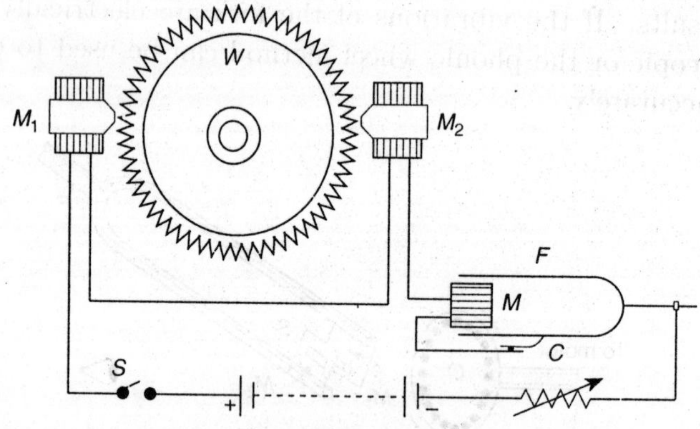

Fig. 13.2 Phonic wheel method

(ii) Rayleigh's phonic wheel method: In this method, a soft iron wheel W having a number of equidistant teeth on the periphery is used (Fig. 13.2). The wheel can rotate about a horizontal axis between the pole pieces of the electromagnets M_1 and M_2 The pole pieces are diametrically opposite and almost touch the teeth of the wheel. The electric current energizing the electromagnets is interrupted by a make-and-break arrangement introduced in the circuit which electrically maintains the vibration of the tuning fork F under test.

When the switch S is closed, the electromagnets are excited by the intermittent current whose frequency is determined by the frequency of vibration of the fork. The wheel is now rotated by hand and then let go. The rotation continues since the iron teeth of the wheel are attracted by the poles of the electromagnets. After some time, the wheel acquires a steady rotating speed. This occurs when a tooth of the wheel occupies the position of its preceding tooth between two consecutive passages of current, i.e., in one complete vibration of the fork. If n is the number of teeth of the wheel and N is its number of revolutions per second in the steady synchronized state, determined by means of a cyclometer, then the frequency of the fork (in Hz) is

$$f = nN. \tag{13.2}$$

This method is more accurate than the stroboscopic method.

Observation

In the arrangement of Fig. 13.2, the vibration of the tuning fork F is *electrically maintained* as follows. An electromagnet M is placed symmetrically between the prongs of the fork F. One end of the coil of the electromagnet M is attached to a metal piece which just touches the contact piece C connected to the nearest prong. When the switch S is closed, the circuit is completed through the contact piece C, and the electromagnet M is energized to attract the prongs inward. Hence the contact is broken at C and the current stops flowing. The electromagnet M is consequently deenergized and cannot attract the prongs any more. So, the prongs fly back and the contact at C is reestablished. The process is repeated, so that the vibration of the fork is maintained by feeding electrical energy, the periodicity of which is determined by the natural frequency of vibration of the fork.

13.2 Helmholtz Resonator

A Helmholtz resonator consists of a hollow metal or glass sphere having a wide cylindrical aperture N, termed the *neck*, and a narrow aperture P at the end of a narrow stem, termed the *pip*, diametrically opposite to the neck (Fig. 13.3). The resonator serves as an air cavity of dimensions small compared to the wavelength of the sound it intends to detect. In the type of the resonator shown in Fig. 13.3(i), the volume of the air cavity is fixed whereas that in Fig. 13.3(ii) is variable.

Sound waves are incident on the neck N and the ear is kept behind the pip P to detect the resonance. If the frequency of the incident tone coincides with the natural frequency of the resonator, the latter responds vigorously. The resonance is very sharp, so that the tone in question can be clearly detected in the midst of a complex note.

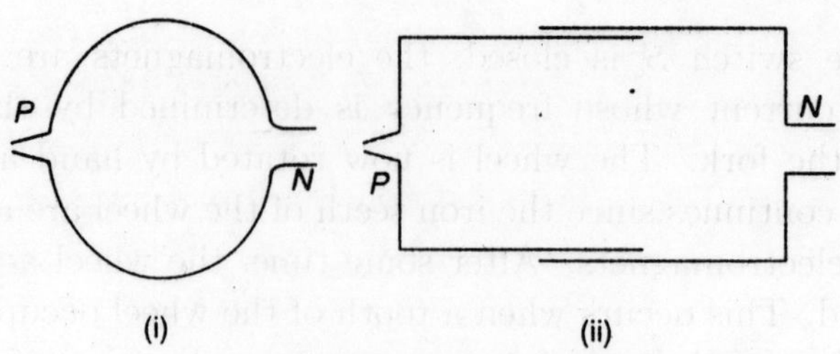

Fig. 13.3 Helmholtz resonators

The air in the neck is first set into vibration by the incident sound wave. This air mass in the neck serves as a piston, whose to-and-fro movement in the neck causes the air inside the vessel of the resonator to suffer adiabatic pressure changes. To find the *natural frequency of vibration* of the resonator, we assume that S is the cross-sectional area of the neck, l is the length of the neck, and V is the volume of the resonator. If ρ is the density of the air, the mass of the air in the neck is $\rho l S$. Let, at any instant of time, x be the outward displacement of the air in the neck due to a slight fall of pressure. The resulting increase in volume of the air in the resonator vessel is $dV = Sx$. As the pressure change in the vessel is adiabatic, we have $d(PV^\gamma) = 0$, where γ is the ratio of the specific heat of air at constant pressure to that at constant volume. Thus, $dP = -\gamma P \frac{dV}{V} = -\frac{\gamma P S x}{V}$. So, the force acting on the air in the neck is $SdP = -\frac{\gamma P S^2 x}{V}$. Neglecting damping, the equation of motion of the air in the neck is

$$\rho l S \frac{d^2x}{dt^2} + \frac{\gamma P S^2 x}{V} = 0,$$

$$\text{or,} \quad \frac{d^2x}{dt^2} + \omega_0^2 x = 0, \tag{13.3}$$

$$\text{where} \quad \omega_0^2 = \frac{\gamma P S}{\rho V l} = \frac{c^2 S}{V l}, \tag{13.4}$$

$c(= \sqrt{\gamma P/\rho})$ being the velocity of sound in air. Equation (13.3) shows that the motion of the air in the neck is simple harmonic, the natural frequency of vibration being

$$f_0 = \frac{\omega_0}{2\pi} = \frac{c}{2\pi}\sqrt{\frac{S}{lV}} \tag{13.5}$$

It is found experimentally that due to the open-end effect, the length l in Eq. (13.5) should be replaced by $l + 0.6r$, where r is the radius of the neck. When the frequency of the incident sound coincides with f_0, resonance occurs yielding a loud sound. In this case, the resonator is said to 'speak'. Its response is very weak at other frequencies.

Forced vibration of the resonator

Let a progressive sound wave be incident at the neck of the Helmholtz resonator from outside. If the incident wave is a simple tone of angular frequency ω, the impressed force on the air in the neck is either $F\cos\omega t$ or $F\sin\omega t$. In complex form, the instantaneous force can be written as $Fe^{j\omega t}$, where $j = \sqrt{-1}$. If x is the instantaneous displacement of the air in the neck, the force of restitution is $-\frac{\gamma PS^2 x}{V}$, where the relevant symbols have been already explained. The dissipative forces in the resonator cause a loss of energy. Another source of energy loss is reradiation from the resonator. These loses can be accounted for by introducing a damping force $-k\frac{dx}{dt}$, where k is the *damping constant.* Hence the equation of motion of the air in the neck is

$$\rho lS\frac{d^2x}{dt^2} = Fe^{j\omega t} - \frac{\gamma PS^2x}{V} - k\frac{dx}{dt}. \tag{13.6}$$

Putting $m = \rho lS$, the mass of the air in the neck, Eq. (13.6) can be rewritten as

$$\frac{d^2x}{dt^2} + 2b\frac{dx}{dt} + \omega_0^2 x = \frac{F}{m}e^{j\omega t}, \tag{13.7}$$

where $2b = \frac{k}{m}$ and $\omega_0^2 = \frac{\gamma PS^2}{mV} = \frac{\gamma PS}{\rho Vl}$.

Since the displacement x ultimately follows the impressed force, we can write $x \sim e^{j\omega t}$, so that $\frac{dx}{dt} = j\omega x$, and $\frac{d^2x}{dt^2} = -\omega^2 x$. Hence Eq. (13.7) gives

$$(-\omega^2 + j2b\omega + \omega_0^2)x = \frac{F}{m}e^{j\omega t}$$

$$\text{or,}\quad x = \frac{(F/m)e^{j\omega t}}{(\omega_0^2 - \omega^2) + j2b\omega} \tag{13.8}$$

$$= \frac{(F/m)e^{j(\omega t-\phi)}}{\left[(\omega_0^2 - \omega^2)^2 + 4b^2\omega^2\right]^{1/2}}, \tag{13.9}$$

where

$$\tan\phi = \frac{2b\omega}{\omega_0^2 - \omega^2}. \tag{13.10}$$

The velocity of the air in the neck, $\frac{dx}{dt}$, can be found from Eq. (13.8)

$$\frac{dx}{dt} = \frac{j\omega(F/m)e^{j\omega t}}{(\omega_0^2 - \omega^2) + j2b\omega}$$

$$= \frac{Fe^{j\omega t}}{k + j\left(\rho lS\omega - \frac{\rho c^2 S^2}{\omega V}\right)} = \frac{F}{Z_m}e^{j\omega t}, \tag{13.11}$$

where

$$Z_m = k + j\left(\omega\rho lS - \frac{\rho c^2 S^2}{\omega V}\right). \tag{13.12}$$

The quantity Z_m is the *acoustical impedance* of the system.

If the force is $F\cos\omega t$, the physical solution for the velocity is given by the real part of Eq. (13.11). Thus

$$\frac{dx}{dt}=\frac{F\cos(\omega t-\theta)}{\left[k^2+\left(\omega\rho lS-\frac{\rho c^2S^2}{\omega V}\right)^2\right]^{1/2}},\tag{13.13}$$

where

$$\tan\theta=\frac{\omega\rho lS-\rho c^2S^2/(\omega V)}{k}.\tag{13.14}$$

At resonance (*velocity resonance*), the amplitude of the velocity is a maximum. This happens when

$$\omega\rho lS-\frac{\rho c^2S^2}{\omega V}=0$$

$$\text{or,}\quad \omega^2=\frac{c^2S}{lV}=\omega_0^2,\tag{13.15}$$

i.e., when the frequency of the impressed force equals the natural frequency of the resonator. The maximum velocity amplitude at resonance is

$$v_m=\frac{F}{k}.\tag{13.16}$$

The smaller the damping constant k, the higher the velocity amplitude at resonance. Also, the sharpness of resonance will increase with decreasing k. So, for the resonator to respond sharply to a note, the damping must be as small as possible.

The instantaneous damping force is $k\frac{dx}{dt}$, so that the instantaneous power absorbed by the resonator is force × velocity $=k\left(\frac{dx}{dt}\right)^2$. The average power absorbed at resonance is

$$P_{\text{av}}=\frac{1}{2\pi}\int_0^{2\pi}k\left(\frac{dx}{dt}\right)^2d(\omega t)=\frac{F^2}{2k}=\frac{1}{2}Fv_m.\tag{13.17}$$

Thus the power absorbed by the resonator is large when the damping is small. The power radiated by the resonator is also large in this case. Although the sound emerging from the resonator through the pip is greatly intensified, there is no violation of the principle of conservation of energy. The sound is intensified at the cost of a greater drainage of energy from the source by the resonator. A vibrating tuning fork will be quickly damped in the presence of the resonator, if its vibrations are not maintained electrically.

Distinction between a pipe and a Helmholtz resonator

(i) The length of a pipe resonator is comparable to the wavelength of the tone to be detected, whereas the dimensions of the cavity of a Helmholtz resonator is small compared to the wavelength.

(ii) In a pipe resonator, the state of condensation varies from point to point, but the state of compression or rarefaction is the same everywhere in the cavity of a Helmholtz resonator.

(iii) The Helmholtz resonator is extremely selective and sharp in response, since its damping is quite small. If the area of the neck is small and the volume of the cavity is large, the vibration of a tuned resonator persists for a long time. In a pipe resonator, the damping is larger, so that it is less selective and its vibration does not persist for a long time.

Uses of a Helmholtz resonator

(i) Analysis of sound: The analysis of sound involves the detection of the different frequencies of the component tones. A number of resonators of different natural frequencies are taken.· Alternatively, a single resonator of continuously variable volume, such as the one with a sliding cylinder [Fig. 13.3(ii)] can be used. The ear is kept behind the pip of the resonator which speaks when its natural frequency happens to coincide with the frequency of a tone in the sound incident on the neck. In this way, the tones are detected or the sound is analysed.

(ii) Amplification of sound: A weak sound is immensely amplified by a Helmholtz resonator if its frequency is the same as that of the resonator.

(iii) Synthesis of sound: Synthesis is opposite to analysis. A Helmholtz resonator can analyse a sound, giving its frequency components. This enables us to synthesize a sound, i.e., to generate the desired sound by combining the appropriate frequency components.

13.3 Acoustic Filter

An acoustic filter significantly attenuates or filters out certain frequencies present in a complex sound. The filter consists of branch pipes (open or closed), branch cavities and tubes of constricted or expanded cross-section connected to the main tube carrying the complex sound. By a suitable arrangement of these pipes, cavities and tubes, the desired filtering action, i.e., attenuation of the required frequencies of the complex sound is achieved.

A *low-pass* filter allows the passage of low-frequency sound waves up to a certain frequency, called the cut-off frequency of the filter. A *high-pass* filter permits the frequencies above a certain frequency, termed the cut-off frequency, to pass through. If medium frequencies above a certain low frequency and below a certain high frequency are required to pass through, a *band-pass* filter is needed. A *band-elimination* filter allows all frequencies except those in a specific band to pass.

The simplest form of the acoustic filter is *Quinck's filter* (Fig. 13.4). A complex sound wave passes through the main tube P fitted with a side tube S. The wave in P resounds with the fundamental or overtones generated in S. Hence the energies of those frequency components of the sound wave are absorbed by S. Those frequency components of the complex sound are thus filtered out. If S is a closed tube, the frequencies which are filtered out or attenuated, are $c/(4l), 3c/(4l), 5c/(4l)$, etc., where c is the speed of sound and l is the length of S.

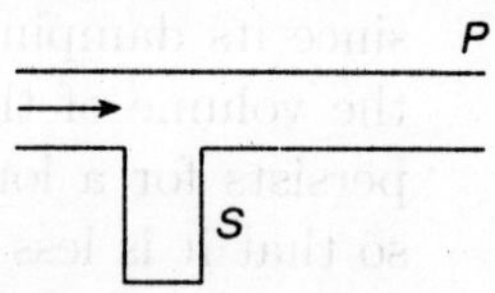

Fig. 13.4 Quinck's acoustic filter

13.4 Measurement of Velocity of Sound

The measurement of the velocity of sound in a gas or a solid rod and several other measurements can be done with the help of a simple device, called *Kundt's tube* (Fig. 13.5). The method employs the stationary waves formed in a closed air or gas column in a tube by the longitudinal vibrations of a solid rod clamped at the centre. A movable piston P_1 is fitted at one end of a horizontal glass tube T about a metre long and 3 cm in diameter. Near the other end of the tube, a light disc P_2 is attached rigidly to a metal or glass rod R which is tightly clamped at its middle point C. The piston P_1 almost fills the entire cross-section of the tube.

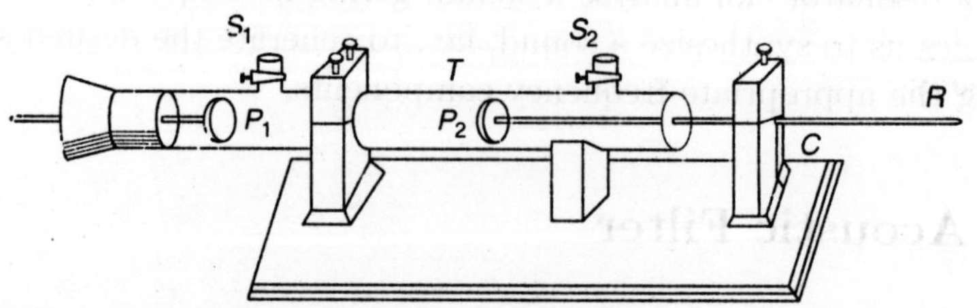

Fig. 13.5 Kundt's tube

The tube is thoroughly cleaned and dried. A thin layer of dry lycopodium powder is spread uniformly inside the tube along its length. Any desired gas can be inserted in the tube through the side tubes S_1 and S_2. The outer portion of the rod R is pulled lengthwise with a resined cloth so that the rod is set into longitudinal vibration, giving out a shrill sound. The longitudinal vibration of the rod causes the disc P_2 to move to-and-fro producing waves of compression and rarefaction in the air column in the tube T. These waves move towards P_1, and after reflection at P_1 superimpose on the direct wave, thus producing stationary waves having nodes (N) and antinodes (A) (Fig. 13.6). Since P_1 is at rest, a node is formed there. As P_2 is in vibration, a node is formed a little away from it. The position of the piston P_2 is adjusted till the air inside the tube exhibits resonance. The powder inside the tube is then violently agitated and finally settles in small heaps at the nodes N where the motion of the air is a minimum. At the antinodes A, the air particles vibrate vigorously and so the powder produces prominent *striations* at the antinodes at right angles to the length of the tube. The lengths of the striations are a maximum at the antinodes and they fall off gradually as one approaches the nodes. The striations are equidistant, the distance depending on the gas pressure, the number and the size of the particles, and the intensity of the sound.

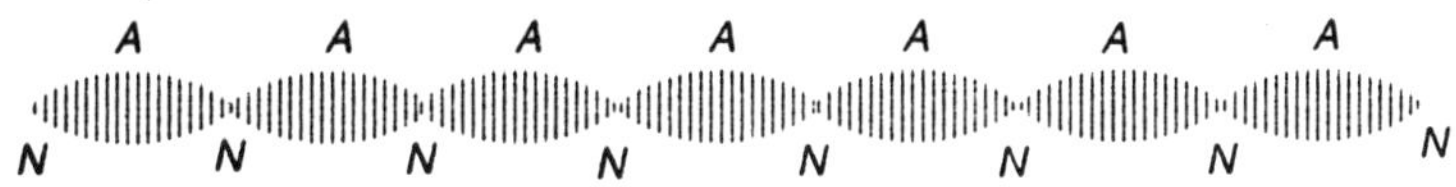

Fig. 13.6 Nodes and antinodes in Kundt's tube

The striations consist of an aggregation of particles in laminae across the length of the tube. Their formation is a hydrodynamical problem, as discussed below. When a flattened body is suspended in an alternating air current, it is acted on by a torque tending to rotate it at right angles to the direction of the air flow. A small sphere, however, does not experience any torque due to its symmetry. If two such bodies are closely spaced, the alternating air flow produces a force which can be attractive or repulsive depending on the arrangement of the bodies. For two particles closely spaced along the length of the tube, the air speed between them is small because of the screening effect of the particles. So the pressure is large, causing a repulsion between the particles. On the other hand, for two particles closely spaced broadside-on to the air stream, the air speed between the particles is large. Hence the pressure is low, causing an attraction between the particles. Thus two particles along the length of the tube are repelled, and if any one of them happens to come near another particle broadside on to the length of the tube, then the two particles are attracted. This accounts for the aggregation of the particles producing striations. The effect is predominant near an antinode since the longitudinal velocity of the air layers is a maximum there.

Let l be the average distance between the successive nodes or antinodes *at resonance*. If f is the frequency of the sound generated by the rod, the speed of the sound in air is

$$c_a = 2lf. \tag{13.18}$$

If l_s is the length of the rod R, the velocity of sound in the rod is (see Sec. 7.2)

$$c_s = 2l_s f. \tag{13.19}$$

If Y is Young's modulus and ρ is the density of the material of the rod, respectively, we have

$$c_s = \sqrt{\frac{Y}{\rho}}. \tag{13.20}$$

In the above, it is assumed that only the *fundamental longitudinal mode of vibration* of the rod is excited.

From Eqs. (13.18) and (13.19) we obtain

$$\frac{c_a}{c_s} = \frac{l}{l_s} \tag{13.21}$$

$$\text{or,}\quad c_s = c_a\frac{l_s}{l}. \tag{13.22}$$

Measuring l_s and l, c_s can be at once determined if c_a is known. Thus *the velocity of sound in a solid rod can be determined in the Kundt's tube experiment.*

Also, if ρ is known, then Y can be found from Eq. (13.20) using the measured value of c_s. Thus, *Young's modulus* of a solid can be determined with the help of a Kundt's tube.

If c_s is known, the *velocity of sound in air* (c_a) can be found from Eq. (13.22). By inserting other gases in the tube the velocity of sound in those gases can be obviously determined in this experiment. By filling the tube with a gas at different pressures, the same arrangement can be employed to study the *effect of pressure* on the velocity of sound in the gas. Furthermore, if the tube is surrounded by a heating coil, then by adjusting the electric current through the coil, the *velocity of sound in the gas at different temperatures* can be found.

The velocity of sound in a gas is given by

$$c_a = \sqrt{\frac{\gamma P}{\rho}}, \tag{13.23}$$

where P and ρ are the pressure and density of the gas, respectively, and γ is the ratio of the specific heat of the gas at constant pressure to that at constant volume. Thus from a knowledge of c_a, P and ρ, *one can determine* γ.

Observations

(i) The velocity of sound in gases in a tube is slightly less than that in the unrestricted medium. This happens due to the viscosity effects and the heat conduction at the walls. However, if the tube diameter is sufficiently large, the effect can be neglected.

(ii) In an improved version of the Kundt's tube, the exciting rod is replaced by an electrically driven loudspeaker. The frequency of the wave in the air column in the tube is the same as that of the current in the loudspeaker at resonance.

13.5 Measurement of Intensity

(i) Rayleigh disc method: Rayleigh observed that a light disc suspended in an air stream tends to set itself at right angles to the direction of the stream. If the stream is alternating as in a sound field, it produces the same effect on the suspended disc as that produced by a steady stream provided the mean square particle velocity of the former is equal to the steady particle velocity of the latter. Assuming that a thin circular disc of radius r is placed in a plane progressive sound wave field, the torque Γ acting on the disc is

$$\Gamma = \frac{4}{3}\rho r^3 u^2 \sin 2\theta, \tag{13.24}$$

where ρ is the density of the fluid medium, u is the rms particle velocity, and θ is the angle between the normal to the disc and the direction of the fluid flow. Clearly, the torque is a maximum when $\theta = 45°$. The diameter of the disc must be *small* compared to the wavelength of sound; otherwise the *diffraction of sound* by the disc will influence the result. The *viscosity effect* of the medium imparts a linear motion in addition to the rotation of the disc. This effect is not negligible for a heavy medium such as a liquid. Here the torque Γ given by Eq. (13.24), has to be multiplied by a factor $(1-\beta)^2$, where β is the ratio of the velocity amplitude of the disc to that of the medium. Hence the effective torque is

$$\Gamma_e = \frac{4}{3}\rho r^3 u^2 (1-\beta)^2 \sin 2\theta. \tag{13.25}$$

In the experiment, the disc is suspended vertically from a fine glass or quartz fibre at $\theta = 45°$. As the sound field is switched on, the disc rotates to a new position. By turning the torsion head from which the disc is suspended, the disc is brought to its original position. If τ is the torsion constant and ϕ is the angle of rotation of the torsion head, then

$$\Gamma = \tau\phi. \tag{13.26}$$

Calculating Γ from Eq. (13.26), one can obtain u^2 from Eq. (13.24). The intensity I of the sound is then found from the relationship

$$I = \rho c u^2, \tag{13.27}$$

where c is the velocity of sound.

Care should be taken to avoid air currents. Furthermore, the *frequency* of the incident wave must be different from the *natural frequency* of vibration of the disc; otherwise, the intensity measurement becomes troublesome.

The experimental arrangement is shown in Fig. 13.7. A thin disc (D) of mica, 1 cm in diameter and fitted with a light mirror on its surface, is suspended at the opening of a Helmholtz resonator by means of a thin quartz fibre about 10 cm long. The disc is set at an angle of 45° to the resonator axis. The suspension fibre in Fig. 13.7 is perpendicular to the plane of the paper, and its torsion head (H) has a circular scale. A pencil of light from a source (S) is reflected from the disc (D) and is incident on the scale (Sc).

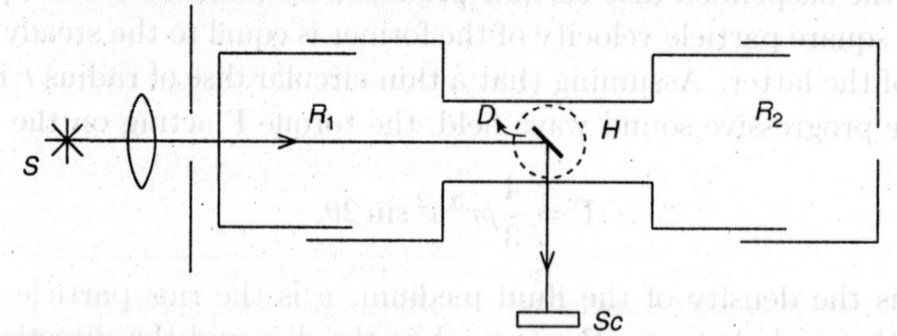

Fig. 13.7 Apparatus for intensity measurement by Rayleigh disc

When the sound field is turned on, the disc is deflected. The torsion head is turned till the spot of light on the scale Sc, and hence the disc comes back to its original position. The angle of rotation is found from the scale of the torsion head. Hence, if the torsion constant τ is known, the torque Γ can be determined from Eq. (13.26).

To achieve an enhanced accuracy, the disc is placed at the neck of the inner resonator of a double resonator R_1, R_2. The open end of R_2 picks up the sound from the source. The lengths of the resonators are adjusted for resonance. To find the intensity of the incident wave, the pressure amplification in the resonator must be taken into account.

If the angle of deflection is small, the intensity of sound is proportional to the deflection of the disc. Thus the intensities of two sound waves, I_1 and I_2, can be compared by measuring the corresponding disc deflections ϕ_1 and ϕ_2, since

$$\frac{I_1}{I_2} = \frac{\phi_1}{\phi_2}. \tag{13.28}$$

(ii) Hot-wire microphone method: The principle of this method is that the air movement due to an incident sound wave cools slightly an electrically heated wire. Consequently, the electrical resistance of the wire decreases, throwing an initially balanced Wheatstone bridge out of balance. The off-balance current is a measure of the intensity of the incident sound wave.

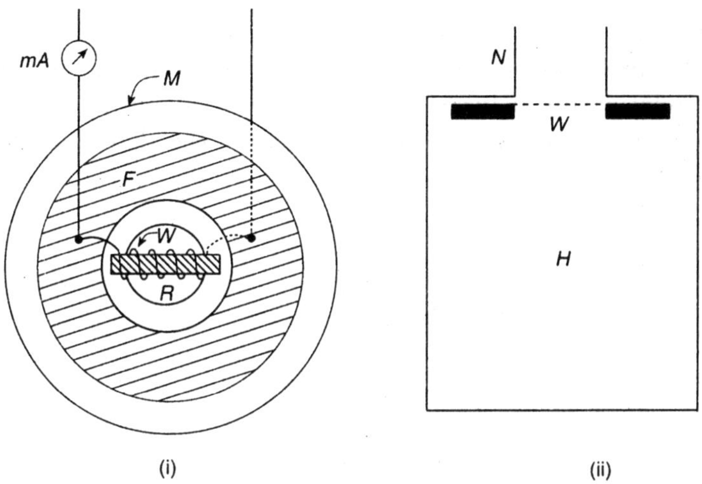

Fig. 13.8 Apparatus for intensity measurement by hot-wire microphone method

The apparatus consists of an electrically heated platinum wire W mounted on a glass rod R that bridges a central hole in a mica disc M (Fig. 13.8(i)]. The ends of the wire are soldered to two annular rings of silver foils F deposited on the opposite sides of the mica disc M.

The assembly is mounted at the neck N of a Helmholtz resonator H [Fig. 13.8(ii)]. The wire W is maintained just below dull red heat by an electric current. The resonator is made of a brass cylinder closed at one end and having several holes on the wall for communication between the inside and the outside air. The purpose of the holes is to prevent the vessel from serving as a resonator for a single compression wave. If the sound of suitable frequency is incident on the resonator, the air in the neck N is set into vibration. The consequent movement of the air particles cools the hot wire.

The wire W forms one arm of a balanced Wheatstone bridge. When the wire is cooled by the incident sound wave, its resistance decreases, thereby off-balancing the bridge. The off-balance current flows through an Einthoven string galvanometer whose response is very quick. The deflection of the string, as observed in a microscope, is found to be proportional to the intensity of the incident sound. If a calibration curve is drawn by plotting the galvanometer deflections against several known sound intensities, then this curve can be used to find the unknown intensity of a sound from the measured galvanometer deflection.

(iii) Sound radiometer: In Sec. 5.14, we found that the sound waves apply a static pressure, called the *radiation pressure*, on the surface on which they fall. The radiation pressure P is given by

$$P = \frac{I}{c}, \tag{13.29}$$

where I is the intensity of the sound and c is the speed of sound. Owing to this simple relationship between P and I, the measurement of the acoustic radiation pressure gives an *absolute method* for the determination of the sound intensity.

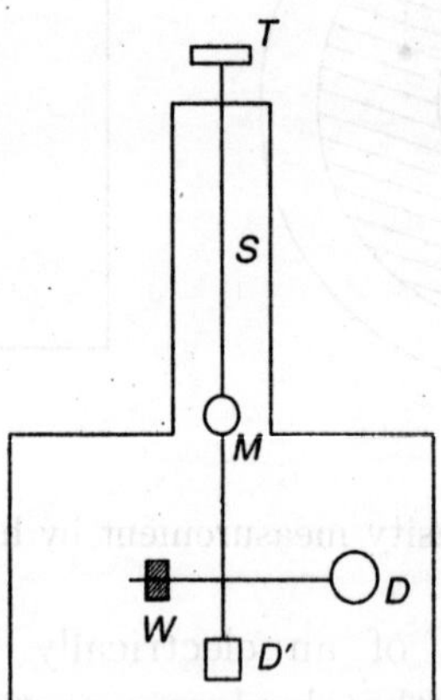

Fig. 13.9 A sound radiometer

A sound radiometer is schematically shown in Fig. 13.9. A light metal disc D closes a hole in a wall through which sound waves are incident on it. The disc D is fixed at one end of the arm of a sensitive torsion pendulum. At the other end of the arm of the torsion pendulum, there is a counterpoise weight W. The system is suspended at its centre of gravity by a torsion fibre S from a torsion head T. There is a damper D' at the bottom of the arrangement. As the radiation pressure deflects the disc, the rotation of the arm is indicated by the movement of a spot of light on a scale, the pencil of light being reflected by a small mirror M attached to the suspension fibre. By turning the torsion head, the spot of light and hence the disc is restored to its original position. If A is the cross-sectional area of the disc, l is half the length of the torsion arm, and P is the radiation pressure, the torque experienced by the disc D is PAl. We have $PAl = \tau\phi$, where τ is the torque per unit twist of the suspension and ϕ is the twist of the torsion head. Hence the intensity of sound is

$$I = Pc = \frac{c\tau\phi}{Al} \tag{13.30}$$

(iv) Optical interferometer method: In this method, an interference pattern is produced by two beams of light. One of these two beams passes transversely through a pipe carrying the sound wave, of which the intensity is to be measured. The other beam passes through the undisturbed air. The sound waves cause a variation of the air pressure, thus changing the refractive index. As a result, the interference pattern changes. Since the pressure and hence the refractive index fluctuates rapidly, the path difference between the interfering beams also changes rapidly. Consequently, the interfering fringes oscillate about their mean positions, and the persistence of vision makes them look wider. By measuring the fringe width, the intensity can be determined.

By inserting in the pipe air at different static pressures and measuring the corresponding shifts of the fringe system, the interferometer can be calibrated.

Observation

By measuring the intensity of sound, the loudness in phon can be determined. For details, refer to Sec. 5.15.

13.6 Determination of Quality

The *quality* of a musical note is the physical cause which distinguishes between the notes of the same pitch and loudness, but generated by different musical instruments. The difference in the quality of different notes arises from the difference in the waveforms of the notes. A musical sound or a note is a mixture of a number of sinusoidal component tones. Each single-frequency component tone is called a *partial.* Usually, the tone of the lowest frequency, i.e., the fundamental tone has the greatest intensity. The frequency of the fundamental determines the *pitch* of the note. The quality of the note is governed by the number, the frequencies, and the relative intensities of the component tones making up the note. Thus, if the frequencies and the relative intensities of the component tones are found, the quality of the musical note can be determined.

The ear of an experienced listner can detect some of the partials making up a musical note and can also roughly estimate their intensities. As an aid to the ear, instruments like Helmholtz resonators, Miller's phonodeik, cathode ray oscillograph, etc. are employed in sound analysis.

(i) Helmholtz resonator: The resonant frequency of a Helmholtz resonator can be varied by varying the volume of the cavity, as discussed in Sec. 13.2. The frequencies of the tone components of a complex sound can be found by observing the resonance frequencies of the Helmholtz resonators. The relative intensities of the tones can also be estimated roughly by these resonators. For a more precise estimate of the relative intensities, a hot-wire microphone can be employed.

(ii) Miller's phonodeik: In Miller's phonodeik, the waveform of the musical note under test is recorded on a film. A Fourier analysis of the waveform gives the frequency components of the note, so that the quality of the note can be studied.

The instrument consists of a conical horn H, the narrow end of which is fitted with a very thin glass diaphragm D, held between soft rubber rings (Fig. 13.10). A very fine wire W is attached to the centre of the diaphragm. The wire passes round a tiny pulley P, and is attached at the other end to a spring S which keeps it taut. The pulley is mounted on a vertical spindle R held on jwelled bearings B. A small mirror M is attached to the spindle R. Light from a source L is reflected by the mirror and is focussed on a vertical film F.

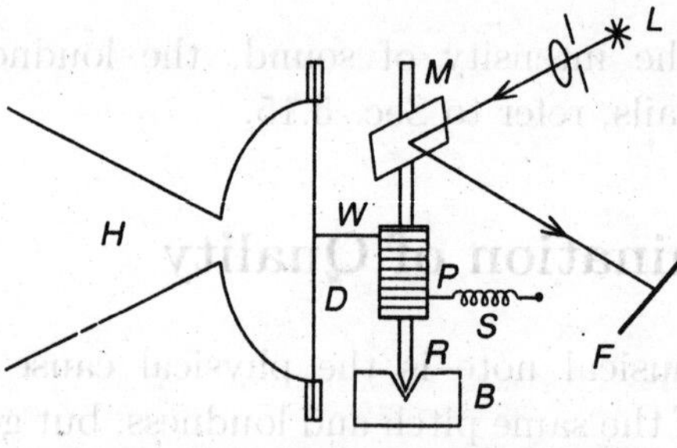

Fig. 13.10 Miller's phonodeik

The sound, collected by the horn H, sets the diaphragm D into vibration. As the diaphragm moves, the wire W is pulled and released, so that the spindle R and hence the mirror M suffer alternating rotation. As a result, the image of the source on the film F is displaced horizontally. The film is moved vertically with a uniform speed, so that the spot of light traces out on it the time-displacement graph of the diaphragm. Thus the pressure variation of the incident sound or its waveform is recorded on the film.

The defect of the phonodeik is that its frequency response is not uniform over the whole range. Resonance occurs if the frequency of any component of the incident sound coincides with any of the natural frequencies of vibration of the diaphragm and the horn. The response shoots up to a large value at these frequencies. To avoid this difficulty, the phonodeik was calibrated by Miller using a number of specially constructed organ pipes.

(iii) Cathode-ray oscillograph: The cathode-ray oscilloscope (CRO) was discussed in Sec. 1.8. The sound under test is picked up by a capacitor microphone whose output is fed to a distortionless amplifier to obtain an alternating voltage having the same waveform as the sound. This voltage is applied to the vertical deflecting plates of a CRO. A sawtooth voltage of negligible flyback time is applied to the horizontal deflecting plates. The horizontal displacement of the light spot on the CRO screen is thus proportional to time, giving the 'time base'. The frequency of the sawtooth voltage is adjusted such that in its one full cycle, the alternating voltage from the microphone performs one or more cycles. A steady pattern is obtained on the screen by proper synchronization. This pattern gives the visual display of the sound waveform. It can be photographed to obtain a permanent record for analysis.

13.7 Solved Problems

Problem 1. *A Helmholtz resonator has a volume* 500 cc. *The radius of the neck is* 1 cm *and the length of the neck is* 0.1 cm. *Calculate the frequency of the resonator. Velocity of sound at room temperature is* 330 m/s. **[C.U. 1986]**

Solution. The natural frequency of the resonator is given by

$$f_0 = \frac{c}{2\pi}\sqrt{\frac{S}{lV}}.$$

Here c = velocity of sound in air = 330 m/s, S = cross-sectional area of the neck $= \pi \times 1^2\,\text{cm}^2 = 3.1416 \times 10^{-4}\,\text{m}^2$, l = length of the neck = 0.1 cm = 10^{-3} m, and V = volume of the resonator = $500\,\text{cm}^2 = 5 \times 10^{-4}\,\text{m}^3$. Hence

$$f_0 = \frac{330}{2 \times 3.1416}\sqrt{\frac{3.1416 \times 10^{-4}}{10^{-3} \times 5 \times 10^{-4}}} = 165\sqrt{\frac{10^3}{3.1416 \times 5}} = 1316\,\text{Hz}.$$

Problem 2. *A Kundt's tube is excited by a solid rod* 120 cm *long and clamped at the middle. At resonance, the powder in the tube settles in loops,* 5 *of which occupy a length of* 60 cm. *If the velocity of sound in air is* 330 m/s, *calculate that in the rod.*

Solution. The length of a loop, i.e., the distance between two consecutive nodes is $l = 60/5 = 12$ cm. If c_s and c_a are the velocities of sound in the rod and in air, respectively, and l_s is the length of the rod we have

$$c_s = c_a\frac{l_s}{l} = 330 \times \frac{120}{12} = 3300\,\text{m/s}.$$

Problem 3. *For what maximum frequency can Kundt's tube method be employed to determine the speed of sound in a solid if the minimum detectable distance between antinodes is* 4 mm? *The speed of sound in air is* 340 m/s. **[Burd. U. 1993]**

Solution. We have $c_a = 2lf$, where l is the distance between successive antinodes, f is the frequency, and c_a is the velocity of sound in air. Here the minimum value of l is 4 mm $= 4 \times 10^{-3}$ m. Hence the maximum frequency is

$$f = \frac{c_a}{2l} = \frac{340}{2 \times 4 \times 10^{-3}}\ \text{Hz} = 42.5\ \text{kHz}.$$

REVIEW QUESTIONS

1. Discuss the principle of the stroboscopic method for the determination of the frequency of a tuning fork whose vibrations are electrically maintained.

2. For an accurate determination of the frequency of a vibrating tuning fork, would you prefer Rayleigh's phonic wheel method over the stroboscopic method? Describe Rayleigh's phonic wheel method.

3. (a) What is a Helmholtz resonator? What measurements can be made with its help? **[Burd. U. 1995]**

   (b) Deduce an expression for the natural frequency of vibration of a Helmholtz resonator. **[C.U. 1988, Burd. U. 1997]**

4. Discuss with necessary deductions the principle of a Helmholtz resonator. In what respect does its action differ from a vibrating air column in a pipe? **[Burd. U. 1990]**

5. Explain the action of Kundt's tube. Mention the various types of acoustic experiments that can be done with its help. **[Burd. U. 1993]**

6. (a) How can the velocity of longitudinal waves in an elastic solid rod be measured using Kundt's tube? **[C.U. 1993]**

   (b) What happens if the rod is narrow?

7. Describe in some detail the Rayleigh disc method for the measurement of intensity of sound. What other methods can be used to determine the sound intensity?

8. Describe the principle of the hot-wire microphone and the sound radiometer.

9. Explain the principle of any method of measuring the intensity of sound. **[N. Beng. U. 2001]**

10. What do you mean by the quality of a musical note? What are the instruments used in the study of the quality of a note?

11. Discuss Miller's phonodeik method for the determination of the waveform of a musical note. How can the waveform be analysed?

12. Write short notes on :

(a) Helmholtz resonator, [**C.U. 1992**]

(b) Rayleigh's disc, [**C.U. 1992**]

(c) Kundt's tube, [**C.U. 1991**]

(d) Hot-wire microphone,

(e) Miller's phonodeik,

(f) Electrically maintained tuning fork, and

(g) Sound waveform analysis by CRO,

(h) Acoustic filter.

PROBLEMS

1. The volume of the vessel of a Helmholtz resonator is 1 litre. The cross-sectional area of its neck is π^2 cm^2 and the length of the neck is 1 mm. Determine the natural frequency of vibration of the resonator, if the speed of sound in air is 340 m/s.

[**Ans.** 1,700 Hz]

2. When a Kundt's tube is filled with air of density 0.00128 g/cm^3, the two successive antinodes are found to be separated by 3.46 cm. When argon of density 0.00178 g/cm^3 replaces the air in the tube, the corresponding separation is found to be 3.16 cm. From these data, what conclusion can you draw regarding the nature of argon?

[**Ans.** $\gamma_{\text{argon}} = 1.62$; hence argon is monatomic]

3. A stroboscopic disc having 30 dots revolves at the rate of 1200 rpm. The dots appear to be stationary when viewed through two slits attached to the prongs of a vibrating tuning fork. What is the frequency of the tuning fork?

[**Ans.** 300 Hz]

Index